ACCELERATED RELIABILITY AND DURABILITY TESTING TECHNOLOGY

WILEY SERIES IN SYSTEMS ENGINEERING AND MANAGEMENT

Andrew P. Sage, Editor

A complete list of the titles in this series appears at the end of this volume.

ACCELERATED RELIABILITY AND DURABILITY TESTING TECHNOLOGY

LEV M. KLYATIS
Professor Emeritus
Habilitated Dr.-Ing., Dr. of Technical Sciences, Ph.D.

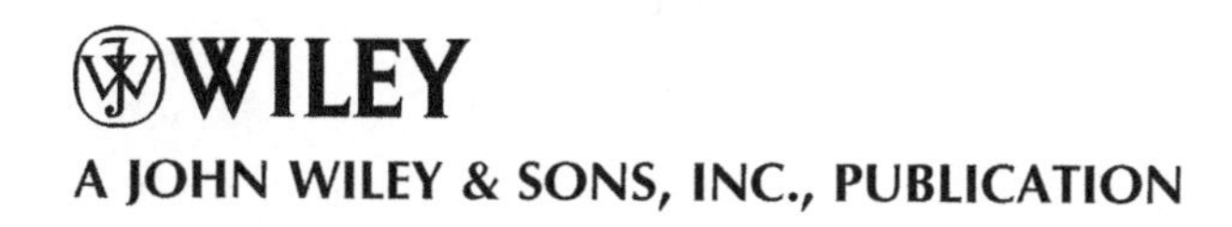

A JOHN WILEY & SONS, INC., PUBLICATION

Published by John Wiley & Sons, Inc., Hoboken, New Jersey.
Published simultaneously in Canada.

Library of Congress Cataloging-in-Publication Data:

Klyatis, Lev M.
Accelerated reliability and durability testing technology / Lev M. Klyatis.
p. cm. – (Wiley series in systems engineering and management)
ISBN 978-0-470-45465-7 (cloth)
1. Reliability (Engineering) 2. Accelerated life testing. I. Title.
TS173.K63 2009
620'.00452–dc22

2009017252

10 9 8 7 6 5 4 3 2 1

To My Wife Nellya Klyatis

Contents

Preface

In 2009–2010, Toyota experienced an increase in global recalls jumping to 8.5 million cars and trucks [1]. Similar situations have occurred and could happen again to other automakers as well as to companies in other industries. Toyota can have similar problems in the future as well. The basic reason behind many recalls and complaints, as well as higher cost and time for maintenance and life cycle cost than was predicted during design and manufacturing is the inaccurate prediction of the product's reliability, durability, and quality during design and manufacturing. Predictions may be inaccurate due to the lack of proper accelerated reliability testing (ART) and accelerated durability testing (ADT) as a source of initial information for the prediction.

The focus of this book is to show multiple applications of the technology (methodology and equipment) that provide physically ART and ADT of an actual product in a way that represents the real world's many product influencing interactions.

Integrated global solutions for many engineering problems in quality, safety, human factors, reliability, maintainability, durability, and serviceability were not available in the past. One of the basic reasons for this design deficiency was the inability to solve a fundamentally important problem—accelerated reliability (durability) testing. ART and ADT provide an integrated solution that will positively influence product development time, cost, quality, design, and effective product/process. The context for this treatise is an industrial product design and development. This approach applies to the development of a large number of products and processes such as, for example, in the consumer goods, industrial, producer, medical, banking, pharmaceutical, teaching, and military factors.

The weakness of prior accelerated testing approaches such as vibration testing, corrosion testing, step-stress testing, thermoshock testing, highly accelerated life testing (HALT), highly accelerated stress screening (HASS), accelerated aging, mechanical crack propagation and growth, and environmental stress screening (ESS) was the result of inappropriately using only a few of the influencing factors and using these factors in isolation. Such testing improved the design in a one-dimensional way but failed to provide the needed global optimum solution formed by combining the many complex factors in a systematically integrated approach. These types of individually conducted tests do not provide sufficient information for the accurate prediction of the interval product/process degradation and failures. The complex interaction of these factors with the multitude of real-world factors is not considered.

Hence, the results are potentially (and usually) misleading. Such suboptimal solutions cannot accurately account for the global impact of complicated interactions that result in delays in development, time to market, and in increased costs related to

- Design time and result
- Customer satisfaction and expense
- Maintenance frequency, cost, and access
- Warranty costs and recalls
- Degradation of product/process over time
- Failures during time intervals and warranty periods
- Quality requirements and indices
- Product safety
- Human factors

Many people seek to artificially increase the value of limited testing by including the word "durability" in the title (e.g., "Vibration Durability Testing"). Vibration testing is not sufficient for the evaluation or prediction of product durability. Vibration testing is only one of many components of mechanical testing as a part of complicated durability testing. Vibration testing alone is not sufficient for accelerated development predictions for reliability, maintainability, and durability improvement or for solving many other related problems.

ART or ADT based on a combination of many types of accelerated tests (field, laboratory, multi-environment, mechanical, electrical, etc.) integrated with safety and human factors helps to solve the identified problems. ART and ADT both use an accurate simulation of a field environment. Chapter 1 describes this phenomenon.

What is technology? Technology is often a generic term to encompass all the technologies people develop and use in their lives. The United Nations Education, Social and Cultural Organization (UNESCO) defines technology as "... the know-how and creative processes that may assist people to utilize tools, resources and systems to solve problems and to enhance control over the natural and made environment in an endeavor to improve the human conditions" [2]. Thus, technology involves the purposeful application of knowledge, creativity, experience, insight, and resources to create processes and products that meet human needs or desires. The needs and wants of people in particular communities coupled with their creativity determine the type of technology that is developed and how it is applied.

The simple definition of ART and ADT technology is a "complex" composed of specific testing methodologies, equipment, and usage that influence accurate prediction and successful accelerated development of product quality, reliability, durability, maintainability, availability, and supportability.

Each type of reliability and durability laboratory testing conducted as a component of ART or ADT consists of many subcomponents. The simulation of multiple inputs individually and simultaneously influences the result. For example, multi-environmental testing consists of a combination of temperature, humidity, chemical pollution, dust pollution, a complex of ultraviolet, infrared, and visible parts of the light spectrum, air pressure, and other influencing factors.

This is the first book on ART and ADT that intends to acquaint the reader with the evolving methodology and equipment necessary to conduct true ART and ADT. To answer the question "How?" the author uses more than 30 years of experience in this field, especially in the area of ART and ADT, as well as drawing from the world's experience in this area.

This work covers new ideas and technologies for accurate ART and ADT that enhance the high correlation between testing and field data. This important testing process is continuously developing and expanding. In the real world, most of the significant factors for design and development are interconnected. Simulation, testing, quality, reliability, maintainability, human–system interaction, safety, and many other factors are interconnected and collectively influence each other. This is also true for the interacting influences of temperature, humidity, air pollution, light exposure, road conditions, input voltage, and many other parameters.

If one ignores these interactions, then one cannot accurately represent the real-world situation in a simulation. Consequently, testing based on a simulation without considering real-world interactions cannot give sufficient information for the accurate prediction of all the quality parameters of interest. This book provides strategies to eliminate these negative aspects and to sufficiently describe ART and ADT.

ART and ADT is an important component of a more complicated problem: accurate prediction of product quality, reliability, durability, and maintainability using accelerated tests. This unique approach shows that ART and ADT represent a significant tool for the integration of multiple interactions generated by other test parameters. This methodology uses a system of systems approach and shows how to use ART and ADT for estimating reliability parameters and related maintainability, durability, and quality parameters. One uses it to get an accurate prediction of optimal maintainability, durability, and a desired level of quality during a given time (warranty period, service life). This approach reduces customer complaints, product recalls, life cycle costs, and "time to market," while facilitating the solution of related problems.

Most publications concentrate on the theoretical aspects of data analysis (including test data), test plans, parameter estimation, and statistics in the area of accelerated testing. The description of the test equipment, test protocol, and its application is seldom available. One can rarely find any information describing the process and equipment required to conduct ART and ADT.

Engineers and managers particularly need to know *how* to correctly perform ART and ADT. The specific advantage this book provides is an explanation of the technology, technique, and equipment sufficient to enable engineers and managers and other service professionals to successfully conduct practical ART and ADT. The book provides the direction to rapidly find causes (with examples) for the degradation and failures in products and processes and to quickly eliminate or mitigate these causes or their effects. This approach demonstrates how to accelerate the processes to provide an accurate prediction during the development of the product's quality, reliability, durability, and maintainability for a given warranty period and service life.

Each individual product needs a specific test plan and testing technique, but the concepts for solving these problems are universal. Therefore, this book will be useful for different types of products in various industries and applications that work on land, at sea, in the air, and in space.

ART/ADT need an initial capital investment in equipment and high-level professionals to manage and conduct it. Only a limited number of industrial companies have appropriate guidance for this type of testing. Many CEOs who make decisions about testing investments do not sufficiently understand that an investment in ART/ADT will typically result in a 10-fold increase in profits. It reduces complaints and recalls, increases reliability, durability, and maintainability, and decreases total life cycle costs. Accelerated reliability (durability) testing is more complicated than many other types of accelerated tests currently in use.

It was not until the late 1950s that professionals began to understand that the myriad interactions among different influences on the product/process could result in overlooking a significant degradation failure mechanism. Originally, testing occurred in series by using one input influence at a time, for example, temperature alone. Then adding another, humidity, for example, was tested. When it became clear that temperature and humidity did not account for all of the failures, then another influence, perhaps vibration, was tested. This serial process continued with the addition of other influencing parameters as necessary to explain the unexpected failures. Although the design was improved to ensure the equipment would be reliable in each individual environment, unexpected failures occurred when the equipment was finally tested in a real-world environment. No assessment had been made of the collective interactions of the many input influences. Eventually, engineers realized that the synergistic effects of the interaction of different environmental factors caused the degradation and failures.

A new approach to testing, where the product simultaneously experiences different input influences, is combined environmental reliability testing (CERT) [3]. CERT required new testing facilities and testing equipment. For example, vibration tables and ovens were combined to create what were humorously referred to as "shake 'n bake" chambers [4]. In fact, such a multi-environmental test only partly reflects the field input influences. Mechanical testing, electrical testing, and other types of testing also exist. The basis for ART and ADT is the simultaneous simulation of all field influences, integrated with safety and human factors, on the product or process.

Each laboratory research study applies different field simulations and testings. This book may improve the quality of this work by enabling professionals to execute research on a higher level.

ART is identical in many aspects to ADT. Therefore, many refer to "ART" as "ADT." Repetition of the term "simultaneous combination" reflects the basic essence of ART and ADT.

The book is for industrial engineers, test engineers, reliability engineers, and managers. It is also for personnel in the service area, maintenance area, engineering researchers, teachers, and students who are involved in quality, reliability, durability, maintainability, simulation, and testing. The author wishes to express his thanks to Y.M. Abdulgalimov, J.M. Shehtman, E.L. Klyatis, V.A. Ivnitsky, and posthumously to D. Lander for help provided in various stages of this project.

LEV M. KLYATIS

About the Author

Dr. Lev Klyatis is Senior Consultant at SoHaR, Inc. and a member of the Board of Directors for the International Association of Arts and Sciences in New York. His scientific/technical expertise is in reliability, durability, and maintainability. He created new approaches for accelerated solution of reliability/durability/maintainability problems, through innovation in the areas of accurate physical simulation of field conditions, accelerated reliability/durability testing, and accurate prediction of quality/reliability/durability/maintainability. He developed a methodology of complaints and recalls reducing. He holds three doctoral degrees: a Ph.D. in Engineering Technology, a high-level East European doctoral degree (Sc.D.) in Engineering Technology, and a high-level West European doctoral degree in Engineering (Habilitated Dr.-Ing.).

He was named a full life professor by the USSR's Highest Examination Board in 1991 and a full professor at Moscow University of Agricultural Engineers in 1990–1992. He came to the United States in 1993. He is the author of over 250 publications, including eight books. His most recent work is *Accelerated Quality and Reliability Solutions* (2006). He holds more than 30 patents in different countries. He is a seminar instructor for the ASQ seminar *Accelerated Testing of Products*. He is a frequent speaker on accurate simulation of field conditions, accelerated testing, and accurate prediction of quality, reliability, durability, and maintainability at national conferences in the United States as well as international conferences, congresses, and symposiums.

Dr. Klyatis has worked in the automotive, farm machinery, and aerospace industries, among others. He was a consultant for Ford, DaimlerChrysler, Thermo King, Black & Dekker, NASA Research Centers, and Karl Schenck (Germany), as well as other industries.

Dr. Klyatis served on the U.S.–USSR Trade and Economic Council, the United Nations Economic Commission for Europe, and the International Electrotechnical Commission as an expert of the United States and an expert of the ISO/IEC Joint Study Group in Safety Aspects of Risk Assessment. He was the research leader and chairman of the state enterprise TESTMASH, and the principal engineer of a governmental test center. He is presently a consultant of ACDI VOCA, a U.S. agency, and a member of the World Quality Council, the Elmer A. Sperry Board of Award, the SAE G-11 Executive and Reliability Committees, the Quality and Robust Design Committee of SAE, and the Governing Board of the SAE International Metropolitan Section.

Chapter 1

Introduction

1.1 THE PURPOSE OF ACCELERATED TESTING (AT)

In an AT, one accelerates the deterioration of the test subject beyond what is expected in an actual normal service environment. AT began many years ago with the development of the necessary methodology and equipment. Development continues into the future. As the knowledge about life and the laws of nature evolves, the requirements for products and technologies have also increased in complexity. Thus, the requirements for AT have and continue to increase in scope. Often, AT methods and equipment that were satisfactory in the past are no longer satisfactory today. Those that are good today will not satisfy the requirements of producers and users in the future. This encourages research and development for AT. This process, reflected in the literature, encourages and directs the research and advancement of test disciplines.

Unfortunately, in real life, people who perform AT for industry and other organizations usually do not have the time, incentive, or the opportunity to write books. Authors of AT books unfortunately often know their subject primarily in theory rather than from an actual application of AT. The situation is not better if an author includes such terms as "practical," "practice," or "practitioner's guide" in the title of the publication. As a result, most books on AT do not demonstrate *how* to conduct testing or identify what type of testing facility and equipment is appropriate, and they also neglect to identify the benefits of one method over another. Publications usually fail to show the long-term advantages and savings accruing from an investment in more

Accelerated Reliability and Durability Testing Technology, First Edition. Lev M. Klyatis.

expensive and advanced testing equipment to increase product quality, reliability, durability, and maintainability while reducing the development time and decreasing a product's time to market. How can one accomplish this? One must provide a combination of practical and theoretical aspects for guidance and use.

The basic purpose of AT is to obtain initial information for issues of quality, reliability, maintainability, supportability, and availability. It is not the final goal. It is accomplished through prediction using the information provided by AT under laboratory (artificial) conditions. The most effective AT of a product design needs to occur under natural (field) conditions. AT design and the selection of appropriate testing parameters, equipment, and facilities for each method or type of equipment to be tested must be coordinated to provide the test inputs and results that are most beneficial for the quality, reliability, or maintainability problems that the test identifies. An AT design is very important in determining how accurate the decision process is in selecting the method and type of equipment to use.

Quality, reliability, durability, and maintainability are factors that are not separable. They are interconnected, have complex interactions, and mutually influence each other. This complex represents the parameters and processes needed to conduct AT and includes simulation, testing, quality, reliability development, maintainability, accurate prediction, life cycle costs, field reliability, quality in use, and other project-relevant parameters and processes. AT is a component of a complex supporting the design, manufacturing, and usage processes, and its benefits depend on how one configures the complex for optimization.

If industrial companies would properly apply this optimization process, then they would choose more carefully among the many popular current test methods and types of equipment such as highly accelerated life testing (HALT), highly accelerated stress screening (HASS), accelerated aging (AA), and others to use them for the accurate prediction of reliability, durability, maintainability, supportability, and availability. It is verifiable that buying simple and inexpensive methods and equipment for testing becomes more expensive over a product's life. It is also true for a simulation as a component of an AT, evaluation, and prediction. A basic premise of this book is that the whole complex needs to be well-thought-out and approached with a globally integrated optimization process.

1.2 THE CURRENT SITUATION IN AT

The following presents three basic approaches for the practical use of AT as shown in Figure 1.1.

1.2.1 The First Approach

The first approach is special field testing with more intensive usage than under a normal use. For example, a car is usually in use for no more than 5–6 h/day.

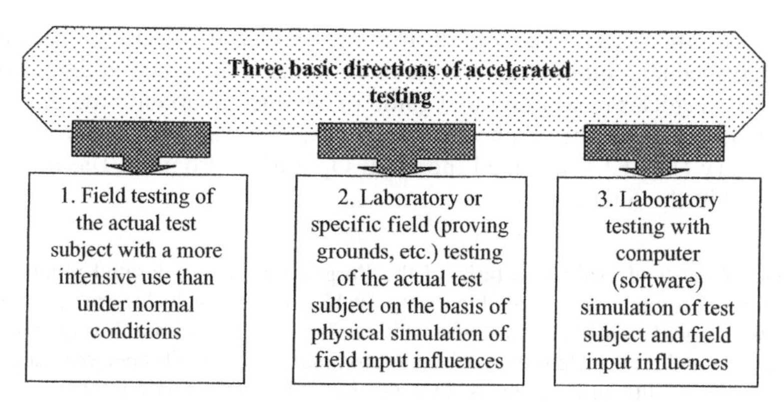

Figure 1.1. The basic directions of accelerated testing.

If one uses this car 18–20 or more hours per day, this represents true AT and provides enhanced durability research of this car's parameters of interest. This is a shorter nonoperating interval than normal (4–6 hours instead of the normal 18–20 hours). The results of this type of test are more accelerated than they would be under normal field conditions.

This type of AT is popular with such world-class known companies as Toyota and Honda; they call it "accelerated reliability testing" (ART). For example, in the report of the U.S. Department of Energy (DOE), INL/EXT 06-01262 [5], it was stated that

> A total of four Honda Civic hybrid electric vehicles (HEVs) have entered fleet and accelerated reliability testing since May 2002 in two fleets in Arizona. Two of the vehicles were driven 25,000 miles each (fleet testing), and the other two were driven approximately 160,000 miles each (accelerated reliability testing). One HEV reached 161,000 miles in February 2005, and the other 164,000 miles in April 2005. These two vehicles will have their fuel efficiencies retested on dynamometers (with and without air conditioning), and their batteries will be capacity tested. Fact sheets and maintenance logs for these vehicles give detailed information, such as miles driven, fuel economy, operations and maintenance requirements, operating costs, life-cycle costs, and any unique driving issues

Another example is cited by Frankfort et al. [6] in the *Final Report of the Field Operations Program Toyota RAV4 (NiMH) Accelerated Reliability Testing.* This field testing took place from June 1998 to June 30, 1999 corresponding to the Field Operation Program established by the U.S. DOE to implement electric vehicle activities dictated by the Electric and Hybrid Vehicle Research, Development, and Demonstration Act of 1976. The program's goals included evaluating electric vehicles in real-world applications and environments, advancing electric vehicle technologies, developing the infrastructure elements necessary to support significant electric vehicle use, and increasing the awareness and acceptance of electric vehicles. The program procedures included specific requirements for the operation, maintenance, and ownership of electric vehicles in addition to a guide to conduct an accelerated reliability test. Personnel of the Idaho National Engineering and Environmental Laboratory (INEEL) managed the Field Operation Program. The following appeared in the final report:

> One of the field evaluation tasks of the Program is the accelerated reliability testing of commercially available electric vehicles. These vehicles are operated with the goal of driving each test vehicle 25,000 miles within 1 year. Since the normal fleet vehicle is only driven approximately 6,000 miles per year, accelerated reliability testing allows an accelerated life-cycle analysis of vehicles. Driving is done on public roads in a random manner that simulates normal operation.

This report summarizes the ART of three nickel metal hydride (NiMH) equipped Toyota RAV4 electric vehicles by the Field Operation Program and its testing partner, Southern California Edison (SCE). The three vehicles were assigned to SCE's Electric Vehicle Technical Center located in Pomona, California. The report adds ". . . To accumulate 25,000 miles within 1 year of testing, SCE assigned the vehicle to employees with long commutes that lived within the vehicles' maximum range. Occasionally, the normal drivers did not use their vehicles because of vacation or business travel. In that case, SCE attempted to find other personnel to continue the test."

A profile of the vehicle's users from Frankfort et al. is presented in Table 1.1. This is a useful work in many areas, but practice shows that this type of field testing is not applicable for an accurate reliability, durability, and maintainability prediction, by this book's definition and methodology, for several reasons:

1. Many years of field testing for several specimens are necessary to gather initial information for an accurate quality, reliability, and maintainability prediction during a given period. This book proposes a methodology and equipment that can accomplish this at a much faster pace and at a lower cost.

TABLE 1.1 Profile of Vehicle Users [2]

Vehicle Number	1	2	3
Normal round-trip commute (miles)	60	120	82
Other daily mileage—lunch, business, and so on (miles)	50 (one to two times per week)	20–30	10–40
Average weekly mileage	410	501	524

2. An industrial company usually changes the design and manufacturing process of its product every few years, not always on a regular basis. In this situation, test results of a previous model's testing have only relative usefulness, but they are not directly applicable.
3. Field testing can only provide incomplete initial information for solving problems related to an integrated system of quality, reliability, and maintainability as will be shown in this book.
4. A combination of laboratory and field testing is more useful for finding a solution to these and many other problems.

These problems show that after describing its field testing, and the tests of the above-mentioned models, Toyota still had many problems in reliability and safety that led to recalls, complaints, degradation, and failures. Consider one more example from Toyota's practice. The report *Hybrid Electric Vehicle End-of-Life Testing on Honda Insight* [7] stated that "Two model year 2004 Toyota Prius hybrid electric vehicles (HEVs) entered ART in one fleet in Arizona during November 2003. Each vehicle will be driven 160,000 miles. After reaching 160,000 miles each, the two Prius HEVs will have their fuel efficiencies retested on dynamometers (with and without air conditioning), and their batteries will be capacity tested. All sheets and maintenance logs for these vehicles give detailed information such as miles driven, fuel economy, operations and maintenance requirements, operating costs, life-cycle costs, and any unique driving issues . . ."

In fact, this was an accelerated field test performed by professional drivers for short periods of time (maximum of 2–3 years). This testing cannot provide the necessary information for an accurate prediction of reliability, life cycle costs, and maintenance requirements during a real service life since it does not take into account the following interactions during the service life of the car:

- The corrosion process and other output parameters, as well as input influences that act during a vehicle's service life
- The effects of the operators' (customers') influences on the vehicle's reliability because it was used by professional drivers during the above testing
- The effects of other real-life problems

Mercedes-Benz calls similar testing "durability testing." For example, the test program for the new Mercedes-Benz C-Class stated in Reference 8 ". . . For the real-life test that involved 280 vehicles they were exposed to a wide range of climatic and topographical conditions. Particularly significant testing was carried out in Finland, Germany, Dubai, and Namibia. The program included tough 'Heide' endurance testing for newly developed cars, equivalent to 300,000 km (186,000 mi) of everyday driving by a typical Mercedes customer. Every kilometer of this endurance test is around 150 times more intensive than normal driving on the road, according to Mercedes. Data gathered are used to control test rigs for chassis durability testing . . ."

A similar situation existed with a Ford Otosan durability testing in 2007. It was stated in the article "LMS Supports Ford Otosan in Developing Accelerated Durability Testing" [9] that "Ford Otosan and LMS engineers developed a compressed durability testing cycle for Ford Otosan's new Cargo truck. LMS engineers performed dedicated data collection, applied extensive load data processing techniques, and developed a 6-to-8-week test track sequence and 4-week accelerated rig test scenario that matched the fatigue damage generated by 1.2 million km of road driving."

Companies specializing in testing areas often find similar situations. For example, in the note about MIRA's (MIRA Ltd.) durability testing [10], it was stated in the Proving Ground Durability Circuits & Features that MIRA's proving ground is used extensively for accelerated durability testing (ADT) on the whole vehicle in addition to these traditional durability surfaces:

- Belgian pave
- Corrugations
- Resonance road
- Stone road

Many other proving ground surfaces and features serve to build up a track base equivalent to real-world road conditions.

Referring to different sources about proving ground testing published 30–40 years ago, in The Nevada Automotive Test Center (NATS) [11], in Kyle and Harrison [12], and in others, one will see that similar proving ground stress testing was used for obtaining initial information for machinery strength and fatigue. Professionals understand that this type of testing cannot offer the information for an accurate prediction of a test subject's durability and reliability because it does not take into account the

- Environmental factors (temperature, humidity, pollution, and sun exposure) and their effect on a product's durability and reliability during its warranty period or service life
- Random character of real input influences that affect a product's performance in the field
- The type of data simulation required for system control is not capable of being simulated during a proving ground test
- Many other real-life tests cannot be simulated on the proving ground

Many authors often ignore the above-mentioned points, especially in publications of companies that design, produce, and use the equipment or methodology for AT. Therefore, this flawed reasoning occurs in many publications that relate to reliability or durability testing.

1.2.2 The Second Approach

The second approach is to use accelerated stress testing (AST). For example, if one conducts research upon or tests the actual car using a simulation of the field input influences with special equipment (vibration test equipment, test chambers, and proving grounds), then the level of the car (or other product) loading is higher than it is in normal usage. In this case, there is a physical simulation of the field inputs on the actual test subject. In most instances, there is a separate simulation of each of the field input influences such as temperature, humidity, sun exposure, pollution, or several of the many field inputs. Therefore, this type of testing does not offer the possibility of obtaining an accurate quality and reliability prediction and of conducting accelerated development.

The level of inaccuracy of this prediction depends upon the level of inaccuracy of the simulation of field input influences, safety problems, and human factors. More details for this situation are provided later in this book.

1.2.3 The Third Approach

This approach relies on using a computer (software) simulation or analytical/statistical methods. A computer simulation is a "computer program that attempts to simulate an abstract model of a particular system" (Wikipedia).

Computer simulations have become a useful part of the mathematical modeling of many natural systems in physics, chemistry, and engineering. They help to gain insight into the operation of those systems. Wikipedia classifies computer models according to several criteria:

- Stochastic or deterministic (and as a special case of deterministic, chaotic)
- Steady-state dynamic
- Continuous or discrete (and as an important special case of discrete, discrete events, or discrete event models)
- Local or distributed

Simulation results are different from actual results.

A simulated test subject is different from the actual test subject, and simulated field input influences are different from the actual field input influences. The results of a reliability and quality prediction and evaluation using computer (software) simulation show a greater difference from the appropriate field results than results from using methods 1 and 2 mentioned earlier. This is attributable to a greater difference from a real field environment. Two examples show the economics of software simulation: "Standish Group, a technology consultancy, estimated, that 30% of all software projects are cancelled, nearly half come in over budget, 60% are considered failures by the organizations that initiated them, and nine out of ten come in late. A 2002 study by the American's National Institute of Standards (NIST), a government research body, found that software errors cost the American economy $59.5 billion annually" [13]. Currently, this approach is in the early stages of development and is more popular with professionals in the software development field.

It is often popular with customers because it is less expensive and less complicated than methods 1 and 2. The following tests are included in computer simulation methods: fixed duration, sequential, test to failure, success test, reliability demonstration, reliability growth/improvement, or others. This book does not address these types of testing. They include qualitative accelerated tests, quantitative accelerated tests, or quantitative time and event compressed testing.

1.2.4 The Second Approach: A More Detailed Review

One common example applies to the following discussion. When Boeing wanted to produce a sensor system for satellites with minimum expenditures, the company specialists decided not to conduct the subsystem testing until they mounted the subsystems with more complicated components to provide testing for the entire block. This approach required more funding than planned. The subsystems that had not been tested had failures that led to the failures of completed blocks, which then had to be dismantled and reassembled [14]. This approach complicates the problem of finding the root cause of failures. Therefore, costs were higher and more time was required to complete testing and reassembly.

There are several approaches to AT, and it is important to differentiate among them because each approach needs its specific techniques and equipment. The effectiveness of these approaches sometimes depends upon the complexity of the product. It sometimes depends on the complexity of the operating conditions for the product, including the need for one or several climatic zones of usage and indoor/outdoor usage. For example, electrode testing requires simpler techniques and equipment than engine testing. Devices or vehicles having indoor applications do not need solar radiation testing. The testing approach for devices that operate for a short period of time needs to be different from the testing approach for devices that operate for a long period of time for greater testing effectiveness. In general, there are three basic methodological concepts to the second approach of AT. Let us briefly describe them:

1. Accelerate the test by reducing the time between work cycles. Many products experience brief usage during a year. Therefore, one can test them by ignoring the time between work cycles and the time with minimal loading that has no influence on the product degradation or failure process. For example, most farm machinery, such as harvesters and fertilizer applicators, have seasonable work schedule. Harvesters work only several weeks during a year. If this work occurs 24 hours a day with an average field loading, one accumulates the equivalent of 8–10 years of

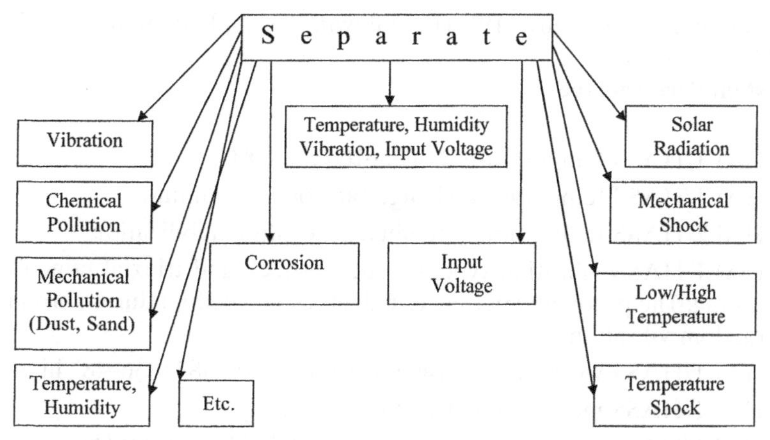

Figure 1.2. Example of separate types of simulation and accelerated stress testing during design and manufacturing.

field operation in several months. The same principle relates to aircraft testing. However, this approach ignores the degradation process during storage time (corrosion and other environmental influences, as well as its contact with mechanical and other factors). Therefore, for reliability/durability testing, one has to take into account stress from the above-mentioned factors.

2. Accelerate the test using stresses. Most industrial companies use this approach to testing (Fig. 1.2).
3. Acceleration through high-level stresses. This method involves increasing the intensity of stress factors. Stress factors accelerate a product's degradation process in comparison with its normal usage. There are many types of higher-level stresses that occur under normal usage: higher loading (tension), higher frequencies and amplitudes of vibration, and a higher rate of change in input influences (temperature, humidity, higher concentration of chemical pollutions and gases, higher air pressure, higher voltage, higher fog, and dew). This approach is often used and is beneficial if the stress does not exceed the given limit. This approach is relatively simple and effective for raw materials and simple components. But often, it applies stresses that are higher than the field stresses and for complicated components or for the entire equipment. The above-mentioned testing approach relates to most types of current AT, including HALT [15], AA [16], and [17], and HASS [18]. Often, these tests are incorrectly called "ADT" or durability testing.

HALT is a process that uses a high-stress approach in order to discover design limitations of products. HALT usually includes two parameters:

vibration and temperature [19]. The following example demonstrates HALT and HASS testing.

System Performance

- HALT/HASS temperature range: –100 to 200°C
- HALT/HASS temperature change rate: 60° per minute
- HALT/HASS temperature stability: ±1°C after stabilization
- HALT/HASS vibration type: repetitive shock and triaxial noncoherent testing: The product experiences 6 degrees of freedom during broadband random vibration
- HALT/HASS working area ranges: 30–48" × 40–48" × 36–48" high
- HALT/HASS maximum vibration power: 60 g
- HALT/HASS frequency ranges: 5–5000 Hz and 5–20,000 Hz

HASS: Apply high stress levels to reduce the reliability stress screening (RSS) time as much as possible. However, do not exceed the specifications of the operational limits of components unless it is a management decision. RSS is a reliability screening process using environmental and/or operational stresses as means of detecting flaws by inducing them as detectable failures.

Combined stresses, combined temperature change, and vibration or bumps are especially efficient for stimulating flaws as failures. Before starting the RSS with its high stress levels, the operational limits for the assemblies must be determined. Furthermore, by repeating the RSS cycle a large number of times, it must be proven that the planned RSS cycles reduce the lifetime of the assemblies to an insignificant degree, even during repeated RSS due to the repairs of induced failures.

Perform the screening process under consideration at the subsystem level of the manufacturing system. The planning includes a number of steps:

Step 1. Specify the maximum allowable fraction of weak assemblies. Perform this step by examining the requirements for the end product including the printed board assembly (PBA) as a subsystem. In this case, no other parts of the end product contribute to early failures. Therefore, the acceptable fraction of weak assemblies that remained after reliability screening is the same for the end product and the PBA.

Step 2. Evaluate the actual fraction of weak assemblies. Calculations in Steps 1 and 2 are required. In this case, there are two rogue component classes: integrated circuits (ICs) and power transistors. It is necessary to reduce the fraction of early failures by an order of magnitude before including the PBA in the end product.

Step 3. Consider the stress conditions. First, identify the flaws that one expects to induce during the assembly process.

For ICs, the following may appear:

- Partial damage of the internal dielectric barriers due to electrostatic discharge (ESD) in the production handling
- Formation of cracks in the plastic encapsulation due to a difficult manual production process

Transistors may appear to have the formation of cracks in the plastic encapsulation due to a difficult manual production process.

Users of the above-mentioned approaches, especially AA, claim that after several days of testing, they can obtain results equivalent to several years of field results. They call these approaches reliability and durability testing. To achieve an accurate field simulation, one has to carefully use and truly understand a high level of acceleration.

Practically, if one wants to obtain accurate initial information for an accurate prediction of product reliability or durability, then one has to take into account that the most current test equipment may only be able to simulate one or a few of the field inputs. But many actual environmental influences such as temperature, humidity, chemical and dust pollution, and sun radiation act simultaneously with many mechanical, electrical, and other influences. Most of the current test equipment simulate these actions (or only a part of the real input complex) separately. Therefore, users tend to implement these parts of the environmental influences separately. As a result, this equipment is not appropriate for accelerated durability, reliability, or environmental testing.

This circumstance applies to the methodology and equipment that simulates not only one type of input influence (e.g., temperature) but also two (temperature and vibration), and three types (temperature, vibration, and humidity) of input influences. The same is true for mechanical and other types of testing. The companies that design and manufacture equipment for AT and especially the users of this equipment should ask themselves, "What can we evaluate or predict after testing? How extensively can we simulate the field environment?" If one wants to cause the product to fail more quickly than in the field, then it is necessary to ask a second question: "How is the product degradation process in the field similar to the degradation process during AT?"

Those who have a high level of professional experience in practical AT for product development and reliability/durability prediction will agree that one cannot accurately predict product durability and reliability if only a part of the field environment is simulated. To solve this problem, some who work in the area of accelerated product development use HALT, AA, and other types of AST with a high level of stress to quickly obtain test results. So, in a few days, one can obtain test results that compare adequately to a few years of use in the field. For example, "When a 10-year life test can be reduced to 4 days, you have time to improve reliability while lowering cost" [16]. This method is simple because the intense stress applied for a short time period is sufficient to determine the results quickly. Also, the equipment is less expensive.

However, what is the quality of these results? The quality of these results is poor. The basic reason for poor results is that by using this approach, one cannot obtain the physics-of-degradation (or the chemistry degradation) mechanism that would be similar to the one obtained in the field physics-of-degradation (or the chemistry degradation) mechanism. Therefore, this approach cannot provide a sufficient correlation between ART/ADT results and field results. If one takes measurements of the time of failure during an AT, it is still impossible to know how accurately these measurements represent the time of failure taken in the field. Moreover, testing may destroy the product during ART/ADT or may show failures in the laboratory that do not occur in the field, because the level of temperature and vibration is higher than in real life.

Today the automotive, aerospace, aircraft, electronic, farm machinery, and many other industries often utilize this approach with minimal success. The accelerated test results (reliability and maintainability) are different from the field results. Consequently, product development, reducing complaints, and recall facilitation need more time, incurring an associated delay in product availability, a decrease in sales numbers, higher production costs, and a decrease in customer satisfaction. Most highly educated professionals in these areas monitor the stress level very carefully and use the physics-of-degradation mechanism as a criterion of simulation. To conduct AT for a unit of electronic equipment, they often combine a minimum of three parameters in the test chambers simultaneously with a minimum level of stress. These parameters include temperature (humidity), multiaxis vibration, and input voltage.

More negative aspects of this approach follow. Those who have practical experience in AT know that one cannot estimate the acceleration coefficient of the whole product (car or computer) or the unit during a test of the whole product or its units (which consist of different assemblies, and each assembly has a different acceleration coefficient). Therefore, if we know the time to failure for different components of the whole product during an AST, we still cannot accurately estimate the time to failure and other reliability parameters of the whole product or unit during real life. This is true because the ratio of the acceleration coefficients (the ratio between the AT time and the field time) for the failures of the test subject elements varies too widely.

This relates to the situation shown in this book when different subunits interact with each other and cause possible failures to disappear. However, additional new subunit failures also appear. Thus, we have nonlinear combinations. For example, for different parts of caterpillar units, the ratio of failure acceleration varied from 17 to 94 [20]. In this case, it is impossible to find the reliability parameters for the whole caterpillar device. One of the research conclusions was that "This confirms the practical impossibility of selecting the regime of AST that will give the ratio of loading of all parts and units of the complete device that will correspond to field" [20].

J.T. Kyle and H.P. Harrison [12] wrote more than 40 years ago that "... Absence of tensions with small field amplitude by AT gives an error in the estimation of the ratio of the number of work hours in the field and on

AST conditions, for evaluation of details under high tension and fluctuation of load." Therefore, this approach to AT is one of the basic reasons why one cannot obtain a sufficient correlation between the AT results and the field results.

Test equipment (chambers) for the automotive industry usually includes a volume from 0.5 to 500 or more cubic meters. Accelerated environmental testing (AET) results of electronics, automobiles, aerospace, aircraft, and other types of products all experience similar negative results.

Specific areas of industry have specific types of AT. For example, AT of farm machinery can be

- In the field
- On special experimental proving grounds
- On special test equipment in the laboratory
- Any combination of the above tests

It can be a complex testing of entire machines or testing of components or combinations of components.

Usually, complex testing of an entire machine occurs in the field and at proving grounds. Components and their combinations are tested at proving grounds and on laboratory equipment. At proving grounds, it moves the whole machine but usually tests only the components of the machine, mostly the body. In the field, one can also test new or modern components that are components of entire machines. Methodological aspects of current AT in the laboratory vary depending on the specifics of test subjects and operating conditions. In general, laboratory AT uses various methods of loading such as

- Periodic and constant amplitude loading
- Block-program stepwise loading
- Maximum stress loading
- Maximum simulation of basic field loads in simultaneous combinations

It is important to consider a load process while analyzing field environments and testing performances, especially when stress testing (AST) is used. One has to identify and evaluate different levels of real-life input influences (loads) and how to account for them when performing an accelerated test. In this case, we classify accelerated tests as constant stress, step stress, cycling stress, or random stress. The highest level is random stress, because it is closer to the real world. In the real world, all loads for mobile equipment, as well as many loads for stationary equipment, have a random character. A field simulation using other types of stress is not accurate, but it has a lower cost. Often people prefer lower cost simulations and tests, but they ignore the consequent increase in costs for the subsequent work during design and manufacturing. If they would take this into account, they would understand that a less expensive test in actuality becomes more expensive and produces more problems and delays during design and manufacturing phases. During testing, one cannot find the real-world degradation and failures and accurately predict field reliability, durability, recalls, time to market, and the cost of maintenance. The least expensive test is the one that uses constant stress, and the constant stress test causes more problems for the subsequent processes.

There are two possible stress loading scenarios: loads in which the stress is time independent and loads in which the stress is time dependent. For a mathematical analysis, models and assumptions vary depending on the relationship between stress and time. Similar to the discussion in the previous paragraph, time-independent stress and loading is the cheapest and simplest to conduct but becomes more expensive and needs more time for subsequent design, research, and manufacturing processes.

In Figure 1.3, one can see the basic reasons that AST often cannot help to accurately predict reliability and durability.

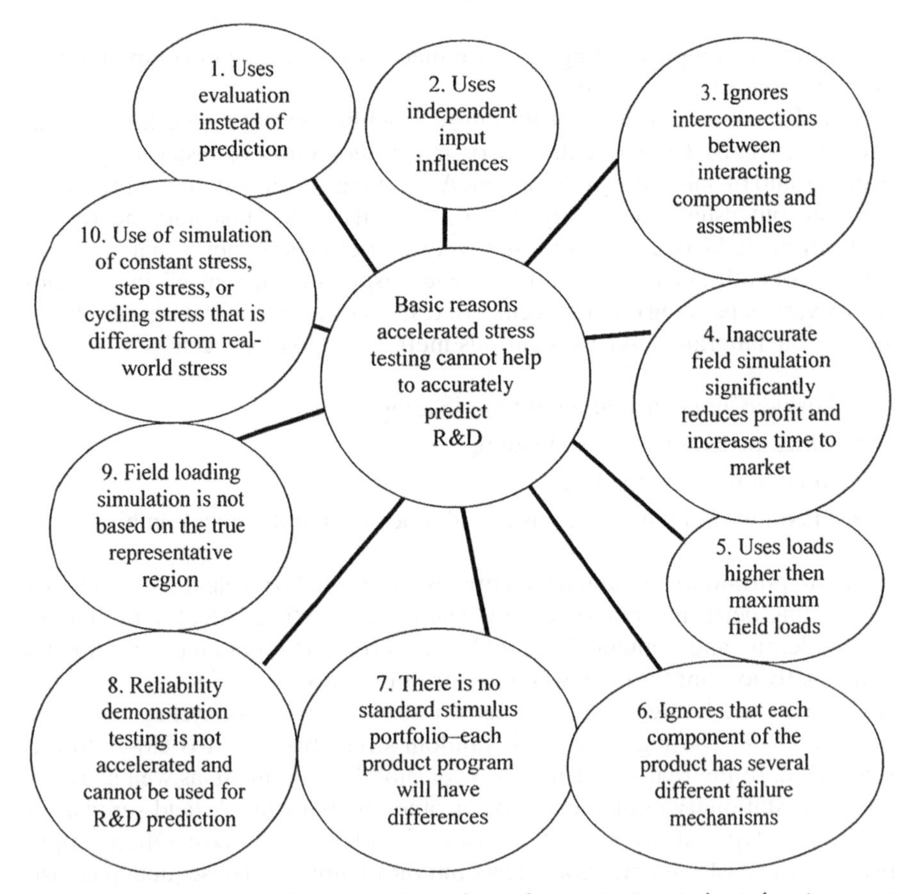

Figure 1.3. Reasons one-dimensional accelerated stress testing makes often incorrect predictions.

AST requires the extensive use of universal and specific test equipment and proving grounds for automobiles, tractors, tanks, farm machinery, and off-highway machinery on concrete and other surfaces. Usually, a number of tracks/surfaces exist in one particular section of the proving ground that is equipped with a drainage system. The procedure for testing under these conditions follows from the principle of a substantial increase in the frequency of application of the maximum working loads. For an accelerated environmental stress test, the increase in temperature, humidity, and/or air pressure makes sense.

The AT of vehicles, tractors, tanks, farm machinery, off-highway vehicles, and other mobile products occurs on specially equipped proving grounds designated for

- Wheeled machine frames by running them under various conditions along a racetrack set with obstacles
- Investigation of the coupling properties of wheeled machines, tool carriers, and wheeled tractors on a concrete track
- Testing of tanks, tractors, agricultural machines, and other mobile vehicles in abrasive media (in bath)

To improve working conditions, in addition to more rationally using the testing time and creating a higher level of testing conditions, one can use an automatic system of control. Usually, this control system includes the following basic components:

- A system to automatically drive a machine along the proving ground track and operate its attachments
- A remote control system for the unit's operating schedule
- A system for the prevention of damage

AST is also applicable to laboratory equipment (universal and specific) found in proving ground test centers. These are different types of vibration equipment, dynamometers, and test chambers. When this equipment is used for an accelerated test for engines of mobile products, the permissible limits of wear on components such as cylinders, pistons, connecting rods, and crank assembles can be determined quickly. Artificially increasing the dust content of the intake air and introducing solid particles into the crankcase oils accelerates engine component wear. These current methods and equipment for AT simulate primarily separate components or subcomponents of field input influences for the real field situation. Chapter 5 describes the testing equipment in more detail.

Some authors have believed accelerated durability testing or durability testing are related to this category of AT. In Reference 21, P. Briskman considered a cycling stress test to be a durability test. This fails to take into account the real-life input influences on the test subject. For example, the Bodycote Testing Group [22] considered environmental testing to be durability testing. Another example, taken from the article "Full Vehicle Durability" from the RGA Research Corporation, shows that "... There are six test tracks with over 30 different types of surfaces available for full vehicle durability tests." But proving ground testing is not accurate durability testing as previously described. In one more example, C.E. Tracy et al. [23] considered ADT of electrochromic windows. They wrote: "... The samples inside the chamber were tested under a matrix of different conditions. These conditions include cycling at different temperatures (65, 85, and 107°C) under irradiance, cycling versus no-cycling under the same irradiance and temperature, testing with different voltage waveforms and duty cycles with the same irradiance and temperature, cycling under various filtered irradiance intensities, and simple thermal exposure with no irradiance or cycling." The above-mentioned citation equating a proving ground test to a durability test is another example of a misconception of what "durability test" or "reliability test" actually means. As one can see from the above, the cycling test is not an "accelerated durability test" because in a field environment, a change of input and output parameters has a random character, but not in cycling.

The use of reliability and durability testing terms in the scientific literature often misleads practical engineers. The term "reliability testing" ("ART," durability testing, ADT) is often blocked out by other words or phrases that change its meaning. This happens most of the time in theoretical discussions, and sometimes it occurs in standards. For example, "accelerated reliability demonstration," "acceleration of quantitative reliability tests," "acceleration reliability compliance or evaluation tests," "acceleration reliability growth tests," and other terms are inappropriate to represent ART.

1.2.5 Durability Testing of Medical Devices

The application of ART/ADT in this book relates not only to mobile equipment but also to stationary equipment. A current situation in AT, especially durability testing, for one group of stationary equipment, medical devices, will be examined. There are many publications [24, 25, 31] that address building state-of-the-art durability and fatigue testing devices for biomaterials, engineered tissues, and medical devices such as stents, grafts, orthopedic joints, and others. Real-time computer control, electrodynamic mover technology, and laser measurement and control options are just some of the features of these test systems. Mechanical testing is a critical step in the development of medical devices. Medical Device Testing (MDT) Services provides a wide variety of mechanical tests from the U.S. Food and Drug Administration (FDA), American Society for Testing and Materials (ASTM) International, and International Organization for Standardization (ISO) medical testing requirements. Some of these tests are

- Stent testing
- Graft testing
- Intravascular medical device testing
- Orthopedic medical device testing
- Dental implant and materials testing
- Biomaterials evaluation testing

However, medical device durability testing has similar negative features as described earlier for the AT of mobile products. Let us begin with durability testing formulation. For example, as the Orthopedic and Rehabilitation Devices Panel of the Medical Devices Advisory Committee [28] described: "The durability testing of medical devices should involve cyclic loading testing several loading models (e.g., flexion/extension, lateral bending, and axial rotation) and involve a maximum of six samples of the worst-case construct out to ten million cycles. This test can incorporate all testing directions into one test or conduct separate tests for each loading mode. Durability testing establishes loading direction, stability of the device, and the potential to cause wear. Clinical justification for the loads and angles chosen should be provided."

As we can see, the above-mentioned test requirements have the same negative features as the description of mobile product tests for farm machinery or the automotive industry product tests that were created 50 or more years ago. In industrial areas, professionals came to the conclusion many years ago that this test could not offer satisfactory initial information for an accurate prediction of durability, reliability, and maintainability in a real-world situation. This is true since this type of testing does not take into account real factors such as

- The speed of change for the real stress processes
- A random character of loading in a real situation
- A simultaneous combination of many factors that influence durability in a real environment

Additionally, this type of testing does not provide an accurate simulation of the whole complex of factors in a real field situation. For example, a heart attack is often the result of a sudden high stress, not step-by-step stresses.

"The 10 million cycles" approach is one used for metals (or other materials). However, it should not be used for devices constructed from these metals because this approach does not take into account local concentrations of tension within these devices.

Consider in more detail some examples of durability testing of vascular stents. The 2005 FDA Guidance document [29] outlines durability testing requirements for vascular stents for use in the United States. According to this document, the primary purpose of testing the materials and structure of a device is to provide accurate initial information to accurately predict the performance of the device during its intended use. Apply finite element analysis (FEA) models to determine regions of high stress or strain in a design under specific boundary conditions. A major variable in properly preparing the finite element model is the accurate measurement of properties of the representative materials, including all processes and treatments on appropriately sized material. Reliable test data on relevant samples of material for the device are critical for a successful FEA model. To properly test the device, it is important to identify potential failure modes for the device during normal use. Typically, an early step in the creation of a new device design is to identify potential failure modes that could occur with the device to determine the effects of these modes. This is the failure modes and effects analysis (FMEA).

For example, a typical FMEA for a balloon-expandable metallic stent may include the following [30]:

- The stent slips off of the delivery catheter prior to inflation.
- The stent snags the vessel during transition to the deployment site.
- The stent exhibits structural failure, that is, breakage of a strut
  - Due to crimping on the balloon
  - Due to expansion of the balloon
  - Due to cyclic distension of the vessel resulting from the pressure change caused by each heartbeat.
- The vessel may close due to insufficient radial strength of the stent.

Additional concerns arise when using the stent in the peripheral vascular system, such as in the femoral artery. This adds additional loading conditions that may seriously affect the performance of the device and may include stent structural failure due to

- Cyclic flexion of the vessel due to regular motion
- Cyclic extension/compression of the vessel due to regular motion
- Compression of the vessel due to regular motion
- Cyclic rotation of the vessel due to regular motion

Once one identifies the failure modes, the next step is to identify the physical tests that are necessary to access the device.

We can see that this is similar to a misconception of a cycling loading test for a durability prediction in engineering. Any surface modification may influence or change corrosion resistance and fatigue life characteristics. Polymeric coatings may crack, tear, slip off, and flake off the stent due to compression of the deployment balloon, deployment expansion, or pulsation distension fatigue. As a result, an embolism and/or thrombosis may occur and the susceptibility to corrosion may also increase. Complete device tests are used to determine the acute failure for durability evaluation of the coated stent during deployment and the chronic failure modes for the pulsating fatigue of the device for the duration of 10 years of equivalent cyclic loading. Most of these tests used stent fatigue testing equipment such as those shown in Figures 1.4 and 1.5.

Accelerated durability testing of a coated stent has to take into account the material limitations of the coating. For example, care is necessary when increasing test frequencies to ensure that the physical properties of the polymer

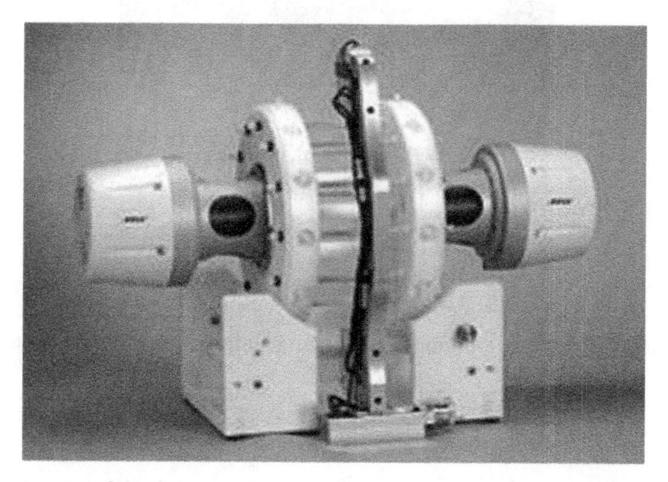

Figure 1.4. Bose® 9100 series accelerated stent durability test instrument [30].

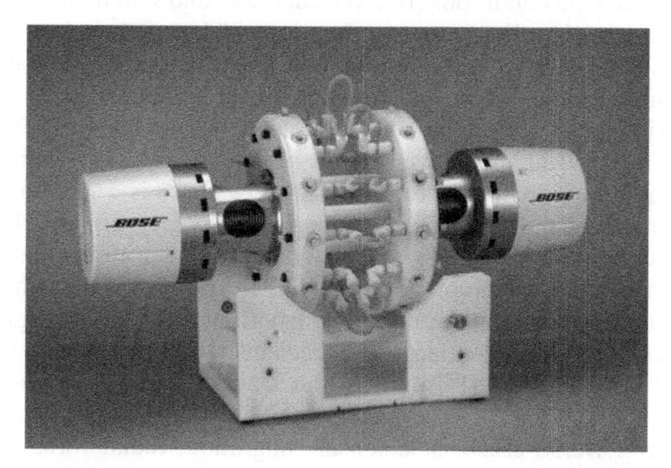

Figure 1.5. ElectroForce 9110-12 stent/graft test instrument (Bose).

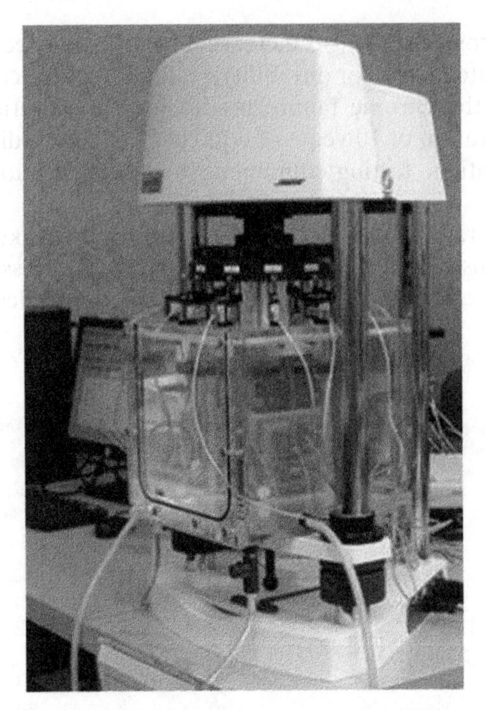

Figure 1.6. ElectroForce® multiple-specimen stent durability testing system [30].

coating do not exceed their glass transition zone. But the basic problem is that this "durability" testing cannot provide sufficient information for an accurate prediction of durability or reliability because the test conditions do not accurately simulate field conditions (random character and simultaneous combination) as shown earlier. The following chapters will demonstrate how to achieve better results by improving test setups. One improvement is the use of multiaxis durability testing equipment, which is shown in Figure 1.6.

Multiaxis fatigue (durability) test systems (such as the Bose 9100 series stent/graft test instruments) have become key components in a design process to bring these devices to market more quickly. There is a growing trend to treat other vascular diseases such as peripheral artery disease (PAD) and carotid artery disease (CAD) [31] with stent and stent–graft testing. The rapid growth of the stent (and its components; Fig. 1.7) and graft markets has created a variety of medical manufacturers needing both stent and graft fatigue testing. MDT Services routinely conduct these mechanical tests using testing methods per FDA 1545, ISO 25539-1, Endovascular Devices, and CE requirements:

- *Strength*: Burst, crush, flex, tensile, migration, radial force, and bond strength

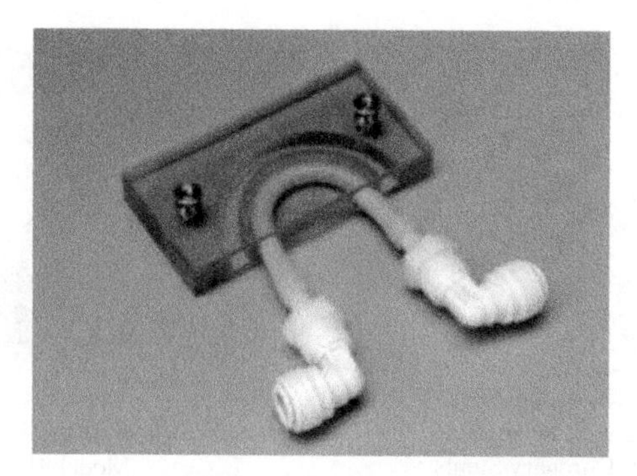

Figure 1.7. Bose fixed bend fixture for stent fatigue testing [25].

- *Stability*: Device pulsating fatigue (durability) testing and material fatigue testing for FEA analysis
- *Fatigue*: Radial, bending, torsion, and compression

Pulse-on-a-Bend Test, Coronary Artery Device [31] Regulatory bodies, including the FDA, have recently requested the performance of pulsating durability tests in a physiologically relevant geometry. For this purpose, MDT Services use Bose ElectroForce Systems (Fig. 1.8) Pulse-on-a-Bend testing equipment. This test can accommodate the following wide range of device sizes and test configurations:

- *Geometry*: Bend radius from 7 to 40 mm
- *Configuration*: 12 tubes, which are able to accommodate either single or overlapped devices
- *Tubes*: Custom dipped to accommodate small bend radii
- *Custom Laser*: Perpendicular measurements along the curve
- *Frequency*: 40–60 Hz, depending on the size of the device
- *Environment*: 37°C saline

Medical device tests provide testing for a wide variety of intravascular devices. Medical device submissions require many tests, including "durability testing to 10-year equivalent cycles." For intravascular devices such as stents or stent–grafts, this translates into 380–400 million fatigue (durability) cycles. Medical devices tests provide testing in the following areas:

- Mechanical test methods per ISO 25539-1
- Heart valve fatigue testing
- Pacemaker lead testing
- Wire fatigue testing
- New device testing

Model 9110-12 with Pulse on a Bend Fixture for Coronary Stents

Model 9130-8 SGT for Vascular Grafts

Model 9120 SGT for Abdominal Devices

Model 9140-20 SGT

Figure 1.8. ElectroForce stent/graft test instruments [26].

Many years ago, practical results of the cycle test method for farm machinery, the automotive industry, and others demonstrated that it alone is not sufficient to obtain the necessary initial information to predict durability accurately because this type of cycle test does not accurately simulate real environments. Dynatek Dalta Scientific Instruments [27] demonstrated that the development of a coating durability tester (CDT) responds to a belief within the industry that developers of drug-eluting and coated devices will soon be required to test the shedding of drug particles as a part of their product's durability evaluation. Dynatek's CDT adds the benefit of real-time evaluation of the device coating for the proven ADT of the small vascular stents prosthesis tester (SVP) and the large vascular stents prosthesis tester (LVP). "Stents have improved the treatment of coronary artery disease, with close to

100 percent penetration in the US, and 50–60 percent in the UK. The market for drug-eluting stents is expected to reach $5.5 billion by the end of 2005 and $6.3 billion by 2007" [27]. But accurate durability prediction for these stents is still not obtainable because their durability testing is an unsolved problem.

Finally, the choice of testing methods and equipment for medical devices has the same principal negative properties as were described for other types of tested equipment. Volunteer standards organizations such as ASTM, International Electrotechnical Commission (IEC), and others have sometimes reflected this issue in their standards. The FDA also based its requirements on these standards. Therefore, we can conclude that different types of AT offer different degrees of accuracy for the results in comparison with field results for the actual product. These tests cannot produce sufficient initial information for the accurate prediction of quality, reliability, durability, and maintainability in the field because test results do not correspond to actual field results.

1.3 FINANCIAL ASSESSMENT OF THE RISKS INVOLVED IN CREATING A TESTING PROGRAM

To optimize the investment in a testing program, a company must consider all financial cost benefits for the program. This is a very complicated process because ART interconnects with quality, reliability and maintainability, accuracy, costs, and many other factors. Therefore, the simplified approach shown here is similar to that described by Dodson and Johnson [32]. Failure costs and significant degradation followed by an abrupt decrease in quality, reliability, and maintainability consist in the following aspects shown in Figure 1.9. Additional costs attributable to early failure include warranty costs, stop shipping costs, recall costs, costs associated with additional complaints, loss of business, loss of good will and reputation, and retrofits. One method to quantify these costs would be to give a score to each element in the testing program

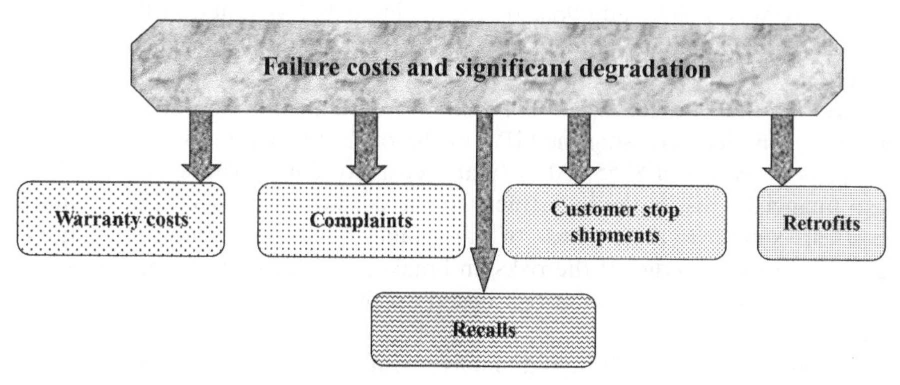

Figure 1.9. Failure cost components.

TABLE 1.2 Example of Test Program Scores [25]

Test Program Item	Score (GPA)
Understanding of customer requirements	B-3
Failure and significant degradation	A-4
FRACAS	C-2
Verification	C-2
Validation	D-1
Manufacturing	B-3
Overall program score	2.33

FRACAS, failure rate analysis and corrective action system.

TABLE 1.3 Example of Field Testing Performance [25]

Testing Performance Item	Cost ($)
Customer returns	8245
Customer stop shipments	0
Retrofits	761,000
Recalls	2420
Overall program costs	1,011,581

then to examine the product's comparative performance in the field. Auditors can perform this task utilizing a grade scale of A through F for each component of the program. Combine the grades for each component into a grade point average (GPA) for the program using four points for an A, three points for a B, and so on. Table 1.2 shows how to score a reliability testing program using this system. Table 1.3 shows how field performance is noted. All laboratory travel costs, paperwork, and research expenses incurred should also be included in the total cost. This can easily be in thousands of dollars. The company's monetary loss from recalls must also be added. It is evident how AT, particularly ART, offers financial benefits. For example, Toyota Motor Corporation, which had a reputation as a high-quality manufacturer, had 1.27 million recalls in 2005 and 986,000 recalls in 2006. (As noted in the Preface, recalls by this company jumped in 2009–2010 to 9 million cars and trucks.)

Figure 1.10 is a scatter chart of results of several programs. The slope of the trend line totals the loss when the testing program is not entirely completed. In this example, increasing the GPA of the overall program by a single point projects a savings of $755,000 in failure cost avoidance. These savings alone can financially justify the investment in the considered project. These failure costs are similar to the expense of water pipes bursting in a house. The homeowner has knowledge of the risks and makes the decision of whether to act on the risk or to tolerate the risk based on the costs involved to rectify the situation.

Another method for bringing management's attention to reliability is presenting the effects on corporate profits using the data shown in Table 1.4. The

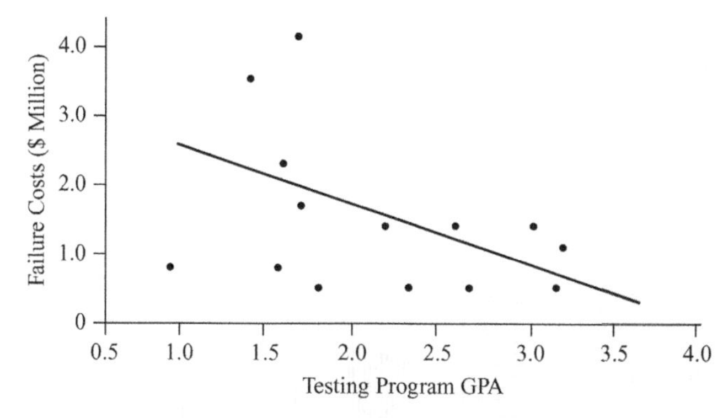

Figure 1.10. Testing program execution score versus failure costs [32].

TABLE 1.4 Vehicle Resale Value [45]

Vehicle (1998 Model Year)	Retail Value as of July 2001 ($)	Consumer Reports Reliability Rating*
A	8430	–45
B	9500	20
C	9725	18
D	11,150	25
E	11,150	30
F	13,315	–5
G	14,365	55
H	15,215	50

*The Consumer Reports scale is from –80 to 80, with –80 being the worst and 80 being the best.

data in Figure 1.11 show an example of the negative impact of an inadequate testing program. Some years ago, money was saved in the short term by taking a chance on a substandard testing program; however, Figure 1.11 reveals that it was not a profitable long-term investment decision. Just as termites can damage a house without the owner's knowledge, hidden low-reliability and low-quality producing programs result in poor decisions that increase losses or damage profits. Losses created by these hidden costs can be much greater than warranty costs. As an example of this concept, consider the automotive industry.

For an average vehicle manufactured by General Motors, Ford, or Chrysler in the 1998 model year, the company had to pay an average of $462 in repairs [33]. These automakers sell approximately 3 million vehicles in North America annually, which results in a total warranty bill of $1.386 billion. This amount may sound like a lot of money, but it is only a very minimal lower bound on the total cost of substandard reliability and quality. Table 1.4 illustrates the retail value of several 1998 model year vehicles sold at prices within a $500 range. As for lease vehicles, the manufacturer absorbs the $5715 difference in resale value between vehicles B and H. For purchased vehicles, the owner of

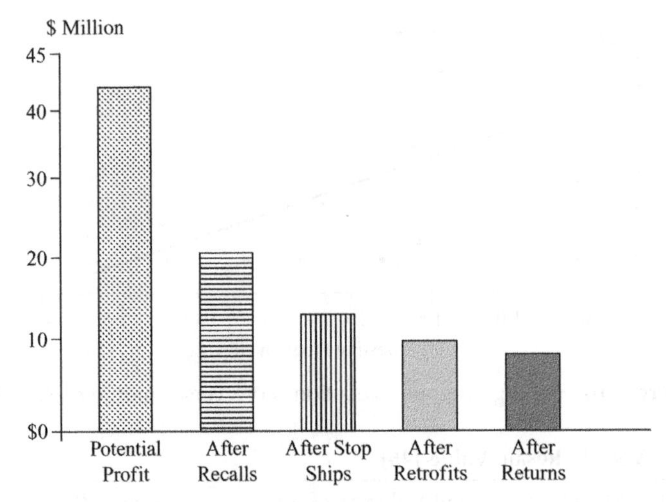

Figure 1.11. Effect of poor reliability and quality on company profits [32].

vehicle B absorbs the cost. But this does not mean that the manufacturer benefits. Reduced retail value becomes evident in the ability of the manufacturer to set prices for new vehicles and sell them. The manufacturer of vehicle H can charge more for new vehicles because its depreciation is slower. The sales for many of these midsize sedans topped 200,000 units, and the $5715 difference in resale value was valued at more than $1 billion annually.

1.4 COMMON PRINCIPLES OF ART AND ADT

Accelerated reliability and durability testing technology is the key factor for the accelerated development and improvement of quality, reliability, durability, maintainability, supportability, and availability for a product/process. This technology offers the possibility for accurate prediction of the above factors as well as a quick method for identifying the reasons for failure and degradation during a given time period (service life and warranty period). One can use this technology to quickly solve many other problems.

1.4.1 The Current Situation

Many engineers and managers use the term reliability testing or durability testing. Few people think that they actually conduct reliability *or* durability testing. Often, in fact, neither of them provides such testing. The basic reason is they do not clearly understand what this type of testing means and how it is different from other types of testing. The literature and practice show that professionals who use vibration testing, thermal shock testing, environmental testing, HALT, HASS [15], or other types of testing [35, 39, 40] often think that they provide reliability or durability testing. Companies from a wide range of industry specialization and size make this error.

TABLE 1.5 Stages, Purposes, and Contents of Reliability Test (from Table 3.1.1 in TOSHIBA, *Testing for Reliability*)

	Stage	Purpose	Content	Test
Semiconductor device development	Verify material, process, and basic design	Determine whether the material, process, and design rules allow the desired quality/reliability objectives and user specifications to be met	A metal electromigration, gate oxide film breakdown voltage, TDDB, MOS transistor hot carrier injection effect, failure rate for medium-and large-scale integrated circuits or products, new package environmental test, and so on.	Process test element groups (TEGs), function block TEGs, and so on.
		Determine whether the product satisfies design quality/reliability objectives and user specifications	Development verification tests (life test, environment test, etc.)	
TEG	Verify product reliability	Determine whether the product quality and reliability are at the prescribed levels	Structural analysis	Products
			Screening and the reliability monitoring (by product family)	Products Reliability TEGs

The following examples demonstrate this:

1. In the book *Testing for Reliability*, TOSHIBA asked [34], "1. What is reliability testing?" And the answer was "Toshiba testing follows the stages shown in Table 3.1.1." Table 1.5 shows no definition of reliability testing, no clear description of their contents, nor how one can provide this (separately or partially simultaneously or all simultaneously). In fact, this is an example of AST, but it is not ART.
2. IBM published the book *Reliability Testing and Product Qualification* [35].

The contents of this book include the following:

- IBM Microelectronics Quality System
- Development Process
  - Technology Feasibility
  - Technology Qualification
  - Product Qualification
- Quality and Customer Satisfaction
- Summary

Reliability testing is only in the title of the book. The "Table of Contents" does not even include reliability testing, what it is, or how to conduct it.

3. Surridge et al. in their paper "Accelerated Reliability Testing of InGaP/GaAs HBTs" wrote [36] "Our standard method of reliability testing is to perform a three temperatures (3T) accelerated test and predict the failure time at maximum junction temperature using an Arrhenius expression." Real-life factors include
   - More environmental factors than temperature only
   - More groups of input influences than environmental

   Therefore, this type of testing will yield a low correlation of test results to field results. The basic reason is that the simulation of the field situation is not accurate. As a result, this test will yield an inaccurate prediction of failures in the field.
4. The paper "What You Should Know about Reliability Testing" [37] consists of only one note on reliability testing: "A final stretch of testing cycles checks for reliability with four different tests: Bare board, flying probe, ICT (in-circuit tests), and functional." In this paper, there is nothing about the definition of reliability testing or how one conducts it.
5. In the book *Reliability of MEMS: Testing of Materials and Devices* [38], the authors compared the mechanical and other properties of thin films to the properties of micro-electro-mechanical systems (MEMS) devices, especially in terms of reliability. In the preface, the authors wrote: "At the present day, industrial products are distributed all over the world and used in a broad range of environments, making reliability evaluation of the products more important than ever." This is true. However, they describe strength and fatigue, as well as the evaluation of elastic properties. So in fact, the authors ignored that fatigue and strength testing is vastly different from reliability testing and cannot provide sufficient initial information for a reliability evaluation or prediction. They mention the standard IEC 62047-3 that specified a standard test piece for thin-film tensile testing for the accuracy and repeatability of a tensile testing machine. But this standard does not consider reliability testing and reliability evaluation.
6. In the paper "Product Reliability Testing and Data" [39], reliability testing is mentioned only in the title.

7. In *LMS Supports Ford Otosan in Developing Accelerated Durability Testing Cycles* [9], in the chapter entitled "Meeting 1.2 Million km Durability Requirements," it states that "... Ford Otosan decided to involve an external engineering partner to establish an accelerated proving ground test scenario that matches the fatigue damage that the truck experiences throughout its lifetime." "They displayed the results in a rain flow matrix format that showed how often events of particular amplitudes occur and extrapolated the data to estimate the damage generated by road testing over the full 1.2 million kilometers. This extrapolation was based on Ford Otosan's targeted weighting mix of 60% highway, 30% local roads, and 10% city driving. The goal was to achieve the full 1.2 million kilometers without any cracks in the major components of the vehicle." Referring to the research on proving ground testing that was published 40–50 years ago [12, 20] and others, one can see that similar techniques were published in those years. But professionals did not call this testing method durability testing, which would be wrong. They called this strength testing or fatigue testing, which is right. Professionals understood that proving ground testing does not take into account the random character of field input influences and environmental influences, such as temperature, humidity, pollution, and light exposure that act on the test subject over the years. A result of their action is corrosion and damage from other sources that the proving ground tests did not address.
8. The brochure *Solar Simulation Systems* reported: "Durability testing, also known as fatigue testing, is used to validate the aircraft's design structural service life (8334 hours) based on a demanding flight spectrum representing expected flight usage. The first such testing was conducted between July 2002 and April 2003. After an inspection through tear-down inspection, the test vehicle was subjected to a second service life testing (an additional 8334 hours) between August 2003 and October 2004" [40].

There are many other examples. Many companies write that they conduct ART or ADT (reliability testing or durability testing) but in fact reveal that they provide the following separately: vibration testing, air-to-air thermal shock testing, temperature cycling, temperature/humidity testing, highly accelerated stress testing (HAST), and high- and low-temperature storage tests (Fig. 1.2).

1.4.2 Improving the Situation

To understand the difference between ART and ADT or AT or reliability testing, one must examine their definitions. AT is a test that quickens the deterioration of the test subject. Reliability testing is testing performed during actual normal service that offers initial information for evaluating the measurement of reliability indicators during the test time.

ART or ADT is testing in which

- The physics (or chemistry)-of-degradation mechanism (or failure mechanism) is similar to this mechanism in the real world using a given criteria
- The measurement of reliability and durability indicators (time to failure, degradation, and service life) have a high correlation with these respective measurements in the real world using a given criteria.

Accelerated reliability and durability testing is connected to the stress process. Higher stress means a higher acceleration coefficient (ratio of time to failures in the field to time to failures during ART) and a lower correlation between field results and ART results. The common layout for accelerated reliability and ADT is in Figure 1.12.

The basic principles of accelerated reliability and ADT are

- A complex of laboratory testing and special field testing as shown in Figure 1.12

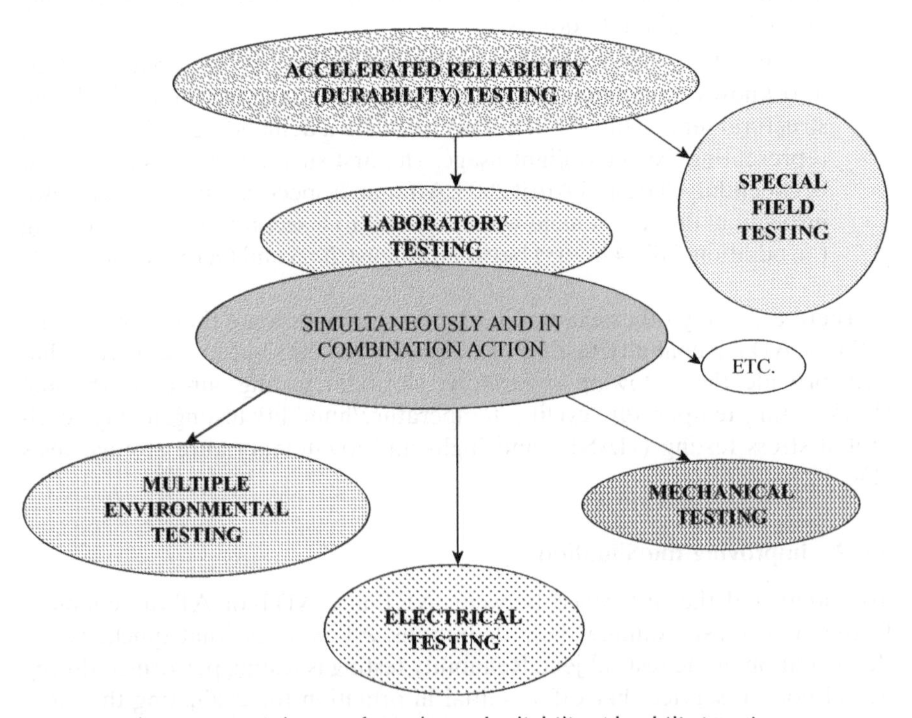

Figure 1.12. Scheme of accelerated reliability (durability) testing.

- That laboratory testing provides a simultaneous combination of a whole complex of multi-environmental tests, mechanical tests, and electrical tests
- That special field testing takes into account the factors that cannot be accurately simulated in the laboratory such as stability of the product's technological process and how the operator's reliability influences the test subject's reliability and durability
- It requires an accurate simulation of the whole complex of field input influences on the product as well as safety and human factors.

Durability is the ability of an object (material, subcomponent, component, or whole machine) to perform a given function under given conditions of use and maintenance until reaching a limiting state. The measurement of durability is its length of time (hours, months, or years) or its volume of work.

ART and ADT have the same basis—an accurate simulation of the field environment. Therefore, if there is no accurate simulation of the field situation, then there is no ART or ADT. The only difference is in the indices of these types of tests and the length of testing. For reliability, it is usually the mean time to failure, time between failures, and other parameters of interest. For durability, it is length of time or volume out of service. ART can be set for different lengths of time, that is, warranty period, 1 year, 2 years, or service life. Accelerated durability testing continues until the test subject is out of service.

A basic goal of the ART/ADT technology is to describe what and how one can rapidly obtain objective and accurate initial information for accurate prediction of quality, reliability, maintainability, availability, and other measurements during the product's design and manufacturing. The basic desirable results of ART are reduction of time and cost for product development and the ability to rapidly find causes for product degradation and failures. The main goal is quick elimination of failure and degradation root causes resulting in a rapid increase in the product's quality, reliability, maintainability, and durability. Currently, the basic causes for seldom conducting ART are the following:

- The knowledge of reliability testing, obtained from the literature, is often poor.
- Many professionals do not understand the specifics of ART or the need to conduct ART.
- The CEOs delegate their responsibilities in quality/reliability to a lower level without delegating the authority and providing the funding to implement ART/ADT.
- CEOs do not understand that it requires an initial investment over a period of time to obtain greater continuing benefits over a longer period of time.

- Lower-level managers involved in quality/reliability are not responsible for expenses and therefore do not have the funding authority to lead the process of reliability testing implementation.
- Not enough accredited professionals in industrial companies can describe to CEOs how to use reliability testing to save money, dramatically decrease recalls, and make the companies more successful in the market.
- The governmental research, development, and engineering centers in the Department of Defense and other federal government departments do not often require reliability and durability testing during acceptance tests. Therefore, they cannot accurately predict the reliability and durability of the tested product and they do not require ART of industrial companies. As a result, U.S. Army personnel in Iraq and Afghanistan often face dangerous situations due to early failures or defective equipment.

1.5 THE LEVEL OF USEFULNESS OF ART AND ADT

Professionals did not begin to understand one of the basic principles of ART until the late 1950s, namely, they did not understand that the interactions of different inputs on the product/process can result in overlooking significant failure (degradation) mechanisms. This new approach to testing, simultaneously subjecting the product to different inputs, was called Combined Environmental Reliability Testing (CERT) (DOD 3235.1H) [3]. MIL-STD-810F made use of this approach at about the same time. In 1981, the DOD CERT workshop confirmed that CERT was ready for implementation.

Then, this approach was undertaken in Japan [41], according to the description of CERT, as a concept and practice of combining effects of environmental factors (especially temperature + humidity, temperature + humidity + vibration + low pressure, and temperature + humidity + isolation).

However, they did not apply this approach for the development of ART using a whole complex of field inputs that interacted with and influenced each other on the test subject. Moreover, this is linked to the development of a more complicated program now known as "system of systems." The simulation of this concerted program in a laboratory will lead to a more accurate simulation of field input influences. It will provide a basis for obtaining, after ART, the initial information for the accurate prediction of quality, reliability, durability, and maintainability. The author's book [42] describes this development. Then, beginning in the 1990s, several other books by the author including *Accelerated Quality and Reliability Solutions* [18] and *Successful Accelerated Testing* [43], several patents, and dozens of articles, papers, and presentations describe this new direction of ART.

Dzekevich [44] from Raytheon wrote that there are some critical questions to answer when planning for reliability testing:

- What is the length of time to market?
- Is this a safety critical product where people's safety or life may be at stake upon a failure?
- What is the life expectancy of the product?
- Does the manufacturing process use accelerated test techniques to find process failures?
- How costly are field failures?
- Are there reliability problems with an existing product?

But if the type of testing is called ART, then it does not necessarily mean that this testing will automatically provide sufficient information for the initial estimation of the quality, reliability, durability, and maintainability parameters as a solution to problems.

ART has different approaches, and many professionals mean different things when they write or talk about reliability testing. The effectiveness of ART depends on the approach taken. With respect to these different approaches, accelerated reliability (durability) tests can be

- Helpful
- Minimally useful or useless
- Harmful

ART is helpful when methods and equipment provide a practical possibility for an accelerated evaluation and prediction of the product reliability and quality with a high degree of accuracy. Accuracy implies the conditions or quality are correct and exact. The level of accuracy of testing results depends on the accuracy of the simulation of the real-life full influences, including temperature, humidity, pollution, fluctuation, pressure, radiation, road conditions, and input voltage, while also including safety and human factors.

One achieves accuracy for field inputs to simulation when the output variables like loading, tensions, output voltage, amplitude and frequency of vibration, and corrosion in the laboratory differ from those under field conditions by no more than a given amount of divergence. The same is true for simulation of safety problems and human factors. If the resulting degradation and failures during ART correlate to those in the field, then accelerated prediction, development, and improvement of reliability and quality occur with minimum expenditure of time and cost. The basic concepts that account for an accurate simulation of field conditions are

- Maximum simulation of field conditions (including all and not only some stresses)
- Simulation of the whole 24-hour day, every day (except the weekend), but not including

  - Idle time (breaks, etc.)
  - Time with such low loading that it does not cause failures
- Accurate simulation and integration of each group of field conditions (full input influences, safety problems, and human factors)
- Accurate simulation of each group of input influences (multi-environmental, mechanical, electrical, etc.)
- Use of the degradation mechanism as a basic criterion for an accurate simulation of field conditions
- Consideration of a system of interacting components as those found in the field while taking into account their cumulative reaction (Fig. 1.13)
- Reproduction of a complete range of field schedules and maintenance (repair)
- Maintaining a proper balance between field and laboratory testing
- Simultaneous simulation of each input influence necessary to accurately replicate field conditions. For example, pollution consisting of chemical air pollution and mechanical (dust and sand) air pollution must be simulated simultaneously.

Current testing methods and equipment are seldom helpful in conducting ART because they do not produce the desired accuracy. Basic problems with implementing the "classical" accelerated life testing (ALT) relate to the second and sometimes to the third group. In Figure 1.14, one can see that ART is minimally useful or useless if

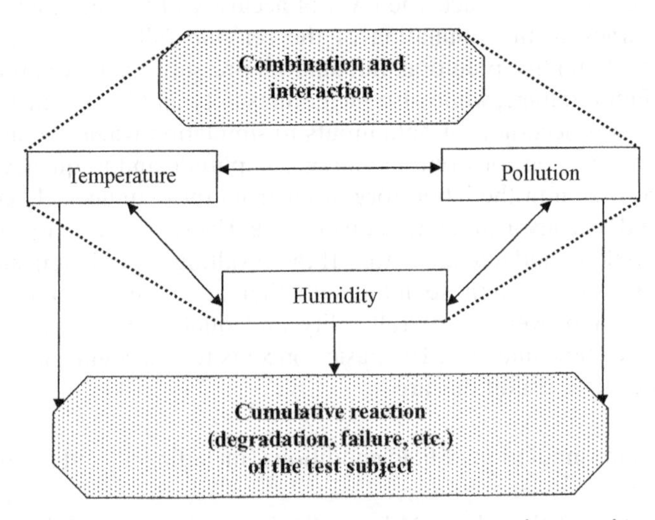

Figure 1.13. Example of the principle of cumulative reaction of a test subject on several input influences.

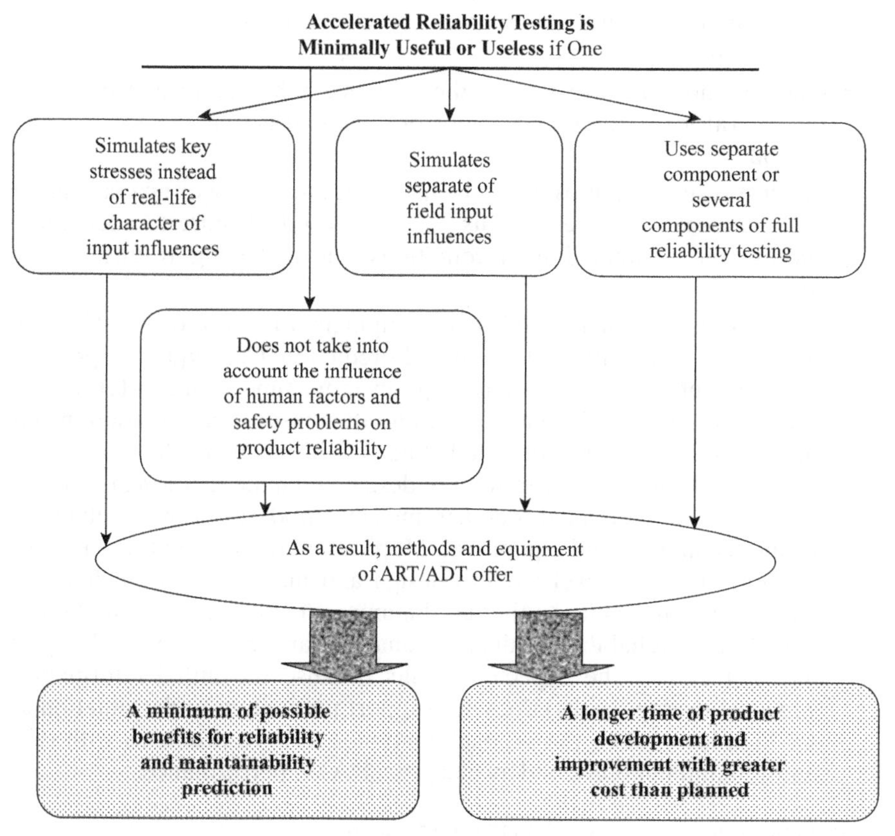

Figure 1.14. Scheme of basic situations when accelerated reliability testing is minimally useful or useless.

- One simulates high stresses instead of real-life input influences. For example, establishing temperatures between –100 and +150°C in the test chamber can change the physics-of-degradation process for many types of test subjects in comparison with the physics-of-degradation process in the field. This does not provide an accurate simulation of real-world and realistic data.
- The test independently simulates one or only a part of full field input influences. Therefore, separate accelerated laboratory testing occurs with either individual input influences or a combination of only a portion of the necessary input influences (temperature/humidity, vibration, corrosion, braking, and electrical) that affect reliability, without a connection and interaction with each other. This contradicts real-life situations where full (or most) of the input influences simultaneously interact with each other.

- One only uses separate components or several components of full ART technology.
- One does not take into account the influence of human factors, including the operator's actions and safety problems affecting the test subject reliability.
- The test results will most likely be incorrect due to low or no correlation between accelerated test results and field results. Failure to adequately address the combined field conditions during testing results in low correlation.
- As a result, current methods and equipment for ART often offer few benefits for reliability evaluation and prediction. Consequently, product development and improvement requires more time with greater expenditures than previously planned because the reasons for degradation and failures in the field are not exactly known. This is one of the reasons for the constant rapid modernization of design and manufacturing technologies for new and updated products and slow modernization of reliability testing techniques and equipment. More often, there is a practical combination of a high-level modern design and manufacturing technology with old testing similar to that used many years ago. As a result, the use of high-level reliability evaluation, analysis, and prediction techniques does not deliver the expected results because the initial information (practical testing results) does not support predictions and evaluation at this high level. Thus, the product continues to have problems with reliability, maintainability, and availability (RAM) performance.

Reliability testing is harmful (Fig. 1.15) when

- It identifies incorrect reasons for field degradation and failures.
- The work for improving the design and manufacturing processes does not lead to an increase in product reliability and quality.
- The work requires excessive time and funds.

Additional basic reasons for the above-mentioned situation include the following:

- Poorly qualified professionals are sometimes involved in ART.
- Following incorrect directions in establishing conditions for ART
- Not enough literature and courses are available for increasing professional knowledge for implementing successful approaches to develop effective ART technology.

These basic negative ART results follow from users selecting or isolating parameters that are different from reality to predict quality, reliability, cost, and time of maintenance. But manufacturing companies also incur losses

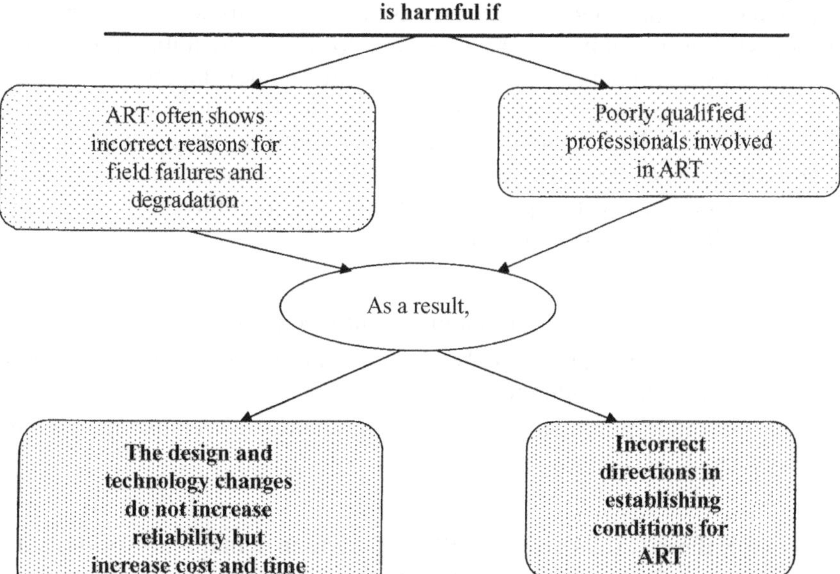
In addition to the scheme of situation shown in figure 1.14, accelerated reliability testing is harmful if
ART often shows incorrect reasons for field failures and degradation
Poorly qualified professionals involved in ART
As a result,
The design and technology changes do not increase reliability but increase cost and time
Incorrect directions in establishing conditions for ART

Figure 1.15. Scheme of basic situations when accelerated reliability testing is harmful.

because they do not make effective use of this situation to achieve better market competitiveness. Without implementing true ART, cost increases probably will not have beneficial results.

How can one eliminate the above-mentioned situation? The basic goal of this book is to introduce principles and descriptions of useful theoretical and experimental technological approaches and practical tools for solving this problem. From the definition of ART, it can be determined that ART enables measurement of reliability indices (time to failures and time between failures) that correlate with the field reliability parameters within the duration of a field test. This makes a direct evaluation and prediction of field reliability and maintainability possible for the service life. ART makes it possible to rapidly increase reliability and improve the robustness of a product. True ART requires an accurate physical simulation of the whole complex of field input influences leading to degradation and failure of the product. As a result, successful identification and resolution of the problems occurring during accelerated development and improvement of the product's reliability and maintainability improved the product. Implementation of this type of ART is rare because management does not choose to simulate the whole complex of real-life input influences with great accuracy in combination with safety and human factors.

One of the basic problems is the cost of AT, especially ART/ADT. Often we read and hear that ART is very expensive. This occurs when professionals consider ART as a goal but not as a resource for obtaining initial information for accurate evaluation, prediction, improvement, and development of product reliability. Therefore, one usually only takes into account the one-time cost of the testing process. But if one also takes into account the design, manufacturing, and usage processes, especially the cost of maintenance, then the cost of ART is not as high. This is especially important to the military.

Many publications and standards

- Include AT as a part of a reliability assurance program
- Describe the strategy for analyzing AT data and other problems based on the information obtained
- Use a single-parameter testing equipment, especially in applied statistics

Of course, these results are also different from an actual real-world results. The standards reflect only past achievement. Therefore, they cannot point the way to improve the current situation of ART. As a result of the poor strategies for testing and predicting the reliability and maintainability of the equipment, many results are several times lower than that predicted due the use of faulty AT during the design and manufacturing phases. This is especially true for military programs and applications. The situation in durability testing is similar.

One can see in Figures 1.16 and 1.17 the way to ART/ADT technology from testing with simulation several or separate field inputs.

The above analysis of reliability/durability testing leads to the following conclusions:

1. There are practical situations involving different testing approaches to ART/ADT that are not intended for accelerated improvement of product quality, reliability, durability, and maintainability.
2. The proposed approach can help to increase the quality of testing, reliability, availability, and maintainability, especially for the military that is interested in increasing equipment RAM and decreasing the total ownership cost. This applies to the design, manufacturing, and utilization phases.
3. It helps to increase the warranty period and especially to predict the RAM and durability performance of systems more accurately. As a result, it helps to dramatically decrease recalls, failures, and complaints.
4. A team of knowledgeable professionals is required to use the above-mentioned approach. These professionals must have knowledge not only of testing and prediction technology but also of advanced technology,

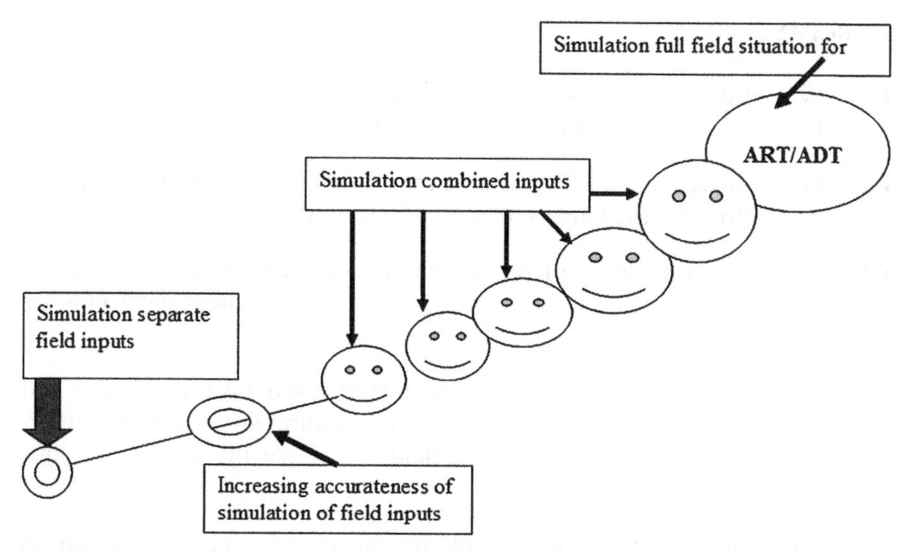

Figure 1.16. The way to accelerated reliability and accelerated durability testing technology.

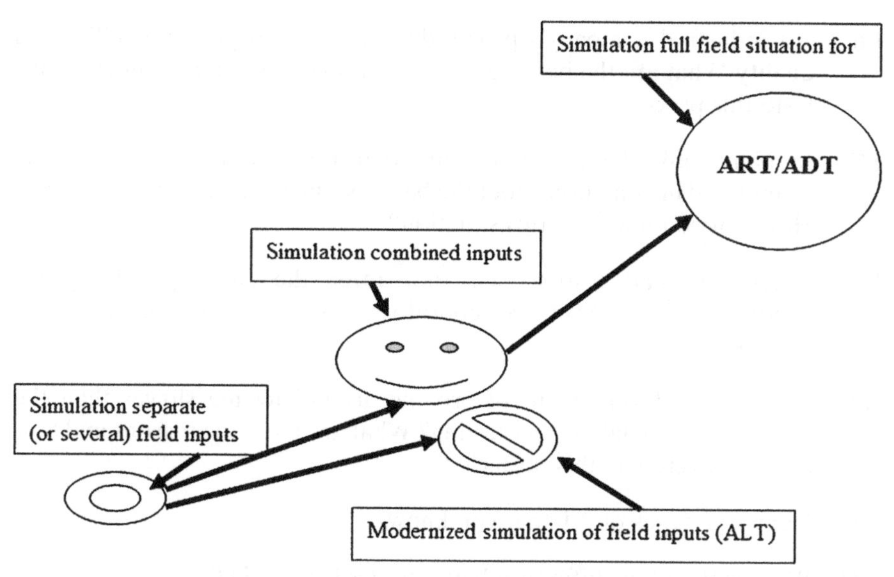

Figure 1.17. The way to accelerated reliability/durability testing technology from traditional accelerated stress testing.

product use, economics, applied statistics, and other areas of mathematics, reliability, fatigue, physics, and chemistry. Requirements for durability testing are similar to those for reliability testing due to interconnection and interdependence of the real world full input influences with safety and human factors.

EXERCISES

1.1 Why are Mercedes-Benz, Honda, and Toyota incorrectly calling the testing that they conduct in the field ART?

1.2 Ford Otosan and LMS said that they conducted durability testing in 2007. How did they incorrectly use the term durability testing?

1.3 Industrial companies often use computer simulations to provide ART. What are the basic reasons they do not simulate field environments accurately?

1.4 Industrial companies are using HALT, HASS, and AA for reliability and durability evaluations. These reliability and durability results are different from those obtained in the field. What are the reasons for this difference?

1.5 Durability testing is not useful for providing accurate predictions of medical device durability. What are the basic reasons for this deficiency?

1.6 An industrial company's profits decrease due to poor reliability and quality. What are the basic causes for this decrease? How can you illustrate this process?

1.7 In many books that mention reliability testing or durability testing, one cannot find information about the basic essence of these types of testing. How can you illustrate this situation?

1.8 Proving ground testing does not offer the same possibility for conducting ART/ADT as laboratory testing. What are the basic reasons?

1.9 Why is the development of proving ground testing slower than the development of laboratory testing? What are the basic reasons and how can you overcome this obstacle?

1.10 Name the types of AT that you know.

1.11 What is the basic difference between ART and ADT?

1.12 What is the difference between AT and ADT?

1.13 What are the components of ART/ADT?

1.14 What are the basic reasons that ART and ADT are seldom used in practice?

1.15 ART can be more or less useful. Determine the basic reasons for this situation and formulate how they differ.

1.16 Why does the reaction of the test subject have a cumulative character?

1.17 What is the difference among basic approaches to current AT methods?

Chapter 2

Accelerated Reliability Testing as a Component of an Interdisciplinary System of Systems Approach

2.1 CURRENT PRACTICE IN RELIABILITY, MAINTAINABILITY, AND QUALITY

Other publications and this text show that technical progress is moving faster in design, manufacturing, and service than in testing. This is demonstrated in Figure 2.1.

Engineering solutions are generally developed in two areas:

- Technological (functional)
- Reliability, maintainability, and supportability during operation

Technical progress consists of new solutions in areas such as physics, chemistry, biology, metallurgy, materials, software, and others. New technology opens the path for advancement. For instance, technological progress in electronics quickly provided new and greatly improved ways of controlling manufacturing and communications systems.

New composite materials made possible great advances in aircraft and aerospace development. Boeing, for example, achieved significant competitive advantages with its 787 aircraft principally through the application of new composite materials in design and manufacturing.

The testing, especially accelerated reliability testing (ART)/accelerated durability testing (ADT) solutions for reliability, maintainability, quality, and supportability, are relatively undeveloped in comparison with technological

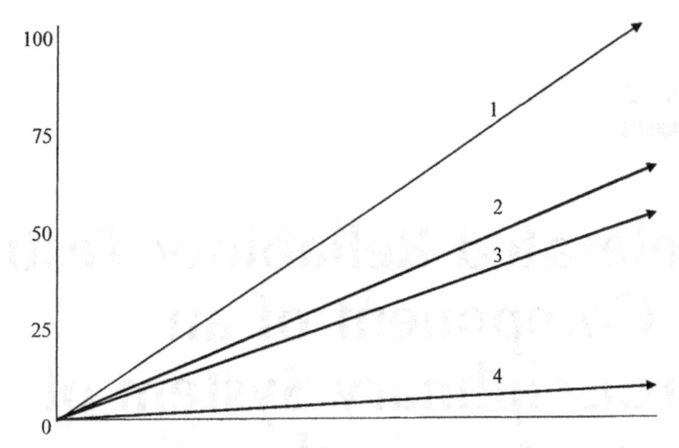

Figure 2.1. Speed of technical progress in different areas of activity (1—design, 2—manufacturing, 3—service, 4—testing).

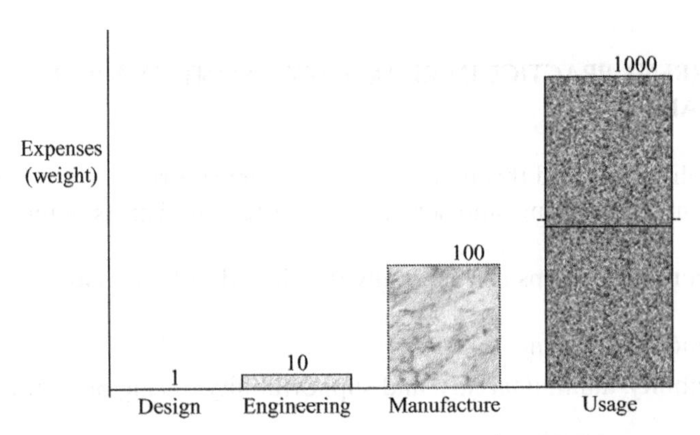

Figure 2.2. Expenses during the life cycle of machinery [78].

(functional) solutions. Additionally, ART/ADT applications are usually more expensive. But alternate solutions based on past and present practice for operations (reliability, maintainability, availability, and supportability) are even more expensive. This leads to a higher consumption cost from usage and labor. One can see the expenses during the life cycle cost of machinery in Figure 2.2.

One basic cost driver in current practice is inaccurate predictions for reliability, quality, maintainability, durability, availability, and supportability during usage. This leads to more (excessive) recalls and more complaints resulting in high cost and time expenditure than predicted to provide corrective maintenance. The problems are mostly a consumer headache. Producers have fewer problems with this situation because their losses are included as part of the product's cost. Loss of market competitiveness is their basic concern. But the associated loss of time, money, and customer "good will" in major recalls are increasing rapidly for the producer. This can also create dangerous situations for those operating in wartime conditions. Negative results of the implementation of new technological solutions occur mostly during operations.

Now let us examine the quality and reliability practices for well-known companies to better understand their product shortfalls.

Example 1. Analysis shows that recalls in Europe mostly affect buyers in the German market.

In 2005, Mercedes recalled 1.3 million vehicles to repair problems with the alternators and batteries [46].

Example 2. In August 2006, Dell and Apple recalled 6 million battery units for Sony notebook computers [47, 48].

Example 3. Automotive vehicle recalls have dramatically increased over the years. For example, annual recalls for 14 years in the U.S. market (1990–2004) increased six times (from 5 million to 30 million). And this process is continued.

Additional customer requirements and increased vehicle volume during the years are two reasons for this increase, but the upward trends are clear. Japanese companies have a high reputation for quality and reliability. Table 2.1 demonstrates the reality of recalls for these high-reputation companies (not all recalls were taken into account).

What about recalls for other automotive companies with a lesser reputation in quality and reliability? For instance, in 2007, Ford Motor recalled 3.8 million vehicles [49].

Example 4. In September 2006, TOSHIBA recalled 340,000 power supply units in Sony notebook computers [50].

Example 5. At the spring 2004 SAE G-11 RMSL Division Meeting, Brigadier General Donald Shenck stated that in the field, the reliability, maintainability, and supportability for military equipment is often at least three to five times

TABLE 2.1 Examples of Recalled Cars by Japanese Companies (see References 45–50, 79–98)

Company	Number of Recalled Cars and Year
Toyota Motor	1 million (2005), 986,000 (2006), 618.5 (2007), 1.4 million (2008)
Honda Motor Co., Inc.	699,000 (2005), 561,000 (2006), 694,000 (2007), 835,000 (2008)
Nissan Motor Co.	2.5 million (until 2003), 1.04 million (1996), 686 (2007), 756,000 (2008)
SUZUKI	157,000 (2004), 120,000 (2006), 456,500 (2007), 497,900 (2008)

lower than what was predicted after testing during the design and manufacturing phases.

Example 6. The army's new Bushmaster troop carrier experience illustrates what can go wrong with defense purchases. The prototype vehicles were 10 times less reliable than the army specified [51].

Example 7. The army deployed a new Stryker transport vehicle in Iraq with many ballistic protection defects that unexpectedly put troops at greater risk from rocket-propelled grenades. This raised questions about the vehicle's development and $11 billion cost. This is reported by a detailed critique in a classified army study obtained by *The Washington Post* [52].

There are many other examples: General Motors Company (GM) recalled 1.5 million vehicles in 2010 for a potential fire hazard; hundreds of thousands of washing machines were recalled; and millions of tainted glasses from McDonald's were recalled.

Global competition in the marketplace has become very complicated in the past few years. Currently, cheaper products do not always sell well. Customers generally prefer a high-quality product that is more economical to use and that requires less maintenance.

Now manufacturers need to more accurately predict quality and reliability over a longer period of time to increase the warranty period and to decrease the number of design-induced failures, complaints, and recalls. The inaccurate prediction of design deficiencies is one of the basic reasons for excessive failures. Overcoming these deficiencies requires accelerated testing to provide accurate solutions for reliability and maintainability problems.

How can one provide adequate solutions?

Examination of some problems in the following example should provide insight.

Currently, the producer controls the market for developing products, not the customer. This is one of the basic reasons for the product design weakness for reliability, maintainability, durability, serviceability, and supportability. For example, Toyota is considered to be a top-quality automotive producer, yet Toyota has recalled around 1 million cars each year. In 2009–2010, the Toyota recalls jumped significantly.

When NASA tested the parachutes for the recovery system on its Orion crew exploration vehicle above the U.S. Army's Yuma Proving Grounds in Arizona [53, 54], The *Status Report*, NASA (2008) stated: "The test proved unsuccessful when a test set-up parachute failed. The failed parachute – called a programmer chute – deployed, but it did not inflate properly and failed to get the test article that simulated the Orion crew module into the correct orientation, attitude and speed for the test, causing the parachute system for the test vehicle to fail." The report also stated that " The programmer chute was designed to stabilize the mock-up before a test of its parachute recovery system, but instead it sent the capsule careening toward the desert floor at the U.S. Army's Yuma Proving Grounds in Arizona" [54]. And importantly, it was added that "... In addition to testing the parachute system, one of the objectives of this test was to demonstrate the testing techniques" [54].

This demonstrates that the reliability of the testing technique was insufficiently developed, causing the simulation of the field situation to be unsuccessful. If one had used the methodology of reliability testing with accurate simulation of the field situation, then one could have avoided the problem and also the experiment's increased time and cost.

Invoking stronger requirements for quality, reliability, maintainability, and durability for equipment can reduce design and prediction problems. Powerful governmental customers such as the Department of Defense or the Department of Transportation can readily impose such requirements. Stronger specifications and standards would dramatically improve the probability of a successful test.

Second, vendor and customer associations need to support the use of strong requirements.

Currently, during the design, development, and improvement of the product and the prediction of the product's parameters, companies use independent solutions for reliability, durability, maintainability, and serviceability. However, in the real world, these processes interact simultaneously as one process: They are interconnected, interdependent, and mutually influence each other. If only reliability problems are considered during research, design, and manufacturing without taking into account these interconnections, then the real-world situation is not being considered. As a result, problems continue to inhibit the successful solution of quality and reliability, resulting in more complaints, recalls, and high life cycle costs. Such problems occur frequently in the automotive, aircraft, aerospace, and other industries.

A similar situation exists within each of these problem areas when the influence of each component is studied separately. For example, during the simulation of field influences on a product or a process, each influencing factor, such as temperature, humidity, pollution, radiation, road surface, air fluctuations, and input voltage, is varied separately.

However, in real life, these factors act on the product simultaneously and interdependently. Studying them separately leads to inaccurate simulations of the real field situation. Hence, research results obtained from independently studying processes in the laboratory or in a short time spent in the field do not reflect the real-life situation. The resulting reliability, maintainability, and durability predictions are mostly inaccurate. This leads to significant problems in applying these results for real-world applications. Similar errors and inaccuracies are found in the results obtained from each independent testing area as well as in the results associated with other applications.

Taking a system of systems (SoS) approach can negate these shortcomings, where the influencing factors are considered independently. In an SoS approach, each case is examined as an interconnected and interdisciplinary problem. Simulation (modeling) is one tool in the SoS approach [55]. One can also find similar solutions to problems of inaccurate prediction for reliability, maintainability, and supportability during design in the book [18].

It is not always possible to model all influencing factors and their interactions. At a minimum, it is essential that individuals working in this field be aware of these interactions. This helps to understand and interpret the observed differences between predicted and field reliability results.

In this chapter, a system to increase quality, reliability, and maintainability through the application of an interdisciplinary SoS approach is outlined. This disciplined approach takes into account the interactions and the interconnections of reliability with quality, safety, human factors, maintainability, and other relevant parameters.

In order to understand the interactions and the interconnections among these influencing factors, one must understand the underlying governing mechanisms. In the following section, reliability and durability, the essential components of the interdisciplinary SoS approach, are outlined.

2.2 A DESCRIPTION OF THE PRODUCT/PROCESS RELIABILITY AND DURABILITY AS THE COMPONENTS OF THE INTERDISCIPLINARY SoS APPROACH

SoSs are composed of components that are systems in their own right (designed separately and capable of independent action) that work together to achieve shared goals [56].

Systems engineering is a discipline concerned with the architecture, design, and integration of elements that, when taken together, comprise a system. Systems engineering is based on an integrated and interdisciplinary approach where components interact with and influence each other. In addition to the technological systems, the systems considered include human and organizational systems where the incorporation of critical human factors with other interacting factors directly affects achieving the enterprise objectives. Simply stated, systems engineering is an interdisciplinary integrated approach that encompasses the composite of people, products, and processes that provide a capability to satisfy customer need.

For the development and improvement of a product or process and the prediction of system parameters, most current approaches rely on separate and independent solutions to the problems of reliability and durability, maintainability, safety, and serviceability. Other influencing factors that are considered separately include quality, human factors, and safety problems.

Since systems engineering processes interact simultaneously in real-world applications, the interacting interdependent components act as one complex. This complex represents the integrated effects of influencing parameters and processes in SoS accelerated testing. The complex may include simulation, reliability development, accelerated testing, maintainability, parameter prediction, life cycle cost development, field reliability prediction, quality in use, and other project-relevant parameters and processes.

Faulty reliability and durability predictions commonly occur in the automotive, aerospace, aircraft, electronics, electrotechnical, medical, and other industries. These seemingly prevalent problems are related to the interconnections of reliability and durability in terms of the following:

- Simulation
- Human factors
- Safety
- Maintainability
- Supportability
- Availability

This integrated plan forms the basis of the interdisciplinary SoS approach. To apply this methodology, an interdisciplinary team consisting of professionals in reliability, safety, human factors, simulation, and testing work together as a coherent group. It is to the benefit of the entire commercial enterprise that management with decision-making authority and funding authority understand, appreciate, and vigorously implement an integrated-team approach.

In order to fully comprehend the use of the above-mentioned interactions and interconnections, it is necessary to understand the governing mechanisms. The following example demonstrates this approach using a step-by-step procedure.

Understanding the field situation for the product or process is the first step. At this step, most components of the problem can be analyzed by using the standard tools of systems engineering. These tools are based on an integrated and interdisciplinary approach that allows the components to interact with and to influence each other [57]. An example from the automotive industry can be seen in Figure 2.3.

Second, one models the product by employing a combination of interconnected actions, beginning with an accurate simulation of the field data and ending with successful reliability and durability development and improvement.

As shown in Figure 2.4, this involves four interconnected steps.

This integrated process is quite complicated and has not been fully demonstrated until now. First, one needs to understand the real-life field situation. Its simulation is the basis for ART. In turn, ART can give initial information for an accurate prediction, and the accelerated development (improvement) of the design reliability, durability, maintainability, supportability, and availability. Thus, one needs to accurately simulate all the basic interconnected influencing complexes simultaneously, considering the cross effects as one

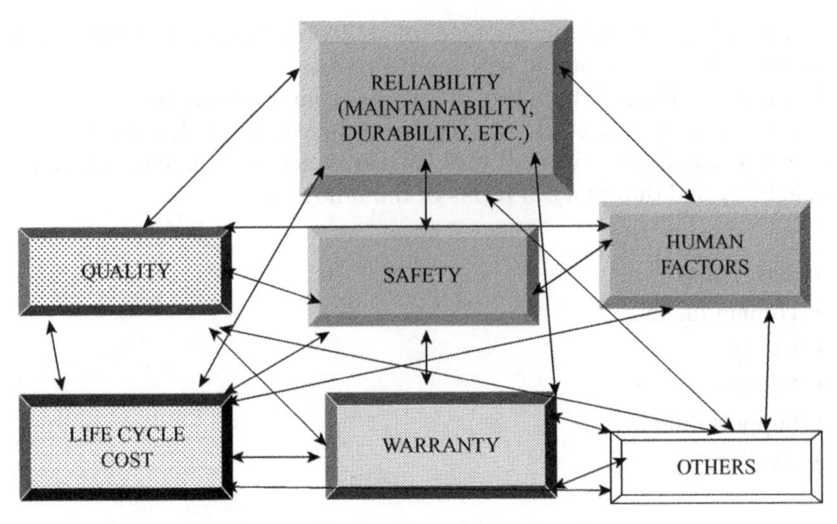

Figure 2.3. Scheme of interaction of field situation components.

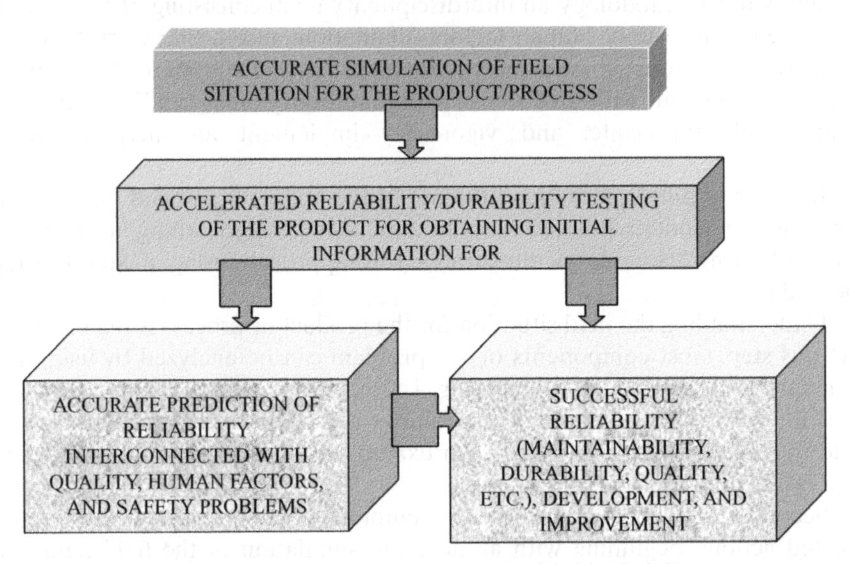

Figure 2.4. Four basic steps for accurate prediction of field reliability (interconnected with quality, human factors, and safety problems) and successful reliability (maintainability, durability, quality, etc.) development and improvement.

complex structure as shown in Figure 2.3. This resultant complex structure is shown in Figure 2.4.

Corresponding to systems engineering approach, ART, accomplished by accurate simulation of the field conditions, is the key driver for the accurate prediction of reliability, maintainability, supportability, availability, durability, and successful design development.

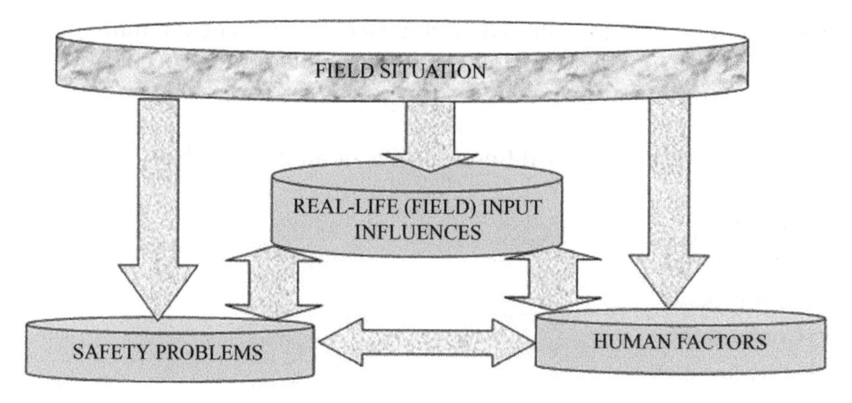

Figure 2.5. Three basic complexes of a field situation.

The basic components/complexes of the field situation, with respect to quality, reliability, maintainability, and other factors, are shown in Figure 2.5. They include the following:

- Interconnected full input influences on the actual product or process
- Safety problems
- Human factors

For an accurate simulation of the field situation, one needs to accurately simulate these interconnected complexes.

2.3 THE COLLECTION AND ANALYSIS OF FAILURE AND USAGE DATA FROM THE FIELD

The collection and analysis of field failure and usage data plays an important role in reliability or durability analysis, analogous to that described in Reference 58. It enables

- Maintenance planning
- Justification of modifications
- Calculation of future resources and spares requirements
- Confirmation of contractual satisfaction
- Assessment of the likelihood of achieving a successful mission
- Feedback for design and manufacturing
- Estimation of the cost of the life cycle and warranty period
- Collection of basic data for possible liability cases

- Collection of usage data to determine a customer's field requirements to provide the supplier's reliability and durability testing specifications and demonstration programs

Data can be collected for different levels of an item, such as

- System
- Equipment
- Unit module
- Component
- Software element

Data can also be collected for the different phases of the life cycle of a product: developmental testing, production, early failures, and operation.

Data collection is often a long-term activity for reliability- or durability-related purposes. Data for many item operations and/or many items may be required before an appropriate analysis can be completed. Data collection should be undertaken as a planned activity and executed with appropriate goals in mind.

In the short term, data collection objectives for dependability-related purposes include the following:

- Identification of the product design shortfalls
- Adjustment of logistic support
- Identification of customer problems for correction and
- Identification of failure modes and their root

Analysis of reliability or durability data requires a clear understanding of the item, its operation, its environment, and its physical properties. A good understanding of the general subject of reliability (durability) and its specific application is also required.

Prior to starting the data collection process, it is important to realize that data collection cannot usually be performed without the cooperation of all the parties involved. This may include item designers, manufacturers, suppliers, repair authorities, and users.

Generally, the life cycle of equipment can be thought of as a three-stage process. These stages are design, manufacturing, and operation. Data collection can be performed at each stage and the information collected can provide feedback to any of the other stages as shown in Figure 2.6.

2.3.1 Quantitative and Qualitative Data Collection

Quantitative data collection is the collection of data that can be stated as a numerical value. Qualitative data collection is the collection of softer informa-

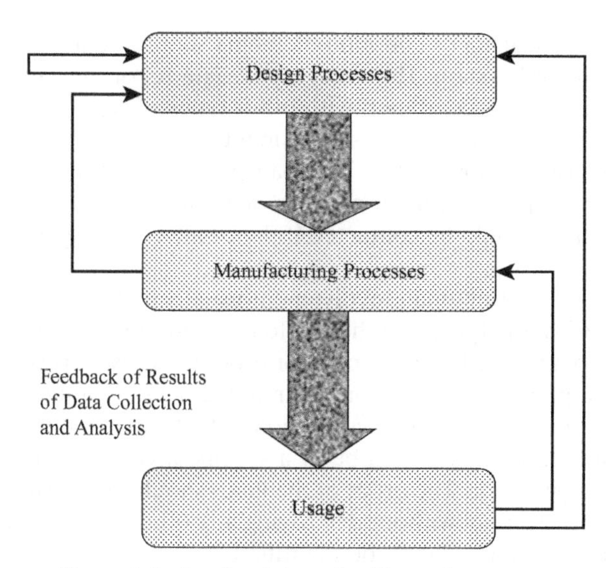

Figure 2.6. Feedback into the life cycle process.

tion, for example, reasons for an event occurring. Both data types are important and support each other. The type of data collected will depend on the sort of questions to be answered by the data.

2.3.1.1 Methods of Data Collection. Reliability (durability) data are often collected over several years and involves many different users and maintenance personnel. This vast amount of data has to be collated by additional staff. Thus, data collection is a large-scale effort with possible sources of data corruption. Accordingly, the data collection, collation and recording process has to emphasize ease of use and error proofing.

Data collection result in a mixture of codes, numerical data and descriptive information. Codes are generally used to identify data within a limited number of previously identified categories to facilitate and limit data entry. This greatly simplifies the subsequent analysis, provided that the categories are appropriate and sufficient for the analysis goals. The categories must be recognizable and understood by the operators. Any ambiguity, or misalignment with actual failure events, will result in corrupt data. A corollary is that the codes and their categories should have clearly published guidelines. When data are collected from numerous sources, it makes sense to harmonize any codes or descriptions across all sources.

This may not be possible if untrained personnel collect the data. Even if codes are used in all mandatory fields, it should always be possible for the individual to give comments. Reliability (durability) and maintenance data share many common elements. Therefore, reliability (durability) data collection should be integrated with the maintenance record system. Whenever possible, all data sharing of common elements should be integrated with the maintenance record system. The accuracy of data reporting can be increased if the reporting form is combined with other types of reporting, for example, economic compensation (spare costs, payment under guarantee, mileage compensation, and time reporting for the repair person). Furthermore, the quality of the reporting improves if the repair person knows how the data will be used. Additionally, they should be contacted if their data reporting is incomplete or ambiguous.

Data collection with human involvement usually involves a series of forms. Forms standardize the format of the written information. The use of standardization [59] provides the first level of error proofing. These forms may be either a paper or computer representation. Human factors that influence the correct completion of a form have to be considered. The form can mirror that layout where a data field is expected to be in a specific layout, including validation masks, when appropriate. The order of the fields should closely reflect the most usual order of collection, without destroying any other natural grouping logic. Mandatory fields should clearly be identified.

The validation of correct data input is essential. When using direct computer data input, it is possible for the computer to simultaneously validate entries against permissible values using more sophisticated checking algorithms.

Finally, one should motivate the person collecting and recording the data. No amount of formatting or field masking or data collection forms will defeat the potential for an unmotivated human to supply corrupt, missing, or incomplete information. The human operator has to feel involved and understand the need for good data. He should feel that his contribution is worthwhile and that he will benefit from the subsequent results.

Data stewardship is one method to motivate data collectors and to help improve data quality. Data stewardship's main objective is the management of the corporation's data assets to improve their reusability, accessibility, and quality. It is the data steward's responsibility to approve the data naming standards of the business, to develop consistent data definitions, to determine data aliases, to develop standard calculations and derivations, to document the business rules of the corporation, to monitor the quality of the data in the data warehouse, and to define security requirements.

Data stewards should have a thorough understanding of how their business works. They have to have the confidence of both the information technology (IT) and the end-user communities in their data collection and analysis processes. Data stewards have to show that they are not creating metadatabases and business rules that are impossible to implement or that run counter to the corporation's policy.

A typical corporate data stewardship function should have one data steward assigned to each major data subject area. These subject areas consist of the critical data entities or subjects, such as customers, orders, products, market segments, employees, the organization, inventory, and other factors.

Usually, there are more than 10 major subject areas in any corporation. The data steward responsible for a subject area usually works with a select group of employees representing all aspects of the company in that subject area. This committee of peers is responsible for resolving integration issues concerning their subject area. The results of the committee's work are forwarded to the data and database administration functions for implementation into the corporate data models, metadata repositories, and the data warehouse.

Just as there is a data architect in most data administration functions, there should be a "lead" data steward responsible for the work of all the individual data stewards. The lead data steward's responsibility is to determine and control the domain of each data steward. These domains can become muddy and unclear, especially where subject areas intersect. Political battles can develop between the data stewards if their domains are not clearly established.

Second, the lead data steward has to ensure that difficult issues are resolved in a reasonable time. If resolution cannot be achieved at the data steward level, then the lead data steward presents the deadlocked and unresolved issues for resolution to the corporate steering committee composed of high-level executives.

Data collection can also be automated or semiautomated by using electronic data-logging devices. Automated facilities reduce errors but may add to the cost, weight, and size disadvantages to the installed platform. The simplest automatic data collection uses bar codes to identify items. The most complex data collection uses built-in electronics to perform the same task.

The choice of an automated system will depend on the complexity, form factor, and cost of the item to be tracked. Automated systems can also be used to track the usage mode, elapsed time, item health parameters, and environmental parameters.

Automated data collection (ADC), also known as automated identification (AutoID) and automated identification and data capture (AIDC), known by many as "bar coding," consists of many technologies including some that have nothing to do with bar codes.

There are two major categories of bar codes: one dimensional (1D) and two dimensional (2D). 1D bar codes are the most familiar and consist of many different symbols or even variations of a specific symbology. Supply chain partners may dictate the symbology used through a standardized compliance label program, or if only used internally, symbology can be based upon the specific application. 2D bar code symbology is capable of storing more data than its 1D counterpart and requires special scanners to read it.

There are primarily two technologies used to read bar codes. Laser scanners use a laser beam that moves back and forth across the bar code reading the light and dark spaces. Laser scanners have been in use for decades and are capable of scanning bar codes at significant distances. Charged-coupled device (CCD) scanners act like a small digital camera and take a digital image of the bar code that is then decoded. CCD scanners offer a lower cost, but they are limited to a shorter scan distance (usually a few inches). However, this technology is advancing quickly, and devices with longer scan distances are becoming available.

Keyboard wedge scanners connect a computer keyboard to the computer and send ASCII data as if the scanner were a keyboard. The advantage is that there is no need for special software or programming on the computer. Keyboard wedge scanners offer a low cost entry into the world of ADC and can provide increasing accuracy and productivity in many stationary data entry applications. There are wireless versions of keyboard wedge scanners available.

Fixed-position scanners are used where a bar code is moved in front of the scanner as opposed to the scanner being moved over the bar code. Applications include automated conveyor systems. Many fixed-position scanners are omnidirectional, which means that the bar code does not have to be oriented any specific way to be read.

Portable terminals come in a vast variety of designs with varying levels of functionality. Batch terminals are used to collect data into files on a device that is later connected to a computer to download the files. Radio frequency (RF) terminals use RF waves to communicate live with the host system or network.

Handheld devices simply use one hand to hold the device, and in most warehouse environments, this is a problem since that hand can no longer be used to handle materials or to operate equipment controls. Handheld terminals generally have very small LCD displays that are usually difficult to read, and very small keypads make it difficult to enter data. Handheld devices often come with integrated bar code scanners that can be used without a scanner or with a separate scanner.

Voice technology is now a very viable and desirable solution in data collection applications. Voice technology is really composed of two technologies: "voice directed," which converts computer data into audible commands, and "speech recognition," which converts voice input into data. The advantages of voice systems are they are hands-free and eyes-free operations.

For many years, optical character recognition (OCR) has been used in sorting mail and document management, but it has seldom been used in warehouse and manufacturing operations primarily because it is not as accurate as bar code technology. The primary advantage of OCR is that it can read the same characters that a human can read. Thus, it eliminates the need for both a bar code and human readable text on labels and documents. This allows the input of data from documents that do not include bar code information.

The results of data collection have to be analyzed. Different types of analyses include correlation, spectral, and regression analysis.

For example, Klyatis and Chernov [60] developed regression models for the analysis of the loading speed for soil-cultivating machinery types . These models are applicable for different soil conditions and different work regimes. This reduced the time required for the experiments. A second-order polynomial was used for a partial description of these processes. It has the following form:

$$Y = b_0 + \sum_{i=1}^{k} b_i x_i + \sum_{i<j}^{k} b_i x_{ij} x_j + \sum_{i=1}^{k} b_i x_j^2$$

where

Y is the response function,
x_i and x_j are factors, and
b_i and b_j are regression coefficients.

2.4 FIELD INPUT INFLUENCES

Field input influences are one of the three basic complexes present in the field situation (Fig. 2.5). The contents of this complex, as well as two others, are common for many types or similar groups of products and processes. The details are more specific for each application. For a car, the field input influences consist of multi-environmental, mechanical, electrical, and electronic groups. Each group is also a complex of lower-level input influences. For this example, a multi-environmental complex of field input influences consists of temperature, humidity, pollution, radiation, wind, snow, fluctuation, and rain. Some basic input influences combine to form a multifaceted complex. For example, chemical pollution and mechanical pollution combine in the pollution complex. The mechanical group of input influences consists of different less complicated components. The specifics of this group of influences depend upon the product or process details and functions. The electrical group of input influences also consists of several different types of simpler influences such as input voltage, electrostatic discharge, and others. Most of these interdependent factors are interconnected and interact simultaneously in combination with each other.

The results of input influence interactions are output variables such as loading, tension, vibration, or output voltage. The result of their actions is a degradation process (see Fig. 2.7): corrosion of metals, wear, crack propagation, decrease of viscosity, decrease of grease, lubricant properties, plastic aging, rubber aging, or other degrading effects.

The final result of degradation is failure (see Fig. 2.7).

For accurate reliability prediction and successful development and improvement, one needs to simulate all the interconnected components that influence one another simultaneously and in combination in the appropriate complex. Independent simulation of separate components as it is usually done in research and design will be inaccurate. The real-life input influences will be considered in more detail in Chapters 3 and 4.

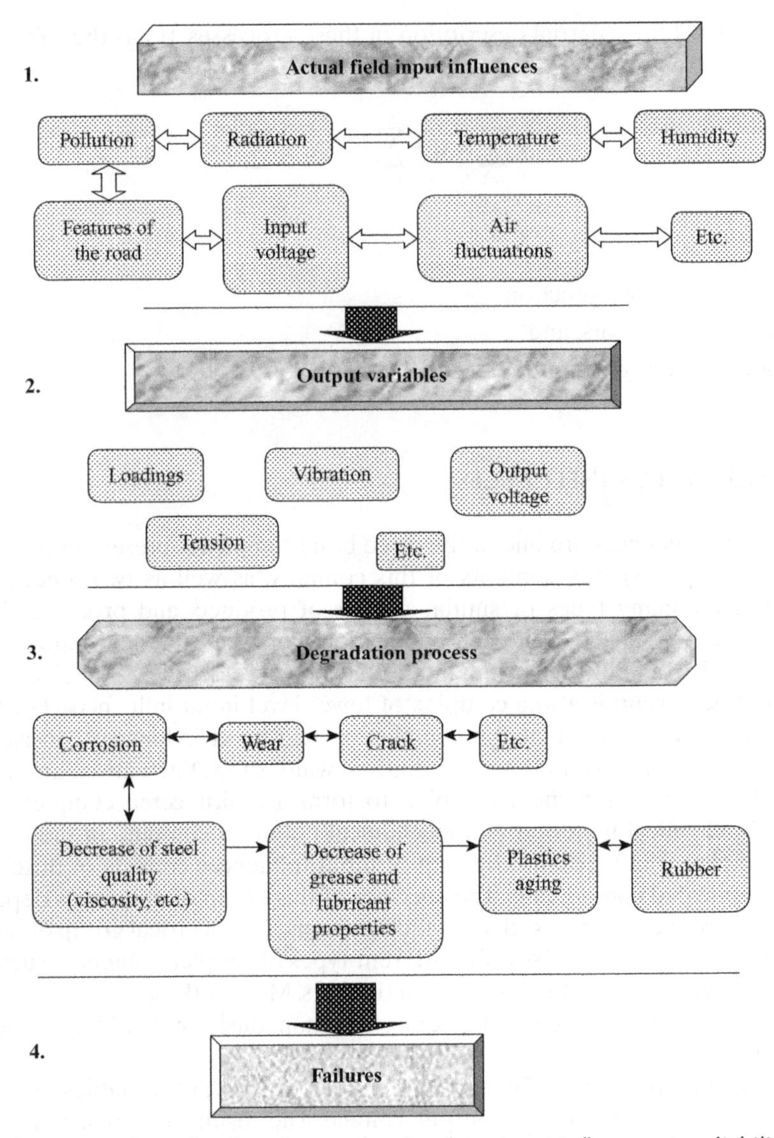

Figure 2.7. Example of a scheme of action from input influences to reliability.

2.5 SAFETY PROBLEMS AS A COMPONENT OF THE FIELD SITUATION

Safety is a combination of two basic components: risk problems and hazard analysis. Both are interconnected with reliability. For example, failure tolerance is one of the basic safety requirements that are used to control hazards. Another example is safety risks that are the result of the hazardous effects of failure functions. Fault tree analysis may be used to establish a systematic link between the system-level hazard and the contributing hazardous event at the subsystem, equipment, or piece-part level.

The same logic applies to the solution of risk problems. There are many standards in reliability and safety that include the interconnection of reliability with safety. Some examples include the updated international standards for space standardization, the standards and guidelines for European Cooperation for Space Standardization (ECSS), and the International Organization for Standardization (ISO), and International Electrotechnical Commission (IEC) standards (ECSS-M-00-03A, ECSS-Q-40B, ECSS-Q-40-02A, IEC 60300-3-7, and ISO/IEC Guide 51).

Failure modes and effects analysis (FMEA) is an example of reliability related to safety. FMEA is a tool used to collect information related to risk identification and management decisions. Documented procedures are available to complete an FMEA with examples of its application in various industries. A higher FMEA score indicates greater risk. Common variables used to quantify risk are the frequency of an activity associated with the defect, the quantity of parts associated with the defect, the ability to detect the defect, the probability of the defect, and the defect severity.

The two basic safety components are a combination of subcomponents. The solution to the risk problem is found in the following subcomponents [18]:

- Risk assessment
- Risk management
- Risk evaluation

Each of these subcomponents consists of sub-subcomponents. For example, risk assessment consists of five subcomponents as shown in Figure 2.8.

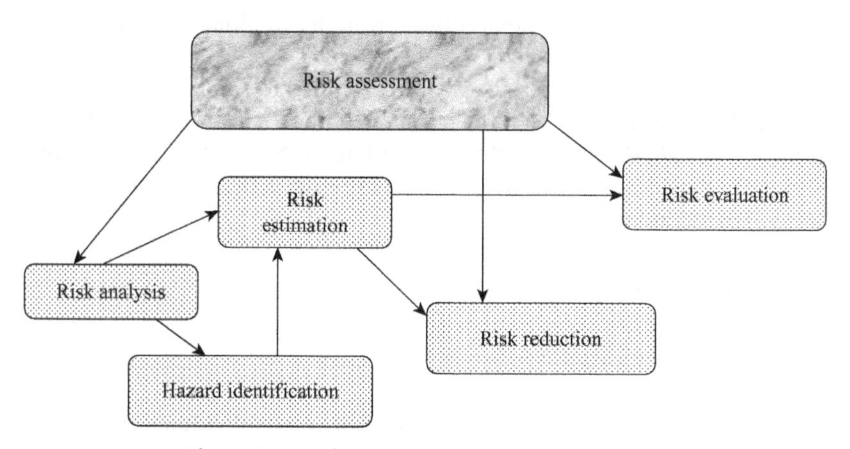

Figure 2.8. Subcomponents of risk assessment.

Solving the safety problem requires the simultaneous study and evaluation of the full complex of these interacting components and subcomponents. This will provide accurate predictions for accelerated development and improvement of the process or product.

To obtain information for risk assessment, one needs to know the following:

- Limits of the machinery
- Accident and incident history
- Requirements for the life phases of the machinery
- Basic design drawings that demonstrate the nature of the machinery
- Statements about damage to health

For risk analysis, one needs

- Identification of hazards
- Methods of setting limits for the machinery
- Risk estimation

For risk reduction, the procedure is based on the assumption that the user has a role to play in risk.

2.6 HUMAN FACTORS AS A COMPONENT OF THE FIELD SITUATION

2.6.1 General Information

Human factors engineering is the scientific discipline dedicated to improving the human–machine interface and human performance through the application of knowledge of human capabilities, strengths, weaknesses, and characteristics [61].

Human factors always interacts with reliability and safety because the reliability of the product has a connection with the operator's reliability and capability. It is called human factors in the United States, whereas in Europe, it is often called ergonomics. This is an umbrella term for several areas of research that include

- Human performance
- Technology
- Human–computer interaction.

The term used to describe the interaction between individuals and the facilities and equipment, and the management systems is human factors. The disci-

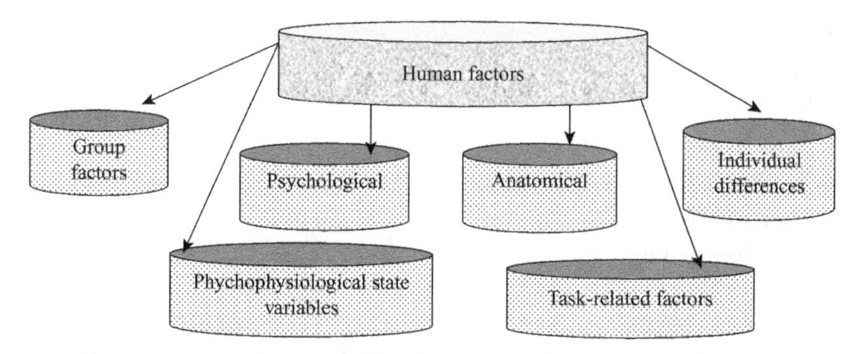

Figure 2.9. Human factors basic components.

pline of human factors seeks to optimize the relationship between technology and humans. Human factors applies information about human characteristics, limitations, perceptions, abilities, and behavior to the design and improvement of objects and facilities used by people [61]. The essential goal of human factors is to analyze how people are likely to utilize a product or process, then to design it in such a way that its use will feel intuitive to them, facilitating a successful operation.

Figure 2.9 shows six basic subcomponents of a human factors complex.

Human factors practitioners come from a variety of backgrounds. They are predominantly psychologists (cognitive, perceptual, and experimental) and engineers. Designers (industrial, interaction, and graphic), anthropologists, technical communication specialists, and computer scientists also contribute to the field. Areas of interest for human factors practitioners usually include

- Workload
- Fatigue
- Situational awareness
- Usability
- User interface
- Ability to learn
- Attention
- Vigilance
- Human performance
- Human reliability
- Human–computer interaction
- Control and display design
- Stress
- Data visualization
- Individual differences

- Aging
- Accessibility
- Safety
- Shift work
- Work in extreme environments
- Human error
- Decision making

A similar term, "human factors engineering," is an extension of the earlier phrase "human engineering." In Europe and in other countries, the term "ergonomics" is more predominant. Ergonomics is made up of the Greek words "argon," meaning work, and "nomos," meaning law.

However, over the last 50 years, ergonomics has been evolving as a unique and independent discipline that focuses on the nature of human artifact interactions. This is viewed from the unified perspective of science, engineering, design, technology, and the management of human-compatible systems. The various dimensions of the ergonomics discipline are shown in Figure 4.1 [61].

The International Ergonomics Association defines ergonomics (human factors) as the scientific discipline concerned with the understanding of the interactions between humans and other elements of a system. This profession applies theory, principles, data, and methods to design in order to optimize human well-being and overall system performance [61].

Corresponding to Salvendy [61], the classification scheme for human factors/ergonomics is much broader than the list mentioned earlier. It includes

- Human characteristics
  - Psychological aspects
  - Anatomical aspects
  - Group factors
  - Individual differences
  - Psychophysiological state variables
  - Task-related factors
- Information presentation and communication
  - Visual communication
  - Auditory and other communication modalities
  - Choice of communication media
  - Person–machine dialogue mode
  - System feedback
  - Error prevention and recovery
  - Design of documents and procedures
  - User control features

  - Language design
  - Database organization and data retrieval
  - Programming, debugging, editing, and programming aids
  - Software performance and evaluation
  - Software design, maintenance, and reliability
- Display and control design
  - Input devices and controls
  - Visual displays
  - Auditory displays and other modality displays
  - Display and control characteristics
- Workplace and equipment design
  - General workplace design and buildings
  - Workstation design
  - Equipment design
- Environment
  - Illumination
  - Noise
  - Vibration
  - Whole body movement
  - Climate
  - Altitude, depth, and space
  - Other environmental issues
- System features and characteristics
- Work design and organization
  - Total system design and evaluation
  - Hours of work
  - Job attitude and job satisfaction
  - Job design
  - Payment systems
  - Selection and screening
  - Training
  - Supervision
  - Use and support of technological and ergonomic change
- Health and safety
  - General health and safety
  - Etiology
  - Injuries and illnesses
  - Prevention of injury

- Social and economic impacts of the system
  - Trade unions
  - Employment, job security, and job sharing
  - Productivity
  - Women and work
  - Organizational design
  - Education
  - Law
  - Privacy
  - Family and home life
  - Quality and working life
  - Political comment and ethical considerations
- Methods, approaches, measures, and techniques

To illustrate how ergonomics (human factors) influence reliability, maintainability, and supportability, consider psychophysiology (PP) as one of the basic components of the human factors in the field. PP needs to be simulated in the laboratory to obtain accelerated solutions of reliability, maintainability, and supportability problems. The PP of professional activities is a part of an interdisciplinary field of theoretical and applied knowledge. Its subject is the investigation of professional activities as a particular form of human behavior. Theoretical investigations are directed at an elucidation of the spiritual, psychical, and physical mechanisms of professional activities. The field of applied PP uses this knowledge as a scientific foundation for the prediction of productivity and for the optimization of intrinsic and extrinsic control of professional activities.

2.6.2 The Influence of PP

One can study the influence of PP on reliability, maintainability, and supportability for a mobile product. Studying the influence of PP requires

- A definition of the admissible interconnected PP parameters of the operator and the established ergonomic characteristics of mobile vehicles
- The likelihood of making both short- and long-term forecasts of the operator's reliability in the system.

The methodological approach to solve this problem must include several considerations. First, the essential value of the quality of the prediction has a quantitative influence on a considerable number of factors and their combinations that depend upon the operator's reliability. The basis of an accurate prediction is a mathematical model of the system that includes the factors influencing the operator. The quality of the prediction will depend on the adequacy and accuracy of the model and its data.

Second, the qualitative prediction of an operator's reliability cannot be evaluated without an accurate goal. It may be formulated as the capacity to execute specific actions with necessary effectiveness during a given time under specific conditions. Prediction of the operator's reliability is problematic. But the prediction of the operator's capacity to execute delineated actions under stated conditions is more practical for specific actions. The important parameter is the length of the time for prediction.

The prediction can be either long term or short term. A short-term prediction is one that is sufficient for executing a specific task during the workday. An example of this type of prediction is preflight tests of aviation pilots or prerun tests of auto drivers. A long-term prediction relates to longer time segments (several days, weeks, months, or years), such as the duration of training required until an operator is capable of performing all tasks required during his shift. An example of this type of prediction is estimating the amount of training required to solve PP problems during an orientation and selection process.

Setting the limits for the various levels of requisite abilities to perform specific work functions is difficult. These tasks combine integral complex characteristics of the functions and the personal qualities of the employee that determine the quality of vehicle controlling actions over a given period of time.

Accordingly, the homeostatic level of a person's basic condition, psychological and emotional tolerance under the influences of complicated and disturbing factors, must be taken into account. This shows that a person may not be able to act favorably under unusual or stressful conditions.

Coping mechanisms are limited and lead to a decline in the ability to adapt under stress. Consequently, the operator's reliability decreases over time. This is especially true for situations that require a high degree of responsibility, such as life-threatening situations or inadmissible psychological or economic consequences of mistakes. Functional mobilization facilitates an operator's reliability for only a limited time before it becomes a dramatic failure.

2.6.3 Accurate Simulation of Human Factors

The accurate simulation of human factors is highly effective and has three advantages:

1. Simulation eliminates risk and the painful consequences of poor decisions. Simulation goes beyond mitigation to completely eliminate costly mistakes.
2. Simulation is able to compress the decision time from years of experience to months or weeks of analysis.
3. Simulation provides the capability to limit the number of variables in a scenario to focus a learner's attention on the desired outcome.

Simulation principles are useful for a wide range of experimental research in engineering, physics, and chemistry. The laboratory researcher needs to simulate the complex parameters of the real-life (field) situation to obtain accurate results. The degree of accuracy depends on the fidelity of the simulation in representing the real world during the research.

For experimental human factors research, a new driving simulator was built at the Toyota Higashifuji Technical Center in Shizouka, Japan. The simulator uses an actual vehicle inside a 23-ft (7-m) diameter dome, with a 360°, concave video screen that simulates a real driving environment [62]. In addition to testing, the car's suspension, brakes, and other mechanical systems, it measures driver behavior under a range of circumstances. A similar driving simulator was built for Mercedes-Benz in the company's Sindelfingen Technology Center.

Mechanical Simulation Corporation's CarSim real-time (RT) software is used to simulate all the critical vehicle dynamics for the system.

Terence Rhoades, President, Mechanical Simulation said: "An installation of this magnitude can allow auto manufacturers to conduct precise tests of their designs, without the time, expense, and hazards of road testing . . . Driver-behavior tests can be created in a repeatable, simulated environment. Without the danger of traffic exposure and unpredictable weather and road conditions, Toyota can simulate driver and vehicle responses to a wide range of environments, as well as driver response to safety technology such as Intelligent Traffic Systems." [62].

The simulator is designed to measure vehicle performance for a wide range of driver characteristics including reduced awareness of one's surroundings, inattentiveness to danger, and impaired driving. The results are used to evaluate the effectiveness of safety technology, including driver-warning and vehicle-control systems plus intelligent traffic systems outside the vehicle.

CarSim uses information derived from driver control inputs (steering angle, throttle, brake) and from the external environment. Its control computer determines how the vehicle will move. It then provides motion instructions to the simulator and scenario generator to create the visual environment.

The 360° concave video screen is positioned more than 14 ft (4.3 m) from the driver and can be manipulated to simulate the sensations of driving, including speed, acceleration, turns, and other maneuvers.

CarSim is Mechanical Simulation's software package for simulating, viewing, and analyzing the dynamic vehicle behavior of cars, light trucks, and utility vehicles using driver, ground, and aerodynamic inputs. CarSim RT, being used by Toyota, features live connections to physical hardware, RT analysis software, and postprocessing capabilities.

Toyota intends to use the simulator for the analysis of the driving characteristics of drivers, research and development of safety technology to reduce traffic accidents, and evaluation and verification of preventive safety technology.

2.7 THE INTERCONNECTION OF QUALITY AND RELIABILITY

In the SoS approach for ART, important common problems are associated with the integration and interaction of

- Reliability (durability and maintainability) with quality during a given time
- Reliability with simulation
- Reliability with safety
- Reliability with human factors

ART/ADT is a highly effective approach to solve these problems through advanced management methods. Practical integration of the interconnected components influencing quality and reliability may be used at every step of research, design, manufacturing, and usage.

How can this be done?

Developing advanced integrated ART/ADT solutions for quality and reliability can accomplish this objective. But this is no easy task for industrial companies that usually solve quality and reliability separately during the design and manufacturing phases. Both are interconnected in the field (Fig. 2.10).

If the solutions for quality or reliability problems are not integrated at an advanced level, then their interconnection will not be effective as can be seen in Figure 2.11. Adopting and implementing this new approach is a difficult strategic problem that most industrial companies have not been able to solve until recently.

To better understand this integration process, let us remember the basic definitions of quality and reliability:

- *Quality* is the *ability* of a product or service to *satisfy the user's needs.*
- *Reliability* is the *ability* of an item to perform a *required function under given conditions for a given time interval.*

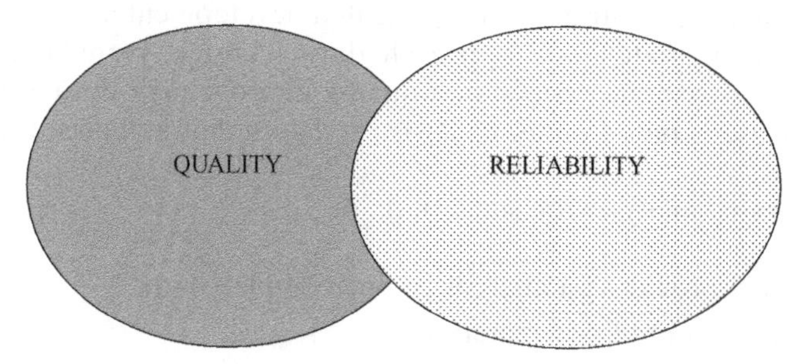

Figure 2.10. General scheme of quality and reliability interconnection.

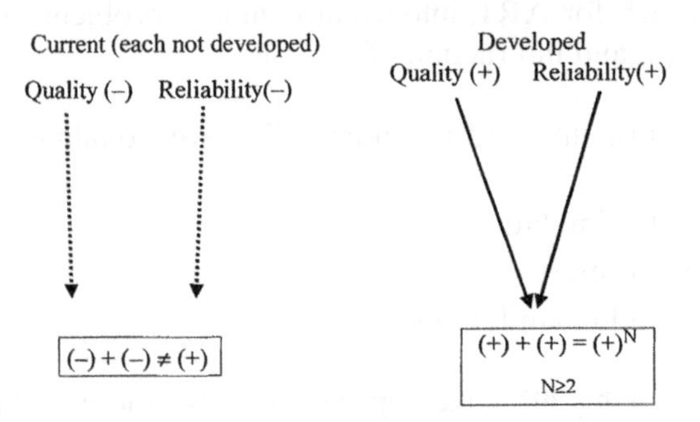

Figure 2.11. Short schematic description of current unsuccessful integration of quality and reliability.

Simulation is an essential tool for the solution of these problems. Without physical simulation, it is currently not possible to effectively conduct laboratory experimental research and testing. Most of the current literature addresses software simulation. Physical simulation of full real world input influences on the actual product is a specific area that is minimally reported in the literature. Accurate physical simulation of field input influences on the actual product is described in Chapter 3.

Industrial companies experience severe problems trying to resolve complaints about their products. They neither rapidly determine the root cause nor quickly resolve the problem. As a result, it is difficult to eliminate complaints during the manufacturing process. It often takes years to increase the product's quality or reliability. Thus, such a product is not adequately competitive in the market. Our methodology [18] shows how one can solve this problem quickly.

A high-quality product can be achieved during design through ART by utilizing integrated physical simulation to accurately predict reliability.

A practical modern strategy of accelerated development and improvement is to use an integrated SoS approach during design, manufacturing, and usage [18]. For example, accelerated quality improvement during the manufacturing phase is based on an analysis of factors that influence the product quality.

The following analyses are suggested:

- One team provides an analysis of one complex in the
  - Procurement of raw materials
  - Procurement of product components
  - Procurement of the final product

- The same team provides an analysis of one complex in
  - Design
  - Technology of manufacturing
  - Quality control and assurance
  - Product use

The following are considered in the analysis of human factors influence

- During design
  - Role, value, scope, and contribution to the field of human factors
  - Including human factors in the planning and execution of product design
  - Digital human modeling for design and engineering
- During manufacturing
  - Design philosophy approach analysis of human factors and their influence on product quality and reliability
  - Simulation of these influences

2.8 THE STRATEGY TO INTEGRATE QUALITY WITH RELIABILITY

2.8.1 The Effects of Engineering Culture

The engineering culture has an important influence on the product's acceptability by directly influencing its quality and reliability. Consider the role of the engineering culture by comparing that of GM with Toyota. Now consider the basic elements of the engineering culture. Alfaro [63] presents three basic elements (see Fig. 2.12): quality, rule/responsibility, and the system. The comparison of the engineering culture at GM and Toyota shows that Toyota pays the most attention to quality, less to the system, and least to rule/responsibility [63]. In part A of Figure 2.12, GM currently pays the most attention to the system, a little less to rule/responsibility, and the least attention to quality. As a result of analyzing this situation [63], GM plans to change its priorities and to give equal attention to all three basic elements.

How can it be more effective?

There is a lot of literature that includes information on the evaluation and development of product quality during its design and manufacturing phases [64–66].

Considerably less literature is published about the accurate evaluation and prediction, accelerated product development, and improvement of durability and reliability. This is especially true of manufacturing companies where the norm is to have no practical methodology and no equipment to use for integrated evaluation and prediction of reliability/durability or for accelerated improvement and development. Absence of accelerated development

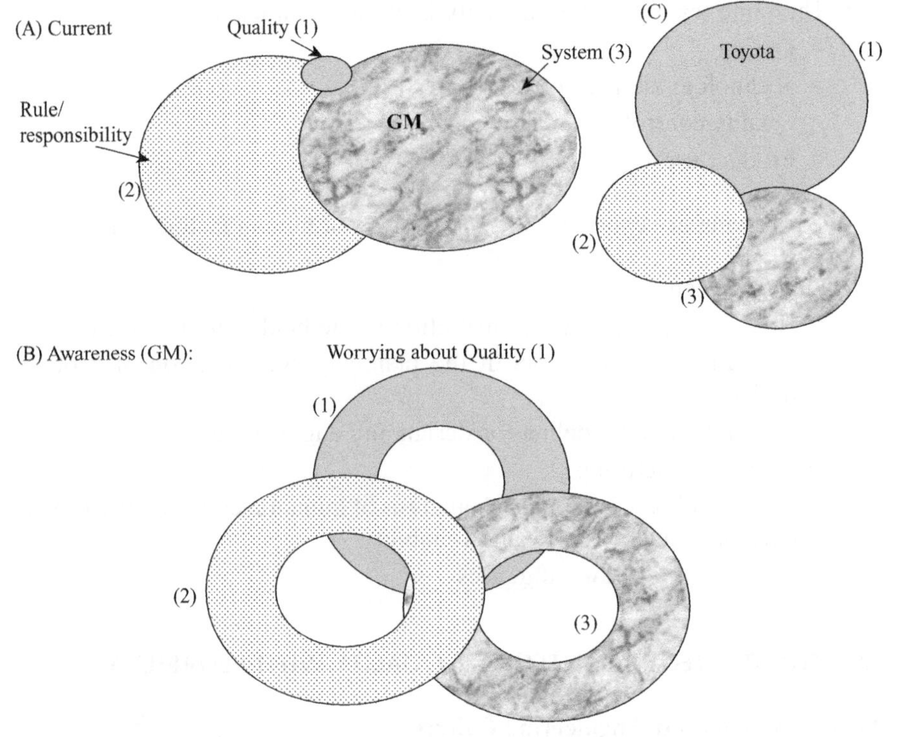

Figure 2.12. Difference in current engineering culture by Toyota (C) and General Motors Company (A—current, B—awareness): (1) is quality; (2) is rule/responsibility; and (3) is system [63].

literature is one of the basic reasons that the reliability and maintainability of equipment is often three to five times lower in operation than what was predicted [67]. Current literature describes mostly mathematical and statistical concepts, but this literature does not address the SoS engineering concepts and necessary methodology and equipment needed to advance industrial implementation.

As a result, the reliability, durability, and maintainability of industrial products are insufficient.

Some companies reduce the negative influences by including incentives to achieve higher quality in their system and culture.

Many of the same companies are not solving the reliability problems and have not achieved their competitive potential. Many companies in different industries do not integrate quality and reliability. This creates market problems for both the customers and the producers including recalls, complaints, warranty repair cost, and loss of goodwill. These companies experience many problems with rework that increase the cost of the product and reduce their competitiveness in the market. This can lead to financial problems.

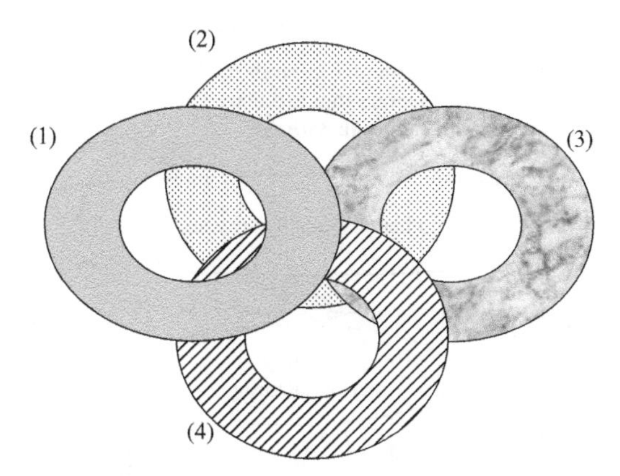

Figure 2.13. Proposed system of integrated quality and reliability (durability/maintainability) with system and rule/responsibility: (1) is quality; (2) is rule/responsibility; (3) is system; and (4) is reliability (durability and maintainability).

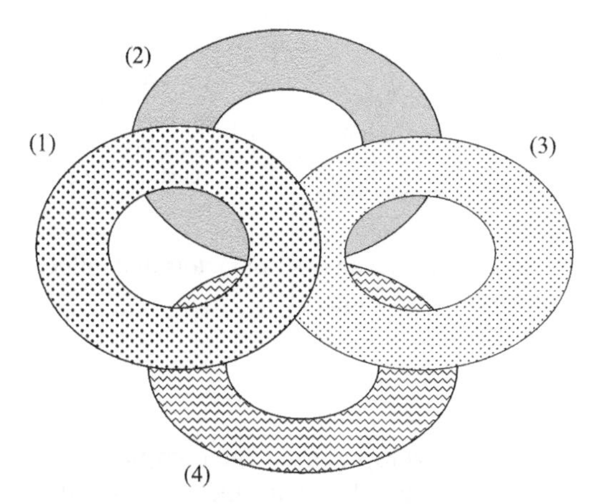

Figure 2.14. Quality, reliability, durability, and maintainability as components of one integrated system: (1) is quality; (2) is reliability; (3) is durability; and (4) is maintainability [18].

Neither GM nor Toyota (Fig. 2.12) adequately considers the repercussions. Both need to change their prioritization. Figure 2.13 shows the approach to an integrated system, where reliability and maintainability are given equal priority with quality, rule/responsibility, and system during the design, manufacturing, and usage phases to achieve better results. Figure 2.14 shows this integration. Using the SoS approach, which integrates the quality, reliability, durability, and maintainability complex, will achieve a more desirable solution.

Industrial companies need this methodology and appropriate equipment to implement this approach.

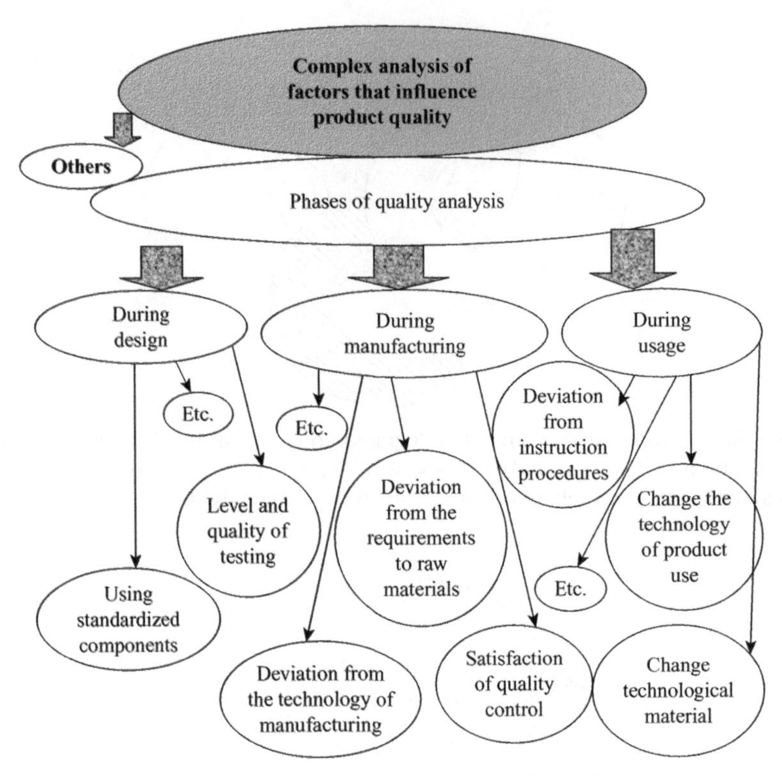

Figure 2.15. The scheme of complex analysis of factors that influences product quality during design, manufacturing, and usage.

How it can be done?

The first element is the development of the complex analysis of factors that influences product quality (Fig. 2.15). This analysis consists of two sections.

The first section includes an analysis in one complex consisting of divergences in the procurement of raw material, in the procurement of product components, in the procurement of the final product, and by the customers. It is based on the author's experience in collaboration with Eugene Klyatis and industrial experience as described in Reference 18. Figure 2.16 shows the current situation in this area.

The second section is shown in Figures 2.14 and 2.17. It includes analysis during the operation, design, and manufacturing phases.

The author and Eugene Klyatis developed the current quality system (Fig. 2.16) that has been successfully used in some companies as shown in Figure 2.17. The quality system was implemented by Iskar, Ltd. (Israel). Warren Buffet later purchased Iskar due to its increased sales and profit attributable to higher product quality. Chapter 7 shows these results and provides substantiation of the improvement.

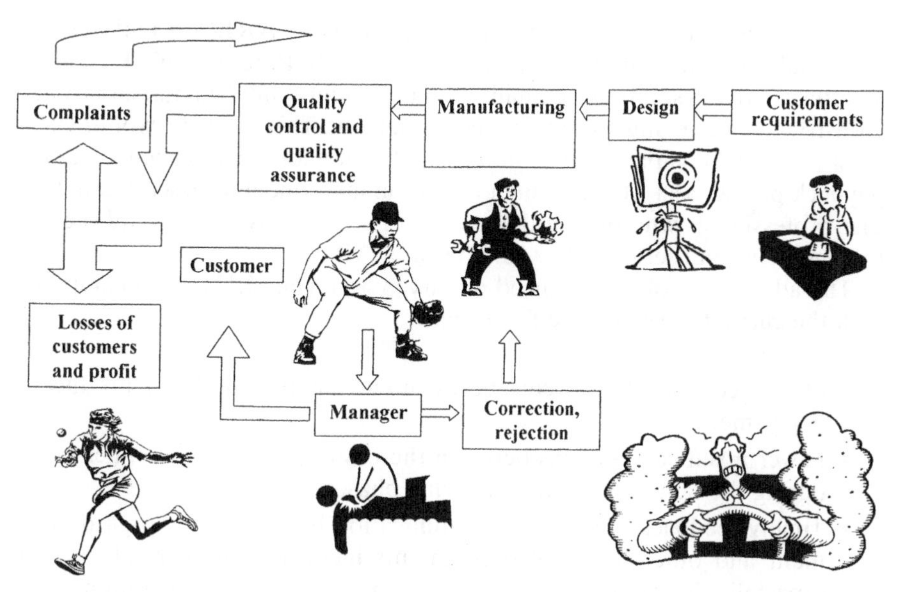

Figure 2.16. Often used current quality system by industrial companies.

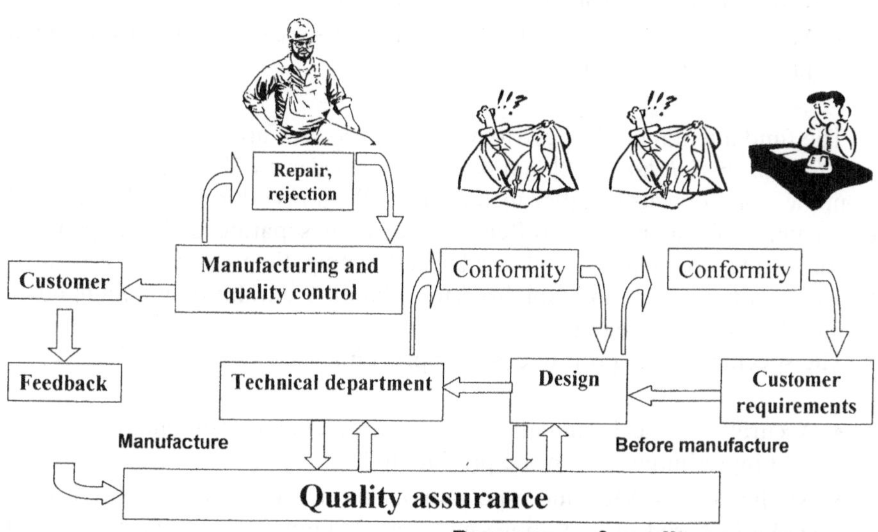

Figure 2.17. The developed quality system for an industrial company.

The second element is the development of useful ART. The author's and industrial experience are both described in detail in Reference 67. To achieve an integrated SoS solution for quality/reliability/durability/maintainability, the industrial company must use this second element. This is a new evolution for most industrial companies. It requires significant additional financing, but the payback period is short with a high return on investment as described in [67]. The professionals who pursue this course of action must understand the difference between useful, useless, and harmful ART.

The advantages of an advanced quality system (Fig. 2.17) when compared with the current norm are the following:

- Online collaboration occurs between customers and all the producer's departments.
- Fewer misunderstandings between the customer and the supplier occur during the design and manufacturing phases.
- Timely feedback and good communication between the design department and other technical departments increase customer satisfaction, especially during the early stages of manufacturing and operation.
- A better management process exists to implement any necessary changes in manufacturing technology.
- Maximum customer participation and satisfaction are achieved through feedback during all stages of design and manufacturing.
- All of the design and production departments are responsible for the high quality of the final product.

The fundamental books describing these processes are *Accelerated Quality and Reliability Solutions* [18] and *Successful Accelerated Testing* [43]. In addition, there are dozens of other technical publications addressing this topic. If one cannot distinguish the difference between separate and independent factor testing versus integrated accelerated testing, then one will not be able to achieve the integration of quality with reliability, durability, maintainability, availability, and serviceability.

The specifics of the proposed SoS approach are

- Accurate physical simulation of the simultaneous combination of the field input influences on the product [68]
- Application of ART and especially useful accelerated reliability testing (UART) based on the above-mentioned simulation (see Section 1.5 of Chapter 1).
- Integration of ART with constant quality improvement [18].

Implementation of this proposed strategy (improved engineering through integrated reliability and quality) should improve the engineering culture of industrial and other types of companies. This strategy has been approved by several industrial companies and is described in some publications [69, 70].

The specifics of the proposed strategy follow:

- Develop integrated reliability, durability, and maintainability solutions that result in quality (Q) improvement during design and manufacturing.
- Conduct accurate physical simulation of full field influences integrated with safety and human factors during design and manufacturing.
- Introduce UART of the actual product based on the above-mentioned simulation.
- Incorporate accurate prediction of quality and reliability on the basis of UART results.
- Develop an integrated quality process for the procurement of materials, components, and vehicles throughout the design, manufacturing, and usage phases. One team of professionals from the design, manufacturing, quality assurance, and marketing departments will provide this integration.
- Incorporate the strategic and tactical aspects of quality/reliability into the process.

Experience shows that this approach can dramatically improve the quality and reliability of a product, reduce recalls, improve the market reputation of a company, and increase sales volume.

The result is an improvement in the company's financial status.

2.9 THE PLACE OF ART/ADT IN HIGH QUALITY, RELIABILITY, MAINTAINABILITY, AND DURABILITY

Industry can develop more accurate predictions for integrated quality and reliability for a given time. This requires the application of accelerated quality and reliability development during the design and manufacturing phases. Normally, the process for accurate prediction of quality and reliability during a given time (warranty period and life cycle) consists of four basic steps (Figure 2.4). The strategy to achieve accurate prediction is presented in Chapter 6.

Clearly, accelerated reliability and durability testing has an important place in product development. In addition, the integrated process provides high quality, reliability, maintainability, and durability.

This system of prediction is accurate if, and only if, the simulation is accurate and ART is possible. Let us analyze this system in steps.

The simulation is the first and very important step of the prediction. One simulates the integrated active field influences and any influences related to any type of research, design, test, and study in the laboratory, or other artificial conditions to provide accelerated reliability and durability testing. One needs to know how effective the research, design, manufacturing, and other solutions will be in the real (field) situation. Current single-dimensional simulation solutions are useful for technology assessment, but they are not practical for reliability, maintainability, supportability, and quality during usage.

The full field input influences on the actual product need to be simulated. In the field, many types of input influences (Fig. 3.2) determine the effects and outcome. They are interconnected with each other and mutually influence each other [42, 70, 71].

If we simulate separate input influences, or just parts of the full complex of these influences, then we will not model the real field situation. As a result, the next steps—studying and testing—will not reflect the real situation. Therefore, such testing and research results cannot give the appropriate information for accurate prediction of quality and reliability during usage time.

The effectiveness of the simulation depends upon its accuracy. An accurate simulation, especially with an accurate prediction, has historically been a key enabling factor for systems and their components, including ART [72–77].

For the successful prediction of product or process quality and reliability, one needs

- Methodology to incorporate all active field influences and interactions
- Accurate initial information from ART/ADT

The possible benefits of using ART/ADT are

- Reducing the time to market of a new product
- Making the product more competitive in the market
- Providing accurate reliability prediction during a given time (warranty period and service life)
- Providing accurate quality prediction during a given time
- Providing accurate maintenance (time and cost) prediction during a given time
- Providing an accurate prediction of the minimal cost service life of the product
- Aiding the rapid identification of the components that limit the product's quality, reliability, and warranty period
- Providing rapid isolation of the reasons for the limitations of the above-mentioned components
- Providing accelerated development and improvement of the product's quality, reliability, maintainability, durability, and supportability
- Reducing the cost of product development and improvement

- Providing a rapid solution to many other problems related to the product's economics, quality, and reliability
- Dramatically decreasing the number of recalls and complaints

This is possible with the implementation of ART/ADT.

EXERCISES

2.1 The technical progress in the operational area is progressing more slowly than in the design, manufacturing, and service areas. Describe why this is the case and illustrate with examples.

2.2 The expenses in the usage phase of the equipment are much greater than in the design, engineering, and manufacturing phases. Give examples of this situation.

2.3 What are the reasons for the increasing number of car recalls? Present examples from Japanese, European, or American companies.

2.4 Are there problems with product reliability and durability? Demonstrate your answer with examples.

2.5 Are reliability and durability problems interconnected with other areas of life? If yes, then identify the interconnecting factors and describe the character of these interconnections.

2.6 Show the scheme of interaction of the field situation components.

2.7 There are several complexes of components that have an influence on the field situation. Determine the basic complexes of these components in general, as well as specifically, for components of cars.

2.8 How many basic components are there in the life cycle of equipment? Describe the basic components.

2.9 What do the reliability and durability processes have in common? What are their differences?

2.10 Input influences act on product reliability and durability. Show these actions step by step.

2.11 What is the role of safety problems on the field situation? How are safety problems interconnected with reliability?

2.12 The process for safety problems is complicated. Identify the basic general components and subcomponents of safety problems.

2.13 Human factors are a complicated process. Determine the basic components of this process and their interconnection.

2.14 What term is used in Europe for "human factors"?

2.15 Show the classification scheme of human factors.

2.16 Demonstrate how the influence of PP can be used in a methodology to solve the reliability problems for a mobile product.

2.17 Demonstrate in a methodological complex how different influences impact an operator's reliability.

2.18 Show the scheme of interconnections of factors that influence an operator's reliability.

2.19 How effective are the human factors in an accurate simulation? Show their advantages.

2.20 Demonstrate and analyze the interconnected factors that affect quality and reliability in all the steps of research, design, manufacturing, service, and operation.

2.21 What are the advantages of the interconnection of quality and reliability solutions and why has this interconnection been a problem for industrial companies?

2.22 Show the basic process for the successful integration of quality and reliability/durability during design and manufacturing.

2.23 Demonstrate schematically why a successful integration of quality with reliability cannot be accomplished by many of the current industrial companies.

2.24 What is the essential aspect for using Exercises 2.21–2.23?

2.25 Describe a reliable way to improve the integration of quality and reliability.

2.26 What is the basic difference in the engineering culture between Toyota and GM?

2.27 Describe the ideal way for GM to dramatically change the engineering culture to improve the role of reliability/durability in this process.

2.28 What strategy is proposed in the book for the solution of Exercises 2.23–2.26?

2.29 Describe the complex analysis of factors that influence the improvement of a product's quality and reliability/durability.

2.30 What specific advantages does an industrial company obtain from a developed quality system?

2.31 Demonstrate schematically how UART influences the accurate prediction of a product's quality, reliability, durability, and maintainability during a long period of time, that is, the warranty period and service life.

2.32 How does ART/ADT impact the quality of research and accelerated improvement of quality, reliability, durability, and maintainability?

2.33 What possible benefits are obtained by using ART/ADT? Schematically describe how to achieve these benefits.

Chapter 3

The Basic Concepts of Accelerated Reliability and Durability Testing

3.1 DEVELOPING AN ACCURATE SIMULATION OF THE FIELD SITUATION AS THE BASIC COMPONENT OF SUCCESSFUL ACCELERATED RELIABILITY TESTING (ART) AND ACCELERATED DURABILITY TESTING (ADT)

3.1.1 General

This chapter describes an approach and methodology for conducting integrated ART/ADT with multiple input influences using simulation. The result should lead to successful ART/ADT. The concept of selecting a representative region for testing is used to extend real-life analysis techniques and to establish the characteristics of the critical input (or output) processes. This leads to the concept of using a representative region that is the most characteristic of the total population representing all anticipated operating regions. Simulation is a tool that uses a representation or model during testing. Simulation is used to evaluate the potential results as a form of testing.

Simulation modeling is used in many contexts to determine how to improve the functioning of natural or human systems, technology, or other products or processes. The key issues in simulation include the acquisition of valid source information for key characteristics and behaviors of the product or process, determination of a method for simplifying approximations and assumptions to build the simulation model, and identification of a methodology to evaluate

Accelerated Reliability and Durability Testing Technology, First Edition. Lev M. Klyatis.

and validate the simulation outcomes. Thus, simulation is an attempt to accurately represent the real situation during research and evaluation.

There are different types of simulations: physical, interactive, computer (software), mathematical, and others. This book considers the physical simulation of the field circumstances applicable to an actual product or process.

Usually, physical simulation refers to a simulation that substitutes physical objects and conditions for the real ones expected during the operation of the product or process. Historically, these physical objects are often selected to be smaller than the actual object or system.

The author considers physical simulation to be the simulation of the actual usage conditions representative of those experienced by the real product or process. These conditions include multiple input influences, human factors, and safety problems. Thus, in the book, the items used in the simulation are typically not smaller or cheaper than those in the real object or system.

3.1.2 Contents

For the ART of a product, the field situation needs to be simulated in the laboratory where artificial input influences are used to model the actual field influences. These input influences are in physical contact and interact with the test subject. For example, for a ground vehicle, it is necessary to provide the physical simulation of the road features (surface, concrete, asphalt, density, pitch, and curvature), environmental factors (temperature, humidity, air pressure, chemical and dust pollution, and radiation), input voltage, electrostatic discharge, and other active influences. Special chambers, shock, vibration equipment, corrosion equipment, and others are utilized for accelerated testing to produce valid results. To achieve the necessary physical simulation of the field input influences in the laboratory, a wide complex of specific special devices and universal devices is used. The physical simulation of the field input influences has to be accurate in order to provide useful ART/ADT results.

The first basic problem is the need to understand what kind of field input influences will be simulated in the laboratory and the purpose of these physical simulation influences.

Understanding how the various types of input influences needed for testing act upon the test subject in the field during its operation and storage (Fig. 3.1) is required. These influences include temperature, humidity, pollution, radiation, road features, air pressure and fluctuations, input voltage, and many others ($X_1 \ldots X_N$).

The results of their action are output variables (vibration, loading, tension, output voltage, and many others ($Y_1 \ldots Y_M$)). The output parameters lead to degradation (deformation, crack, corrosion, and overheating) and failures of the product.

A detailed description of input influences and output variables is provided in the examples [18]. The following practical examples demonstrate how

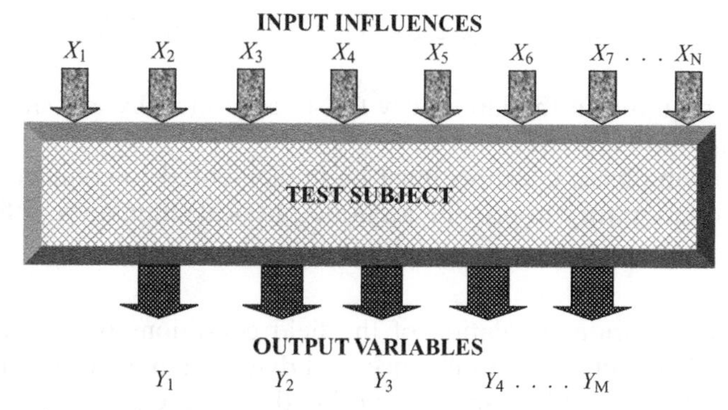

Figure 3.1. Scheme of input influences and output variables of the actual product.

TABLE 3.1 Tensions from Different Sensors of a Car's Trailer in Different Fields

Sensor No.	First Field			Second Field			Third Field		
	m	σ	*f*	*m*	σ	*f*	*m*	σ	*f*
2	6.75	45.0	20	−14.5	44.5	21	−6.5	33.0	—
4	−40	51	18.4	−121	38.0	22.8	−113	35	15.2
5	−450	89.5	11.5	−374	58.0	13.1	−178.0	63.0	14.6
6	−179	103.0	12.7	−46.4	78.0	18.9	−124.0	65.0	16.2
7	91.5	54.0	23.0	89.0	34.5	22.0	13.1	32.0	—

m, mathematical expectation of tension (kg/cm^2); σ, standard deviation (kg^2/m^4); *f*, mean of frequency (1/s).

output variables (tensions) change in different fields (regions). Table 3.1 shows how tension readings can vary for a car's trailer on the same type of road.

Always simulate the full range of input influences ($X_1 \ldots X_N$) in the laboratory when testing the product. ART/ADT offers the initial information for accurate prediction of reliability, quality, maintainability, durability, and other parameters.

An accurate physical simulation occurs when the physical state of output variables in the laboratory differs from those in the field by no more than the allowable limit of divergence. The literature shows that the simulation is accurate if the mean of output variables (vibration, loading, tensions, voltage, amplitude, and frequency of vibration) during testing differs from the mean of the same output variables in the field by no more than a given limit, for example, 3%.

This means that the output variables

$$Y_{1\,\mathrm{FIELD}} - Y_{1\,\mathrm{LAB}} \leq \text{given limit (e.g., 1, 2, 3, and 5\%)}.$$

$$Y_{M\,\mathrm{FIELD}} - Y_{M\,\mathrm{LAB}} \leq \text{given limit } (1, 2, 3, \text{ and } 5\%).$$

What is required in the laboratory for an accurate physical simulation of the field input influences?

The following basic concepts must be implemented to conduct ART and ADT in the laboratory to provide accurate physical simulation of field input influences:

- Provide accurate simulation of the field conditions (quantity of influences, character of each influence, and dynamics of changes of all influences), safety, and human factors using the given criteria.
- Conduct simulation testing 24 hours a day, every day, but not including
  - Idle time (breaks or interruptions) or
  - Time operating at minimum loading that does not cause failure.
- Accurately conduct simulation of each group of field input influences (multi-environment, electrical, and mechanical) in simultaneous combinations. For example, the multi-environmental group is a complex simultaneous combination of pollution, radiation, temperature, humidity, air fluctuation, and other environmental factors.
- Use a complex system to model each interacting type of field influences. For example, pollution is a complex system that consists of chemical air pollution + mechanical (dust) air pollution, and both types of this pollution must be simulated simultaneously.
- Simulate the whole range of each type of field influences and their characteristics. For example, if one simulates a temperature, then one must simulate the whole range of this temperature from the minimum to maximum, its rate of change, and the random character of this temperature if it changes randomly.
- Use the physics-of-degradation process as a basic criterion for an accurate simulation of the field conditions.
- Treat the system as interconnected using a systems of systems approach.
- Consider the interaction of components (of the test subject) within the system.
- Conduct laboratory testing, in combination with special field testing, as a basic component of ART/ADT.
- Reproduce the complete range of field schedules and maintenance or repair actions.
- Maintain a proper balance between field and laboratory conditions.
- Correct the simulation system after an analysis of the degradation and failures in the field and during ART/ADT.

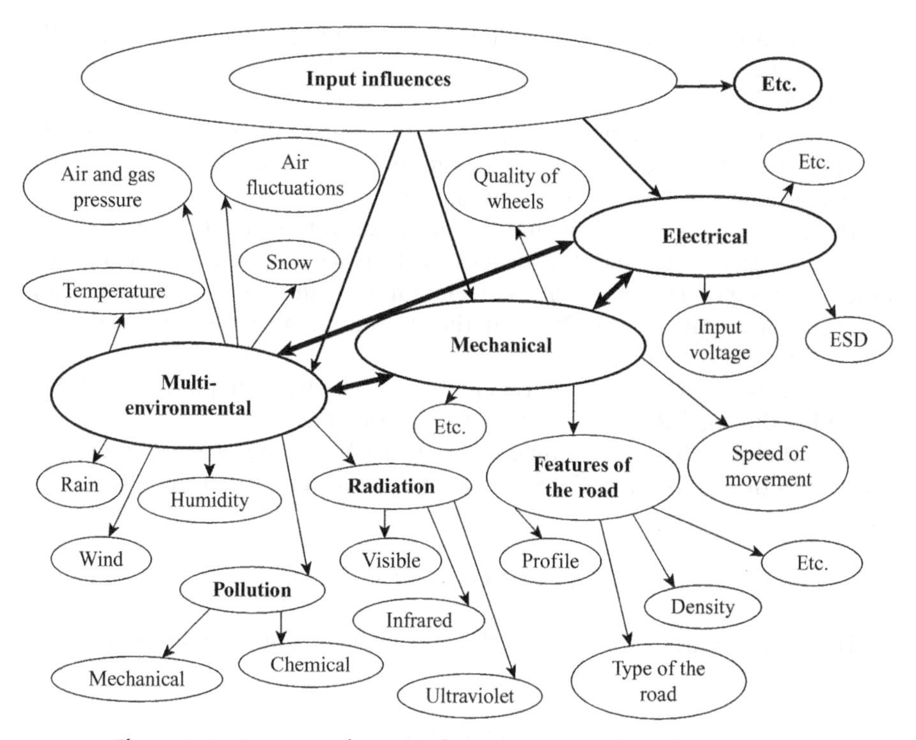

Figure 3.2. Contents of input influences on actual product/process.

These concepts are demonstrated in more detail in References 18 and 43 and in other authors' publications. Simulation of a comprehensive combination of input influences is necessary for accurate simulation.

What does it mean to simultaneously simulate a combination of input influences?

Consider the actual simultaneous combination of influences in the field for automobiles (Fig. 3.2). Most of the field influences act and interact simultaneously. Therefore, for an accurate simulation of the field input influences (the job of appropriate testing equipment), the influences have to act simultaneously and in mutual combination.

Current accelerated testing usually fails to include these interactions. For example, engineers typically simulate the environmental, mechanical, electrical, and other input influences independently. Consequently, they conduct the simulation of chemical and mechanical (dust) pollution separately.

It is very important to understand and identify all groups of interconnected input influences and their many complicated interactions. These interactions have to be simulated in the laboratory for useful ART/ADT.

Likewise, radiation simulation is conducted separately from the simulation of other environmental influences. As a result, one does not take into account the interconnections among different environmental influences.

The methodology for accurate simulation of input influences during ART is demonstrated in Chapter 4. The equipment for this effort is shown in Chapter 5.

The second basic problem is to accurately simulate the range for each type of input influence. The input influences are part of a complicated process, and changing them usually produces randomness.

The author's experience is that the most accurate simulation of the input processes occurs when the statistical characteristics (i.e., mathematical expectation μ, variance D, normalizing correlation $\rho(\tau)$, and power spectrum $S(\omega)$) of all the input influences differ from the measurements under operating conditions by no more than 10% [42].

The following statistical criteria can be useful when comparing reliability/durability during ART/ADT and operating conditions.

The reliability distribution function after ART is termed $F_a(x)$. Under operating conditions, it is $F_0(x)$. The measure of the difference is given in Reference 42 as

$$\eta[F_a(x), F_0(x)] = F_a(x) - F_0(x).$$

The function $\eta[F_a(x), F_0(x)]$ has the limit η_A (maximum of difference). When

$$\eta[F_a(x), F_0(x)] \leq \eta_A,$$

it is possible to determine its reliability using the ART results. But if

$$\eta[F_a(x), F_0(x)] > \eta_A,$$

using the ART results is not recommended to predict reliability.

Often in practice, the values of the functions $F_a(x)$ and $F_0(x)$ are not known, but it is possible to construct graphs of the experimental data for $F_a(x)$ and $F_0(x)$ and to determine the difference

$$D_{m,n} = \max[F_{ae}(x) - F_{oe}(x)],$$

where F_{ae} and F_{oe} are empirical distributions of the reliability function observed by testing the product under operating conditions using accelerated testing.

One can also find $D_{m,n}$ from graphs of $F_{ac}(x)$ and $F_{oc}(x)$.

To do this, it is necessary to determine the value of

$$\lambda_0 = \frac{\sqrt{mn}}{(m+n)(D_{m,n} - \eta_A)},$$

where n is the number of failures in operating conditions and m is the number of failures in the laboratory.

The correspondence between comparable distribution functions is evaluated using the probability

$$P[\sqrt{mn}/(m+n)(D_{m,n}-\eta_A)\geq\lambda_0]<1-F(\lambda_0).$$

If $[1 - F(\lambda_0)]$ is small (not more than 0.1 typically), then

$$\max[F_{ac}(x)-F_{oc}(x)]>\eta_A.$$

An analogous approach has to be used for human factors and safety.

Then, the distribution function for ART/ADT does not correspond to the distribution function in the field. Thus, it is probably not possible to develop an appropriate system for the physical simulation of field input influences in the laboratory (methods, equipment, or both).

The first conclusion from the above-mentioned criterion is that it is useful to compare reliability and durability changes during service life in the field. The second conclusion is that this information may help to determine the limitation on the reliability function between laboratory testing and testing under operating conditions (field). The level of accuracy of the reliability evaluation depends on this limitation.

The choice of influences for ART/ADT is determined by the specific requirements for the test subject under operating conditions. Therefore, it is very important to evaluate the influence of the statistical characteristics of the operating conditions in a region that approximates operating conditions. This is necessary to carry out the requisite physical simulation in a laboratory during ART or experimental research. The point is to represent characteristics that are as close as possible to the statistical characteristics for field operations.

The solution to this problem is very complicated, and it requires new theoretical research. The principal steps are described next.

The accurate simulation of input processes provides a means to calculate the acceleration coefficient. This result enables an accurate calculation to predict the dynamics of changes in reliability, durability, and maintainability during operation in the field after ART/ADT.

How can one evaluate the accuracy of physical simulation? To determine the accuracy, different types of criteria have to be used.

Professionals often use the correlation of ART/ADT results to field-testing results. Reliability indices such as time to failures, types of failures, and the distribution of failures are typically used. Considerable test time is usually required to develop these measures, especially if one wants to evaluate a product's durability.

Another negative aspect of this method is that it typically leads to the destruction of the product. Also, professionals need early knowledge about how accurate the field simulation is to be sure that the criterion was used correctly, or to see whether an earlier correction of the criterion is necessary.

To accomplish this requires the comparison of output variables (vibration loading and tension) in the field and in the laboratory. Sensors that measure tension and load are frequently used. They are easy to install, have a short and simple output, and are frequently used. Using this method does not require a long time, but experience shows that this method often indicates when the primary output parameters (tensions and loads) are not sufficient to characterize the simulation process. A good correlation can exist between tensions and loads during field and laboratory testing, but an insufficient correlation may still occur between the time to failure and other reliability indices.

Experience shows that a comparison of the degradation mechanism or time to failure during ART/ADT or experimental research during field testing is a better method.

The destructive testing and appropriate destructive failure analysis of components of the system are not useful for reliability/durability testing and accurate prediction.

The physical (chemical, mechanical) mechanism of the degradation approach is the second and basic criterion for an accurate simulation of real-life processes during ART. It compares the results of field and laboratory testing and enables quick correction for an incorrect simulation process.

This approach assumes that the degradation mechanism during ART/ADT must be approximately similar to the degradation mechanism during field testing [43]:

$$D_L \approx D_F,$$

where

D_L is the degradation mechanism during ART and

D_F is the degradation mechanism during field testing.

The degradation mechanism may provide a stress limit. When the stress is over that fixed limit, the degradation mechanism during laboratory testing is different from the degradation mechanism in the field. This relates to highly stress and test approaches.

The "physics-of-degradation" result is not enough for using ART.

For example, chemical degradation can be chemical corrosion as well as a change in the chemical structure of a plastic that has a different degradation mechanism. This is also true for mechanical degradation or electrical degradation. Therefore, the type of degradation (physical, chemical, or electrical) depends on the type of process and on the subject of testing (research).

The degradation mechanism can be estimated through its degradation parameter during product testing (Fig. 3.3).

Each product has different parameters for the degradation mechanism. For an accurate simulation, one must evaluate the limits for each parameter.

Accelerated testing usually uses only a stress-based simulation. However, in this case, it is necessary to remember the following axiom of stress testing:

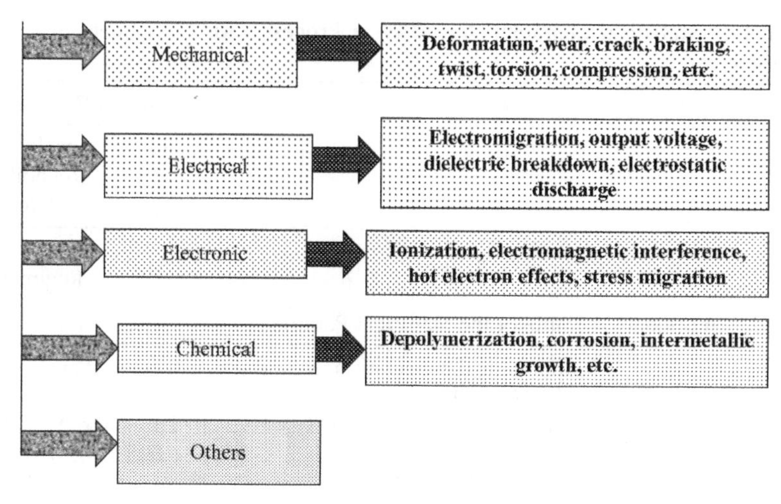

Figure 3.3. Types of degradation mechanisms and their parameters.

More stress means greater acceleration and a lower correlation of accelerated testing results with the field results. This is true when performing ART/ADT.

In real life, the mechanical, chemical, and physical types of degradation mechanisms often interact with each other. The degradation parameters for mechanical degradation are deformation, wear or cracks, and other changes. The degradation parameters for chemical degradation are corrosion, depolymerization, intermetallic growth, and other changes.

The degradation parameters for electrical degradation are electromigration, electrostatic discharge, dielectric breakdown, and other changes. The degradation parameters for electronic degradation are ionization, electromagnetic interference, hot electron effects, beta degradation, stress migration, and other changes.

If sensors can be employed to evaluate these basic parameters, then one can measure their rate of change during a particular time to compare field and laboratory testing, and determine how similar the ART conditions are to real-life conditions. If these processes are similar, then there may be a sufficient correlation between the ART/ADT results and the field-testing results.

Consider the comparison between the parameters of the degradation mechanism for wear of the common drive shaft of a harvester after 600 hours of field testing and the appropriate time for ART. This parameter (wear) has a probabilistic character during the test time and can be evaluated by three statistical characteristics: the mathematical expectation $\bar{A}$, the standard deviation σ^2, and the time of correlation τ_k.

The results are presented in Table 3.2. The difference between these testing results (degradation) is no more than 10%. Therefore, ART offers accurate accelerated information about wear in the field.

TABLE 3.2 Statistical Characteristics of Wear Processes for Component of Harvester

Studied Component	Name of the Statistical Parameter	Value of Parameters: Field	Accelerated Reliability Testing, Number of Regimens: 1	2
Common drive shaft	Mathematical expectation $(H \cdot M) \times 10^{-1}$	20.2	19.3	21.2
	Standard deviation σ^2 $(H \cdot M) \times 10^{-2}$	181.1	186.0	191.0
	Time of correlation, τ_k, c	0.09	0.10	0.08

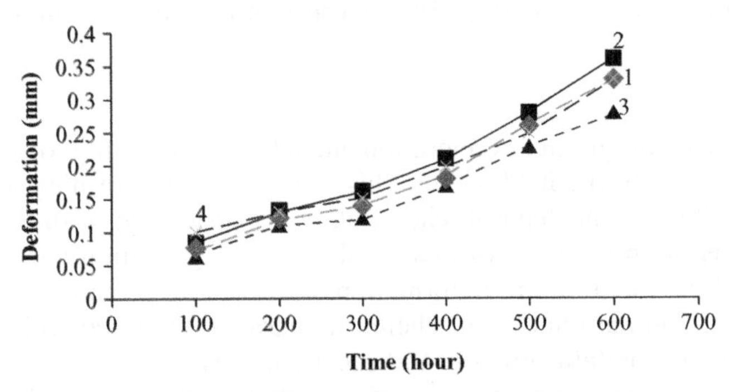

Figure 3.4. Deformation of a metallic sample during the time [43].

How is a practical comparison between the parameters of the degradation mechanism in the field and in the laboratory conducted? Consider the metallic sample deformation in Figure 3.4. The value of confidence interval is ±10%. The mean of the deformation in laboratory testing is within the ±10% confidence interval (between the upper and lower confidence limit) of the field deformation. This means that there is a sufficient correlation between the accelerated testing results and the field results. The confidence interval of ±10% is a good measure in practice. The comparison process of the field and laboratory results has a statistical character.

What is the connection between the physical simulation (its result, the degradation) and ART? It is a complicated process, because we need to take into account many components of the test subject and appropriate degradations. The degradation and failures in the field and in the laboratory (if there is an accurate simulation of the field influences in the laboratory) are a result of the combined complex of factors (multi-environmental, mechanical, electrical, etc.) influencing the product's reliability.

3.2 CONCEPTUAL METHODOLOGY FOR THE SUBSTANTIATION OF A REPRESENTATIVE REGION FOR AN ACCURATE SIMULATION OF THE FIELD CONDITIONS

The test subject may experience different field conditions and stresses in operation due to physical location. The field conditions may include different climate zones where the test subject is being used, as well as other varying field situations. The regions where the product will be used experience different environments, different mechanical influences, road surfaces, loads, and electrical influences, among others. The human factors and safety problems are also variable. In this case, for accurate simulation of all these different conditions (factors) in ART/ADT, it is useful to select a region that is representative of the multitudes of field conditions. The author presents a methodology to determine a representative input region for an accurate simulation of the field conditions [18]. This process can be used for a wide range of parameters. Each region has a different influence on the product's quality, durability, reliability, maintainability, and other factors. This is especially true for mobile products. Therefore, a region representative of multiple field conditions needs to be simulated to provide ART/ADT. The region can be characterized by input influences, output parameters, safety, or human factors, that is, the parameters that best represent the multiple characteristics of the applicable regions. The author developed a special methodology for the selection of a representative region. The full methodology is published in Reference 99. The basic components are shown in the subsequent example using the results of sensor data. There, data can be the result of vehicle load measurements that reflect the field condition in the region. We identify there data by the code name "oscillogram." The methodology is summarized in the algorithm presented next.

3.2.1 Algorithm

The algorithm for selecting a representative region for an accelerated test is the following:

Step 1. Identify the type of process exhibiting a random rate of change and establish whether it is a stationary process through the following steps:

A. Characterize the process in terms of its mathematical expectation, standard deviation, correlation, and power spectrum.

B. Evaluate whether the process is ergodic by assessing whether the important correlation tends to zero as time tends to infinity.

C. Use the Pearson criterion, or equivalent, to assess the normality of the process.

Step 2. Measure the divergence between the basic characteristics of the process evaluated in different regions.

Step 3. Select and measure a characteristic (usually length) in a representative region.

Step 4. Identify the representative region that has minimal divergence.

3.2.2 Selection of the Process and Its Characterization

Assume that there is an interval $[0, T]$ in the written characteristics of the measurement of a process. Name this an oscillogram that characterizes the changes in loading or wear (or another physical parameter, X) of a product during use. Next, divide the interval $[0, T]$ of the oscillogram into the smaller intervals of length T_p and evaluate the mathematical expectation (μ_X), correlation function $\rho_X(\tau)$, power spectrum $S_X(\omega)$, and their accuracy. For example, one can use the Lourie method [100]. The ergodicity of the process can be checked to ensure that it appropriately uses one interval realization of the process as being representative of all realizations of the same duration. The ergodicity of the process can be assessed formally by checking that the correlation function tends to zero as $\tau \to 0$. It is appropriate to combine this information with informed judgment relating to the particular application and the physical conditions of the process being studied.

To check the hypothesis that the process exhibits normality, assume that using one realization of the specified interval length T_p is sufficient to characterize the entire process because it is ergodic. From the correlation function, identify the value of τ for which the correlation is (statistically) zero, that is, $\rho_X(\tau_0)$. For the interval $(0, \tau_0)$ in the oscillogram, measure the values at equidistant points $t_1, t_2, t_3 \ldots t_n$. The usual goodness-of-fit tests, such as Pearson's, can be used to assess whether these values are normally distributed.

3.2.3 The Length of the Representative Region T_R

If it is established that the process is normal, then begin to evaluate the length of the representative region T_R that has to be large enough to evaluate the correlation function ($\rho_X(\tau)$) and the mathematical expectation because of the accuracy of μ_X.

The interval representing T_R is divided into n equally spaced subintervals of length,

$$\Delta t = \frac{T_R}{n},$$

and the observed values of the oscillogram at these times are denoted by $x(t_i)$. Hence, the correlation function can be estimated using

$$\hat{\rho} = \frac{1}{m-n} \sum_{t_i=1}^{n-m} [x(t_i + m) - \mu_X][x(t_i) - \mu_X]. \tag{3.1}$$

To calculate the preliminary values of T_R, it is necessary to know the frequency ranges in the oscillogram where the parameters are changing. The row of full low-frequency periods, denoted by T_1, T_2, $T_3 \ldots T_n$, each having the random rate, is marked at equal intervals from each other. The mean, denoted by T_M, can be evaluated by

$$T_M = \frac{T_1 + T_2 + \ldots + T_n}{n},$$

where n ≥ 3.

The lowest frequency of the process rate is given by $f_L = 1/T_M$ and the value of τ_{max} can be selected using the condition $\tau_{max} \approx 1/f_L = T_M$ [76]. The value of T_R can be established from τ_{max}. Typically, $T_R = 10\tau_{max}$ because the error associated with the estimation of μ_X and $\rho_X(\tau)$ should be no more than 5%.

To estimate Δt, consider the highest frequency changes of the process, which are denoted by f_m and, in practice, $\Delta t = 1/0.6 f_m$.

3.2.4 Comparison of Values between Different Regions of the Oscillogram

Stationary normal processes can be characterized by their expectation (μ), variance (σ^2), and normalized correlation structure ($\rho_N(\tau)$). Evaluate the entire oscillogram over the range [0, T] and for k different regions of the oscillogram of length T_R, where $k = T/T_R$. For example, denote the vector of the characteristics of the ith region as $[\hat{\mu}_i, \hat{\sigma}_i, c|\rho_{Ni}(\tau)|]$, where c is the normalization constant. For the entire oscillogram, the equivalent vector will be written as $[\hat{\mu}, \hat{\sigma}, c|\rho_N(\tau)|]$, where

$$\hat{\mu} = \frac{\sum_{i=1}^{k} \hat{\mu}_i}{k}, \quad \hat{\sigma} = \frac{\sum_{i=1}^{k} \hat{\sigma}_i}{k}, \quad \hat{\rho}_N(\tau) = \frac{\sum_{i=1}^{k} \hat{\rho}_{Ni}}{k}, \quad k = \frac{T}{T_R}. \tag{3.2}$$

The divergence between any two regions, $\underline{a_i}$ and $\underline{a_j}$, may be estimated as a weighted linear combination of the squared differences between their expectations, variance, and correlations. For example,

$$\Delta(\underline{a_i}, \underline{a_j}) = p_1(\hat{\mu}_i - \hat{\mu}_j)^2 + p_2(\hat{\sigma}_i - \hat{\sigma}_j)^2 + p_3 \max[\hat{\rho}_{Ni}(\tau) - \hat{\rho}_{Nj}(\tau)]^2, \tag{3.3}$$

where the weights p_1, p_2, and p_3 sum to one and are selected using the method of least squares to minimize the sum of squared deviations between the corresponding components of the two vectors. Hence, higher weights will be given to the larger differences between the corresponding statistics. If there is no difference between any of the components within the vector, then all components will be equally weighted. Using the maximum deviation between the correlations as a measure of the most extreme difference between the two correlations is recommended. Only a selection of information in the correlation structure is utilized. Using the estimated spectral power function ($\hat{S}(\omega)$) instead of the correlation function may be preferred. In this case, the deviations between the two regions are written as

$$\Delta(\underline{a_i}, \underline{a_j}) = p_1(\hat{\mu}_i - \hat{\mu}_j)^2 + p_2(\hat{\sigma}_i - \hat{\sigma}_j)^2 + p_3\left[\hat{S}_i(\omega) - \hat{S}_j(\omega)\right]^2, \tag{3.4}$$

where $p_1 + p_2 + p_3 = 1$.

3.2.5 Choice of the Representative Region

Select the region that has minimum divergence among the whole range of regions. Use it as the representative region. The representative region will be denoted by the vector $\underline{a_R}$ and minimizes the following function written first for the correlation structure and then for the power spectrum:

$$\min \Delta_1(\underline{a_R}, \underline{a}) = \min\left\{p_1(\hat{\mu}_R - \hat{\mu})^2 + p_2(\hat{\sigma}_R - \hat{\sigma})^2 + p_3 \max_{\tau}[\hat{\rho}_{NR}(\tau) - \hat{\rho}_N(\tau)]^2\right\}$$

or

$$\min \Delta_2(\underline{a_R}, \underline{a}) = \min\left\{p_1(\hat{\mu}_R - \hat{\mu})^2 + p_2(\hat{\sigma}_R - \hat{\sigma})^2 + p_3 \max_{\omega}\left[\left(\hat{S}_R(\omega) - \hat{S}(\omega)\right)\right]^2\right\}. \tag{3.5}$$

Thus, the region of length T_R that satisfies the above-mentioned condition is selected for use in simulation during accelerated testing. It provides a representative selection of the oscillogram of reduced length, that is, $(0, T_R)$ rather than $(0, T)$, which will facilitate a more efficient testing.

3.2.6 An Example of Selecting a Representative Region

The following example illustrates the application of the proposed methodology in an actual accelerated reliability test. The data to be analyzed come from the oscillogram that was written by a tension sensor on a truck with a mobile computer. The curve of the oscillogram consists of 20 separate regions, each of equal length 800 mm, and reflects the influence of the field on the sensor. The estimates of the mean and standard deviation over all regions of the oscillogram are $\hat{\mu} = 1.58$ mm and $\hat{\sigma} = 5.98$ mm. The statistical characteristics of the tension changing in 19 regions are summarized in Table 3.3. Note that the values of the estimated parameters are shown in millimeters (1 mm = 2 MПa). The data for region 15 have been excluded since they exhibited very different characteristics from the other regions. It represented an atypical field experience. Figure 3.5 shows the correlation function for regions 10–12.

To illustrate the process used to assess the normality of the data, consider the data for region 6. The first time the estimated correlation tends to zero is 0.1 second, set $\tau_0 = 0.1$ second. In the region $(0, \tau_0)$, 158 observed values occur-

TABLE 3.3 Statistical Characteristics of Variable Tension Process [98]

Number of Regions	$\hat{\mu}_i$	$\hat{\sigma}_i^2$	$\hat{\sigma}_i$	$\lvert\hat{\mu}_i - \hat{\mu}\rvert$	$\lvert\hat{\sigma}_i - \hat{\sigma}\rvert$	$\max_{\tau}\lvert\hat{\rho}_{Ni}(\tau) - \hat{\rho}_N(\tau)\rvert$	$\max_{\omega}\lvert\hat{S}_i(\omega) - \hat{S}(\omega)\rvert$
1	−0.72	68.3	8.26	2.3	2.28	0.23	0.024
2	2.22	45.3	6.73	0.64	0.75	0.26	0.024
3	2.79	35.6	5.9	1.21	0.08	0.11	0.005
4	1.12	21.6	4.65	0.46	1.33	0.13	0.08
5	1.04	33.2	5.76	0.54	0.22	0.15	0.007
6	3.66	30.2	5.49	2.08	0.39	0.18	0.023
7	−1.21	42.7	6.5	2.79	0.52	0.15	0.011
8	3.48	27.8	5.27	1.9	0.71	0.24	0.008
9	1.47	56.4	7.5	0.11	1.52	0.15	0.012
10	2.21	26.6	5.16	0.63	0.82	0.17	0.015
11	1.5	51.4	7.17	0.32	1.19	0.24	0.012
12	1.71	39.4	6.8	0.13	0.82	0.16	0.010
13	0.08	24.8	4.98	1.5	1.00	0.19	0.023
14	−0.19	30.8	5.55	1.77	0.43	0.15	0.013
16	1.03	34.0	5.84	0.55	0.14	0.23	0.020
17	5.0	22.2	4.7	3.42	1.28	0.27	0.027
18	4.6	18	4.24	3.02	1.74	0.20	0.012
19	−0.95	35	5.92	2.53	0.06	0.19	0.003
20	0.82	38.4	6.2	0.76	0.22	0.18	0.018

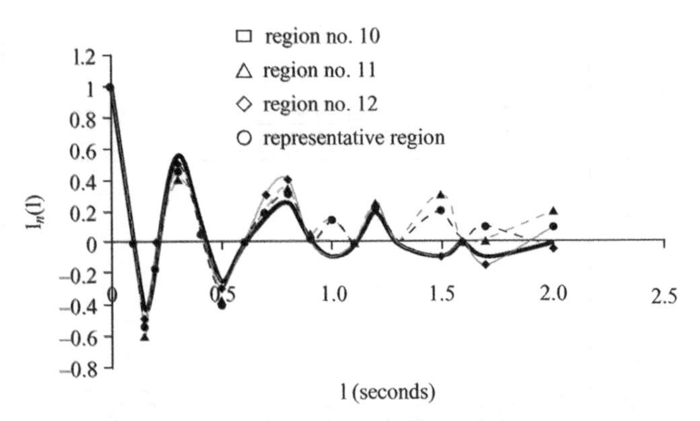

Figure 3.5. The correlation functions of regions 10–12 and the representative region.

TABLE 3.4 Frequencies and Probabilities for the Groups in Region 6

Groups	1	2	3	4	5	6	7	8	9	10	11
Frequency	1/225	5/225	18/225	25/225	35/225	57/225	47/225	22/225	9/225	3/225	3/225
Probability	0.0078	0.0238	0.0628	0.1227	0.1875	0.2127	0.1816	0.1167	0.0564	0.0205	0.0067

TABLE 3.5 Frequencies and Probabilities for the Groups in Region 8

Groups	1	2	3	4	5	6	7	8	9	10
Frequency	5/225	9/225	27/225	44/225	47/225	55/225	45/225	8/225	3/225	3/225
Probability	0.0221	0.0493	0.1146	0.1878	0.2242	0.1952	0.1958	0.0579	0.0189	0.0046

ring at a random rate were recorded. These data values are 33, 30, 55, 46, 55, 42, 29, 38, 42, 48, 40, 41, 38, 37, 35, 32, 32, 40, 43, 46, 28, 20, 31, 48, 44, 36, 30, 30, 37, 38, 40, 32, 34, 36, 36, 36, 38, 34, 25, 33, 41, 39, 34, 28, 38, 44, 54, 41, 25, 33, 49, 45, 40, 37, 37, 37, 36, 34, 31, 26, 36, 35, 43, 33, 25, 30, 33, 39, 33, 38, 29, 30, 32, 40, 42, 44, 39, 36, 26, 27, 46, 39, 30, 31, 34, 38, 33, 36, 44, 40, 28, 28, 36, 38, 36, 39, 35, 38, 38, 36, 35, 37, 39, 38, 31, 32, 34, 39, 38, 42, 39, 28, 31, 44, 42, 39, 37, 36, 37, 27, 30, 43, 42, 36, 35, 38, 37, 32, 36, 40, 36, 38, 37, 30, 37, 40, 38, 34, 31, 33, 32, 31, 43, 38, 24, 30, 35, 40, 37, 34, 44, 39, 28, 28, 36, 36, 38, and 37.

The estimated mean and standard deviation for region 6 are 36 and 5.56, respectively. Dividing the range of the data into 11 equal classes, compare the observed frequency with that expected under a normal distribution as shown in Table 3.4. The chi-squared goodness of fit [100] gives a test statistic of $\chi^2 = 9.14$, and since this statistic has a chi-squared distribution with 8 degrees of freedom, the P-value is found to be 0.34. This suggests that the normal distribution is a suitable model for the data.

Check the normality of data from region 8. Again, 158 sample values were analyzed. Table 3.5 shows the probabilities observed over 10 equal-sized classes, and the mean and standard deviation for the data set are 36.68 and 5.25, respectively. The chi-squared goodness-of-fit test statistic is 6.9 and has a chi-squared distribution with 7 degrees of freedom, giving a P-value of 0.43.

Again this provides evidence that the normal distribution is a suitable model for the data. This is the case for the analysis of all 19 regions.

Additional checks on the expectation and variance of the process suggest that it is not stationary. For example, consider the results presented in Table 3.3 for the mean, variance, and standard deviation. There is a considerable spread in the estimates. In contrast, the estimate's correlation across regions shows that the maximum deviations in the normalized correlation are almost equal. This suggests the process is not stationary.

Analysis of the power spectrum over the interval of ω from 9 to 25 s^{-1} covers most of the spectral density as shown in Figure 3.5. The correlation function for regions 10–12, shown in Figure 3.5, appears to have a periodical component. Therefore, the correlation should take in a larger interval, such as from 0 to 10 seconds.

The modulus of mathematical expectations, standard deviations, correlations and spectral functions, and the divergences from the different regions are based upon the characteristics of the whole oscillogram. The divergences presented in Table 3.3 are substantial. An alternative way of examining the deviations between each region and the whole oscillogram is to superimpose the corresponding functions on the same graph. Taking the small influence of the tension expectation value into account gives estimates of $p_1 = 0.1, p_2 = 0.5$, and $p_3 = 0.4$ and allows calculation of the divergence of Δ_1 and Δ_2. The results are presented in Table 3.6. Regions 3 and 19 are closer in their characteristics to the whole oscillogram than any of the other regions. Criterion Δ_1 is closer to region 3, and criterion Δ_2 is closer for region 19. The power spectrum shows that frequency is the most important factor that characterizes the dispersion distribution (amplitude of oscillation). Frequency is a very important technical characteristic. Therefore, region 19 provides the best input influences to represent all regions.

The associated procedures are measures of the representative region with a minimal divergence resulting from the multitude of input influences (or output variables) under field conditions. The whole process is illustrated by a practical example. The solution is shown for normal processes by the use of the Pierson criterion, but it could also be shown using other types of random rates.

This methodology is useful in substantiation of ART regimes. It leads to successful ART. This helps to obtain accurate preliminary information for reliability prediction, for predictive maintenance, and for the solution of other problems of reliability.

3.3 BASIC PROCEDURES OF ART AND ADT

The following procedures demonstrate how to provide ART/ADT.

Many industrial companies use separate field conditions or combine several field conditions. They seldom use fully integrated multidimensional ART technology. This is one of the basic reasons that their methods, test equipment, and processes used for accelerated testing often give erroneous results and cannot properly predict operational performance.

TABLE 3.6 Divergences Measuring Between Characteristics of 19 Regions

Divergence	R	E	G	I	O	N	S			R	E	G	I	O	N	S			
Measuring	1	2	3	4	5	6	7	8	9	10	11	12	13	14	16	17	18	19	20
$\Delta_1(\underline{a}_i, \underline{a})$	12.55	4.85	0.96	7.27	1.75	2.37	3.48	4.76	8.21	4.84	6.94	4.75	5.91	2.93	1.68	6.82	9.8	1.31	1.9
$\Delta_2(\underline{a}_i, \underline{a})$	12.59	4.93	0.72	7.07	1.45	2.57	3.32	4.12	8.09	4.76	6.46	4.51	6.07	2.85	1.56	6.82	9.48	0.17	0.49

The following components of the basic procedures of ART/ADT technology can help to dramatically improve the accelerated testing, then accurate prediction process.

In 1999, the author first published the step-by-step ART/ADT process that follows [101]. This technology consist of the following 11 procedures.

3.3.1 Procedure 1: Collection of the Initial Information from the Field

This step investigates the kind of input influences that act on the specific product in the field. Select the input influences that must be simulated in the laboratory to provide accurate ART/ADT. This information is necessary to substantiate accurate ART parameters and regimes.

It is necessary to obtain actual field input influences that affect product reliability, durability, and maintainability during operation, that is, the range, character, speed, and limiting range of values for temperature, air pollution, air fluctuations, humidity, solar radiation (ultraviolet, infrared, and visible), input voltage, air pressure, and road features (type, surface, density, and others), for the various field conditions where the product will operate.

It is necessary to study all input influences, in the example for temperature (Fig. 3.6),

- On the product in single steps and in simultaneous combinations including their interaction in the field and
- On the entire system in the full hierarchy as it is shown in Figure 3.7.

The full hierarchy required is the complete product that is used as a test subject (system) in real life consisting of N subsystems that act in a series with interactions. The subsystems also consist of lower-level K sub-subsystems that also act in a series with interactions. Therefore, in real life, each subsystem and

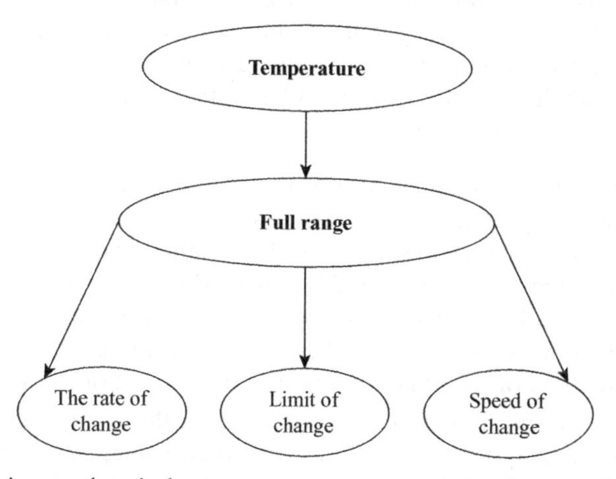

Figure 3.6. Scheme of study for temperature as an example of input influence on the test subject.

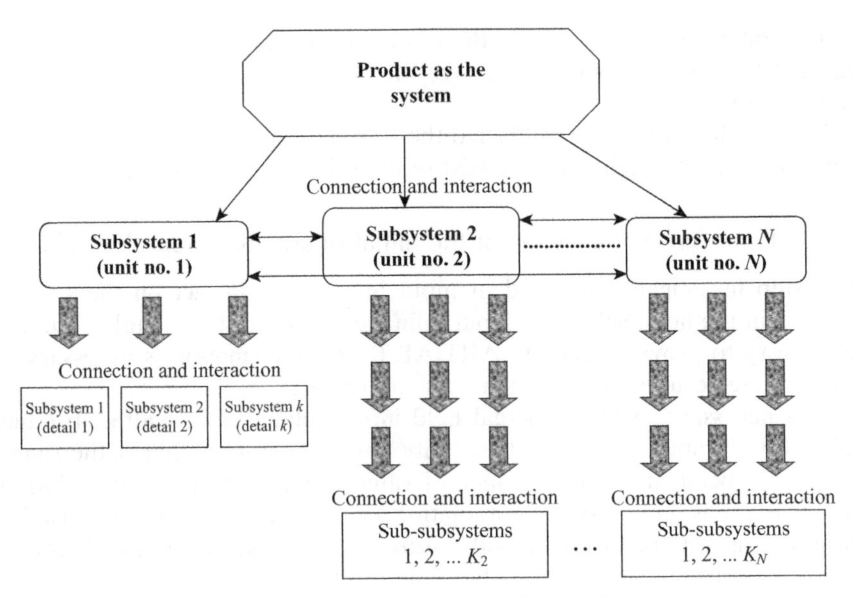

Figure 3.7. The full hierarchy of the complete product and its components as a test subject.

sub-subsystem may act in series with connections to and interactions with all other subsystems of the complete product.

To simulate real-life influences on the product accurately in the laboratory during ART, it is necessary to accurately simulate the full hierarchy of connections and interactions. Simulating only the field input influences for separate subsystems or sub-subsystems to provide accelerated reliability does not take into account the connections and interactions among subsystems and sub-subsystems. This results in an inaccurate simulation that leads to an inaccurate ART with results that will be different from the field results. The outcome is an inaccurate prediction of the test subject's life cycle cost (LCC), reliability, durability, maintainability, and quality in the field. Elements of this inaccurate prediction include

- Output variables such as vibration, corrosion, loading, tension, wear, resistance, output voltage, the decrease of protection after deterioration, the range, character, and speed that change under various field conditions, including climatic areas (Fig. 2.7). Using data from the prototypes can reduce the study time for a developed product. Study variations may exist depending on the facilities of the researchers, the goals of the research, the conditions of the experiments, and the subject of the study.
- The mechanism of degradation observed for the components or the test subject (the parameters of degradation, their value, speed, and the statistical characteristics of these parameters, which change during the time of usage) [18].

- Input influences and output parameters of the data collection and analysis. This includes the types of input influences that impact sufficiently on the degradation and failure process. If it is known with assurance that one or several influences do not impact on the product degradation (failure), then those influences can be eliminated from consideration. This possibility is described in detail for electronic equipment [102].
- The percentage of time the test subject is working under different usage conditions. This is important because most products are used under different conditions with different loads. These conditions depend upon the changing values of each output parameter. The results lead to the distributions that are needed for the programming and understanding of the ART results. An example of these distributions for fertilizer applicators is shown in Reference 42. If the fertilizer applicator is used in different climatic regions, this will probably alter the output distributions.
- The testing scheme developed for the test subject as a system consisting of hierarchical subsystems (units) and sub-subsystems (details) with their connections and interactions shown in Figure 3.7.

Appropriate analysis has to consider safety and human factors.

3.3.2 Procedure 2: Analysis of the Initial Field Information as a Random Process

For a product, especially for a mobile product, there is a randomness into field input influences and outputs.

It is necessary to evaluate the statistical characteristics (the mean, standard deviation, normalized correlation, and the power spectrum) of the studied parameters to estimate their influence. It is also necessary to look at the distribution of input influences and the output parameters. Figure 3.8 shows an example of ensembles for experimental normalized correlation and power spectra of tension data registered by sensor 1 for the car's frame point under different field conditions. The probability distribution can also be utilized for a more detailed evaluation.

3.3.3 Procedure 3: Establishing Concepts for the Physical Simulation of the Field Conditions on the Product

The field input influences safety and human factors must be accurately simulated in the laboratory to achieve high correlation between the accelerated reliability/durability test results and the field results. This leads to the accurate prediction of reliability, durability, and maintainability in the field after ART/ADT. This practice shows that the most accurate physical simulation of the input influences occurs when each statistical characteristic [μ, D, $\rho(\tau)$, $S(\omega)$] for all input influences differs from the field condition measurements by no more than 10%.

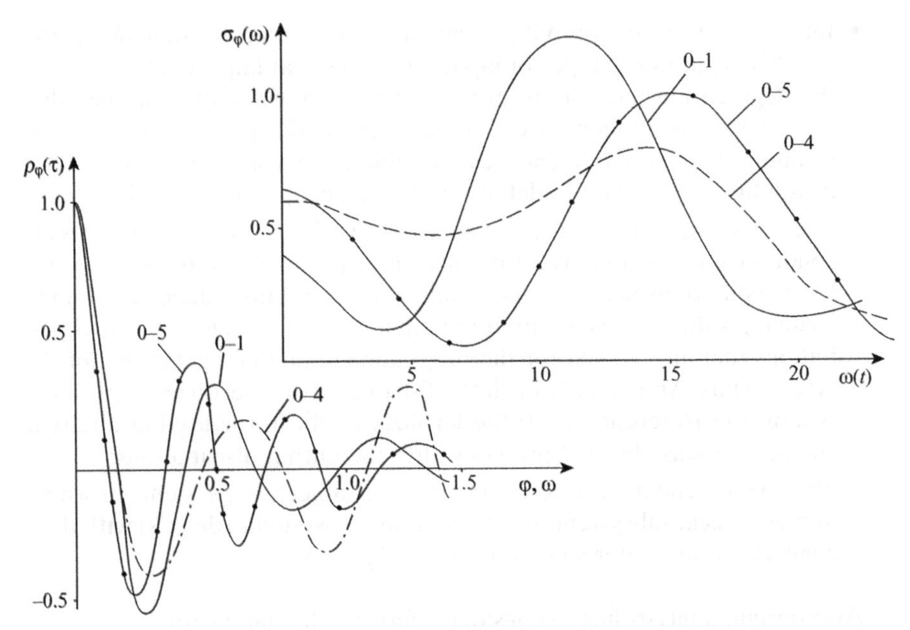

Figure 3.8. Ensembles of normalized correlation and power spectrum of frame tension data of the car's trailer in different field conditions.

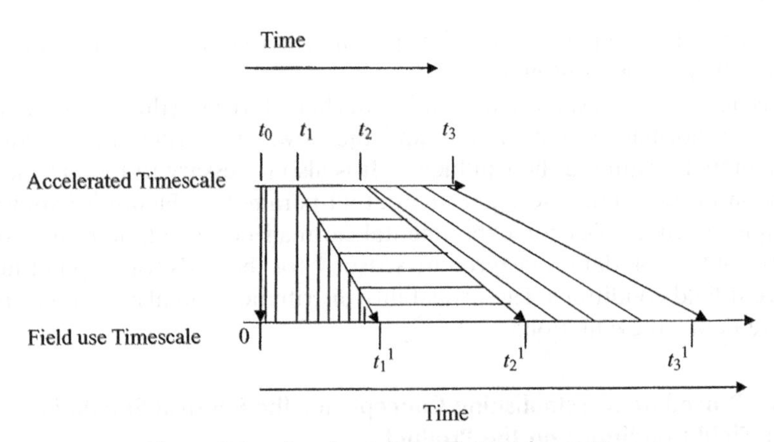

Figure 3.9. Accelerated coefficient description.

For each specific situation it is necessary to use these statistical criteria to compare the reliability measured in the ART/ADT with field reliability.

The ART/ADT methods and other accelerated testing methods have an acceleration coefficient. The similarity of the degradation process in the field and in the laboratory will determine the practical limit of stress (acceleration coefficient) (Fig. 3.9).

The most accurate method of ART/ADT is based on minimizing breaks between work times. This type of testing can be accelerated if the product is tested 24 hours a day excluding idle time and time with minimum loading. This method is based on the principle of reproducing the complete range of operating conditions and of maintaining the correct proportion of heavy and light loads. The author's experience shows that this method is especially effective for a product that is used with long storage periods.

This method has the following basic advantages:

- A maximum correlation occurs between the field and ART/ADT results.
- One hour of pure work performed by the product that faithfully reproduces the stress schedule is identical in its destructive effect to 1 hour of pure work under normal operating conditions.
- There is no need to increase the pace of testing in terms of the size and proportion of stress.
- The results of ART/ADT are obtained 10–18 and more times faster than in the field. This method provides the successful correlation of ART/ADT results and field results.

To evaluate the particular accelerated coefficient, one needs to divide the accelerated timescale by the field use timescale, that is, t_1/t_1^1 (Fig. 3.9).

For storage time acceleration, one can use the stress of the environmental parameters (temperature, humidity, and pollution) for acceleration. However, the calculation of the correlation between the laboratory results and the field results will be more complicated.

More stress means less correlation. Less correlation is less accuracy for the simulation results and greater problems in obtaining an accurate prediction of the product's reliability and durability in the field.

Therefore, the limit of the acceleration coefficient is very important (Fig. 3.10).

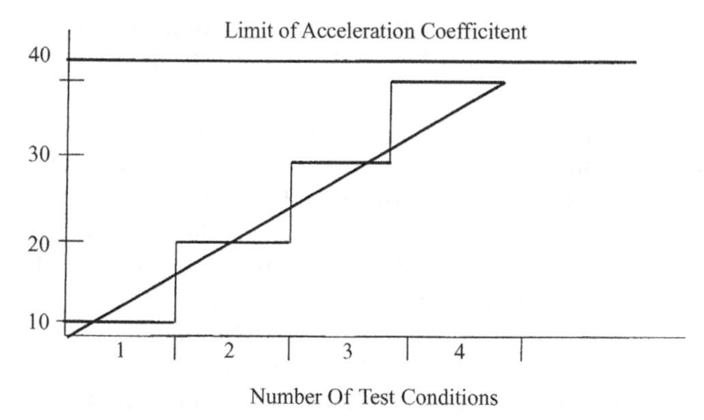

Figure 3.10. Example of stress limit (acceleration coefficient limit).

Less correlation leads to more problems in determining the reasons for failures (degradation) and how to eliminate them.

When using the accelerated coefficient number, it is necessary to decide which way will be more effective for each product.

Accelerated laboratory testing for ART is conducted as a simultaneous combination of different groups of testing (mechanical, multi-environment, and electrical). If only vibration testing of the product is conducted, then it is not possible to accurately predict the reliability or durability of the product. The same is true for the simulation of temperature and humidity testing or both. This is not environmental testing because only one or two parameters out of the combination of environmental field influences are simulated.

Acceleration (stress) factors are factors that accelerate the product degradation process in comparison with normal usage. There are many types of acceleration factors including the following:

- Increasing concentrations of chemical pollutions and gases
- Increasing air pressure
- Increasing temperature, fog, and dew
- Increasing the rate of change of input influences

One widely used testing method employs simulation with only a minimum number of field combinations of input influences. For example, a temperature/humidity environmental chamber is used for environmental testing. It is known that these are only a few of the many environmental influences acting on the product.

Care is required when using a high-level range of acceleration or a combination of the field influences with ART/ADT.

3.3.4 Procedure 4: The Development and Use of Test Equipment that Simulate the Field Input Influences on the Actual Product

Test design is impacted by the result of the analysis of the combination of field input influences on the product ($X_1^1 \ldots X_M^1$) (Fig. 3.1) and how they are simulated in the laboratory.

Test equipment selections are made based on both universal and specific application test equipment. The basic reason to develop test equipment is to provide an appropriate simulation of the field situation.

Universal test equipment can be used on many different types of products. Design and manufacturing companies typically use universal test equipment for specific applications. For example, all types of mobile products and many parts of stationary machinery vibrate in the field. Companies conduct vibration testing using vibration equipment. This is also true for test chambers that simulate the environmental influences and their combinations.

The selection and characteristics of the test equipment are very important in controlling how stress is applied to the product and affects the ART/ADT results. For example, single-axis and multiaxis vibration test equipment (VTEs) have different applications. For many types of stationary and for all mobile products, single-axis VTE cannot simulate real-life vibration. The solution is to use multiaxis VTE.

Vibration input can be generated in up to 3 or 6 degrees of freedom. Vibration is simultaneously generated for one to three linear axes (vertical, lateral, and longitudinal) and for one to three angular (rotational) axes (pitch, roll, and yaw).

Test chambers are another type of universal test equipment. There are many types of test chambers and other types of universal testing equipment available (Chapter 5). ART/ADT requires test chambers for multiple environment testing.

Every industrial company has many types of specific testing equipment that can be used as components to provide ART/ADT. The type of specific testing equipment depends on the type of product manufactured. Usually, industrial companies design, or specify the design for, the required test equipment.

If it is desired to conduct ART/ADT, then it is necessary to use both universal and specific testing equipment.

The above analysis relates to safety and human factors too.

3.3.5 Procedure 5: Determining the Number and Types of Test Parameters for Analysis during ART/ADT

The basic objective is to determine the minimum number of test parameters that are sufficient for the comparison of the ART/ADT and field-testing results to provide an accurate prediction of reliability, durability, and maintainability.

Let us consider the field input influences.

Thus, it is necessary to establish the applicable area of each influence that can be introduced into all varieties of operating conditions to set acceptable test conditions. This is

$$E > N,$$

where

E is the number of field input influences, $X_1 \ldots X_a$ (Fig. 3.1), and
N is the number of simulated input influences, $X_1^1 \ldots X_b^1$ (Fig. 3.1).

And the allowable error for simulation input influences is

$$M_1(t) = X_1(t) - X_1^1(t),$$

where

$X_1(t)$ are input influences of the field and
$X_1^1(t)$ are simulated input influences.

The basic steps for choosing the area of influences that introduce all the basic variations of the field situation are the following:

1. Establish the type of random process to be studied. For example, a stationary process is determined by the normalized correlation from the difference of the variables.
2. Establish the basic characteristics of the process. For a stationary random process, use the mean, the standard deviation, the normalized correlation, and the power spectrum.
3. Define an area's ergodicity, that is, the possibility to make judgments about the process. This occurs when the correlation approaches zero as the time $\tau \rightarrow \infty$.
4. Check the hypothesis that the process is normal. Use Pearson's criterion or another criterion.
5. Calculate the length of the influence area.
6. Select the size of the divergence between basic characteristics in different areas.
7. Minimize the selected measure of divergence and find the area of influence that introduces all the possibilities of the field.

The result of the number and types of field input influences used in the analysis determines the impact on the product's degradation (failure) mechanism.

3.3.6 Procedure 6: Selecting a Representative Input Region for ART

To execute this step, it is necessary to identify one representative short region from the multitude of input influences (or output variables) under specified field conditions with a minimum of divergence. Next, one needs to simulate this representative region's characteristics in performing ART. The solution is shown for normal processes by the use of the Pearson criterion, but other types of statistical criteria can be used.

3.3.7 Procedure 7: The Procedure for ART/ADT Preparation

The preparation for ART/ADT should consider the following:

- The schedule applied to the product's technological processes, including all areas and conditions of use (time of work, storage, and maintenance) for each condition
- The conditions of influences on safety and human factors
- Sensor disposition
- The kind of input influences and output variables that occur under each field condition analyzed

- The measurement regimes used for speed, productivity, and output rate
- The typical measurements of conditions and regimes in the field and their simulation by ART/ADT, as compared to the field
- Establish the value of the testing
- Execute the test
- Determine the schedule for checking the testing regimes
- Obtain the test results and the analysis of the data.

3.3.8 Procedure 8: Use of Statistical Criteria for Comparison of ART/ADT Results and Field Results

The degradation mechanism is the criterion for comparing the field results to the ART/ADT test results. To calculate this comparison, it is necessary to use statistical criteria comparing the ART/ADT results to the field results.

The use of statistical criteria can help determine whether to use the current ART/ADT technique or to continue to develop this technique until the difference between the reliability/durability function distribution during ART/ADT and during field usage is not more than the fixed limit.

The statistical criteria shown in this book must be used in these three stages of testing:

- During ART/ADT for a comparison of the output variables or the laboratory degradation process with the output variables or the degradation process in the field respectively
- During ART/ADT for a comparison of the time required for the maintenance process and the cost of maintenance with the indices determined during normal field conditions
- After completing ART/ADT, to conduct a comparison of the reliability indices (failure modes and effects analysis [FMEA], time to failure, failure intensity) and maintainability indices with field reliability and maintainability indices. Additional information could be generated in this step by a failure analysis of the events classified as failed and not failed test subjects.

The difference between the field and ART/ADT results should not be more than a fixed limit. The limit for fixed parametric differences in the field and in the laboratory depends on the desired level of accuracy, such as 3, 5, or 10% as a maximum difference.

Another criterion can be used for correlating results of ART/ADT with field testing. It can include any or all of the following:

$$[(C/N)_L - (C/N)_f] \leq \Delta_1$$

$$[(D/N)_L - (D/N)_f] \leq \Delta_2$$

$$[(V/N)_L - (V/N)_f] \leq \Delta_3$$

$$\cdots\cdots\cdots\cdots$$

$$[(F/N)_L - (F/N)_f] \leq \Delta_i,$$

where

$\Delta_1, \Delta_2, \Delta_3 \ldots \Delta_i$ are divergences calculated from the results of the laboratory and field testing;

L and f are laboratory and field conditions;

C, D, V, and F are measures of output parameters: corrosion (C); destruction of polymers, rubber, and wood (D); and vibration (V) or tension failures (F);

N is the number of equivalent years (months, hours, or cycles) of exposure in the field (f) or laboratory (L).

Figure 3.11, for example, shows the normalized correlation data for $\rho(\tau)$ and the power spectrum $S(\omega)$ of a car's trailer frame tension data in the field and ART/ADT after using the above-mentioned criteria.

The implementation of criteria for correlating the ART/ADT results and the field results is described in References 42, 103, and 104.

The results of the implementation of these criteria for a car's trailer are also shown in Tables 3.7 and 3.8. These results are also presented in Chapter 4 of this book.

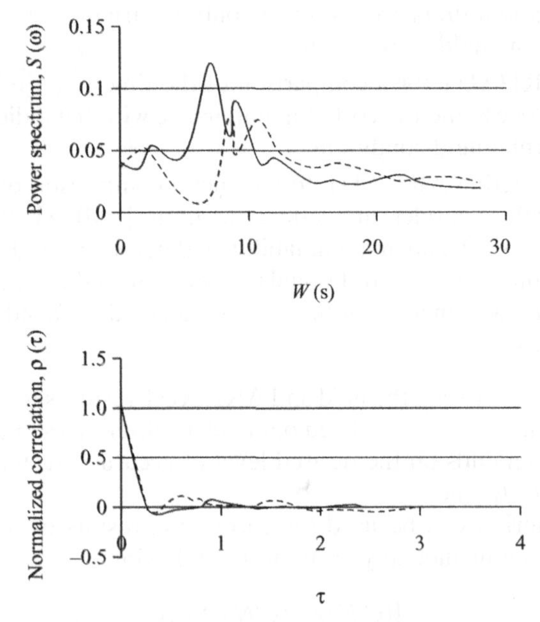

Figure 3.11. Normalized correlation $\rho(\tau)$ and power spectrum $S(\omega)$ of the car's trailer frame tension data [18]: ——— field; ------------ ART/ADT.

TABLE 3.7 Results of the Calculation of Tensions on the Wheel Axles of a Fertilizer Applicator [14]

	Class Frequency		Class Accumulated Frequency		Accumulated Relative Frequency		Modulus of Congruence Accumulated Relative Frequency Difference
Class Number	N_i	N_j	ΣN_i	ΣN_j	$P_i = \frac{\Sigma N_i}{N_i}$	$P_j = \frac{\Sigma N_j}{N_j}$	Max $\lvert D \rvert$
1	10	16	10	16	0.020	0.040	0.020
2	18	10	28	26	0.070	0.065	0.05
3	20	14	48	40	0.121	0.100	0.021
4	32	32	80	72	0.202	0.180	0.022
5	39	50	119	122	0.303	0.305	0.002
6	84	68	203	190	0.512	0.476	0.037
7	68	72	271	262	0.680	0.655	0.025
8	49	63	320	325	0.808	0.812	0.004
9	30	36	350	361	0.884	0.900	0.016
10	24	16	374	377	0.994	0.940	0.004
11	10	12	394	389	0.969	0.971	0.002
12	12	11	396	400	1000	1000	0.000

Where i is during the field and j is during the ART/ADT.

TABLE 3.8 Part of Assembly of Fertilizer Applicator's Tension Data [39]

Digit Number	Field n_i	Laboratory Testing n_j	Difference $n_i - n_j$	Definition from Average $[(n_i - n_j)x]$	Definition from Average in Square $[(n_i - n_j)x]^2$
1	1	1	0	−32.5	1056.25
2	5	4	−1	−33.5	1122.25
3	18	8	−10	−42.5	1808.25
4	20	25	−5	−27.5	756.25
5	121	66	−55	−87.5	7658.25
6	223	174	−49	−81.5	6642.25
7	270	201	−59	−101.5	10,302.25
8	217	107	−110	−142.5	20,302.25
9	105	40	−55	−97.5	9506.25
10	37	11	−26	−58.5	3422.25
11	13	4	−9	−41.5	1722.25
12	3	1	−2	−34.5	1190.25

$\Sigma n_i = 1033$; $\Sigma n_j = 642$; $\Sigma n_i - n_j = 391$; $\Sigma[(n_i - n_j) - x]^2 = 65{,}489.5$.

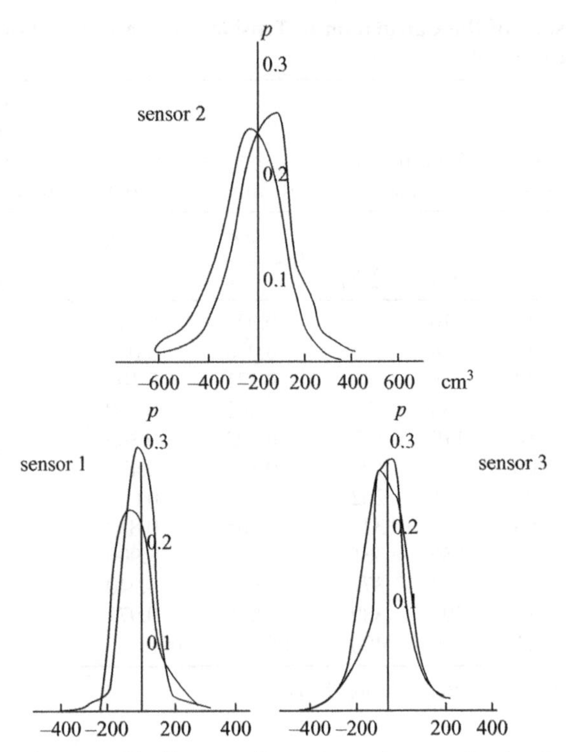

Figure 3.12. Distribution of fertilizer applicator's tension amplitudes during field testing and ART/ADT (-x- is the field; - is the ART/ADT) [18].

The experimental results can be used to find the correlation. Figure 3.12 shows the experimental distribution of tension amplitudes or frequencies on a fertilizer applicator axle (sensor 1), the frame (sensor 2), and the carrier system (tension 3) during fertilizer distribution under field conditions and during ART/ADT (Fig. 3.12).

The results of the calculation based on sensor 1 data are given in Table 3.7. These comparison data show a very small deviation:

$$\lambda = \max / D / \cdot \sqrt{\frac{\Sigma N_i \cdot \Sigma N_j}{\Sigma i + N_j}}.$$

If λ is less than 1.36, then adopt the hypothesis that both samples belong to one statistical population; that is, the loading regimes in the field and in the laboratory are closely tied to each other. Here, $\lambda = 1.36$ is the value of Smirnov's criterion at the 5% level.

Table 3.8 shows an example of the use of Student's t-distribution for the evaluation of the mean.

The following results were obtained by sensor 2. The average is equal to 1.4. Therefore, $1.4 < 1.8$, where 1.8 comes from the tables of the Student t-distribution at the 5% level of degree of freedom.

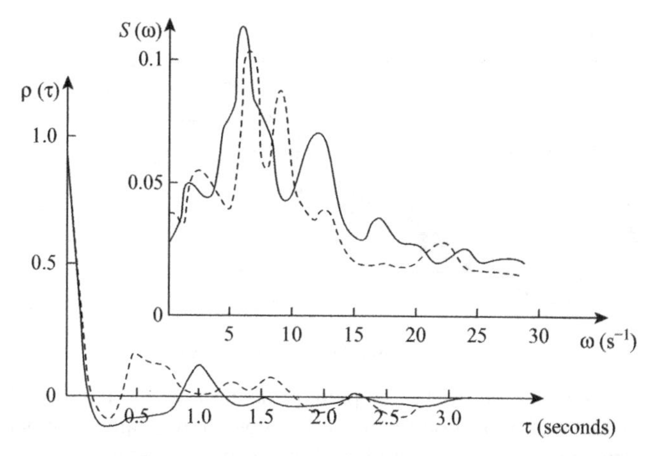

Figure 3.13. Normalized correlations ρ(τ) and power spectrum S(ω) of car trailer's frame tension data [18]: ———— field; - - - - - - accelerated (laboratory) testing.

Therefore, our hypothesis is true for this example.

The results of the comparison of random tension data in the field and in ART/ADT are illustrated in Figure 3.13. The time of correlation is between 0.10 and 0.12; the time of attenuation is the same; the maximum of a power spectrum equals 8–12 seconds; and the interval of frequencies is substantially the same (from 0 to 16–18 seconds). The maxima of the power spectrum have small changes in velocity.

Overall, the test regimen for the carrier and the running gear systems of a fertilizer applicator in the laboratory is closely related to the test regimen in the field.

Sections of Chapters 4 and 5 also cover vibration testing as a component of ART/ADT.

3.3.9 Procedure 9: Collection, Calculation, and Statistical Analysis of ART/ADT Data

Data are the foundation for reliability, quality, and maintainability analyses, models, and simulation.

Data include total operating time, number of failures, and the chronological time of each subsystem failure, component failure, and the times to each component failure, time for maintenance (repair), how failures are displayed, the reasons failure causes, and the mechanisms of failure and degradation. The system for collecting and using degradation and failure data from testing is often called "failure reporting, analysis, and corrective action system (FRACAS)." This step concerns test data collection during the test time, statistical analysis of this data based on the failure (degradation) type and test regimens, and the reasons for deterioration leading to the ultimate failure of the test subject, including the accelerated coefficient.

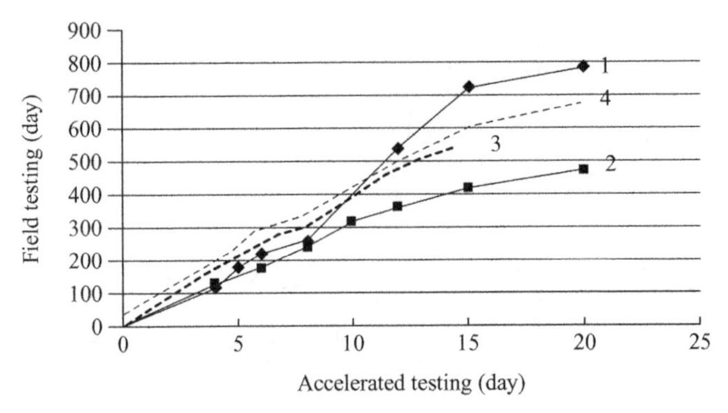

Figure 3.14. Accelerated destruction of paint protection in test chambers (two types of paint) [43]. First type of protection (paint A): 1—protection quality; 2—impact strength; 3—bending strength. Second type of protection (paint B): 4—impact of strength.

Figure 3.14 shows the accelerated loss of paint protection in the test chamber. The acceleration coefficient can be easily found as shown next.

The loss of the protection quality of paint A (curve 1) for 5 days of accelerated testing is the same as for 90 days in the field. The accelerated coefficient is 18.

Since the paint deterioration data show the importance of collecting and analyzing many data points, a computer monitors most tests and the data are automatically collected. The computer is also used to ensure that the test conditions are maintained or to monitor deviations from the desired conditions. The importance of this data collection step is discussed in Reference 102.

3.3.10 Procedure 10: Prediction of the Dynamics of the Test Subject's Reliability, Durability, and Maintainability during Its Service Life

ART is not a final goal; it is a source of initial information for quality, reliability, maintainability solutions, and for the solution of other problems. This initial information is used in the evaluation of the above-mentioned problems for testing conditions or in predicting these problems for use in field conditions. Chapter 6 shows how these conditions can be predicted.

3.3.11 Procedure 11: Using ART/ADT Results for Rapid and Cost-Effective Test Subject Development and Improvement

When the ART/ADT results have sufficient correlation with the field results, it is possible to rapidly find the reasons for test subject degradation and failure. These reasons can be determined by analyzing test subject degradation during the time of usage through the location of initial degradation and the development process of degradation. Next, it is necessary to rapidly eliminate the reasons for degradation. This is the author's approach.

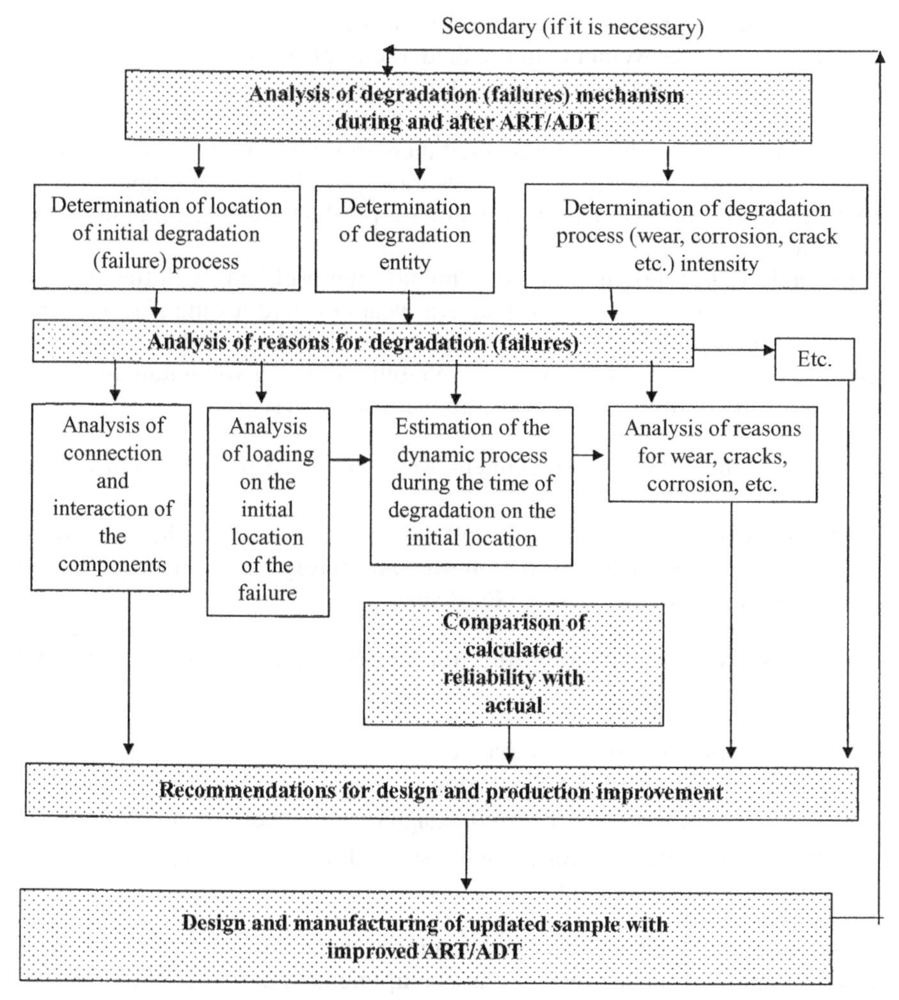

Figure 3.15. The scheme of the updated process of rapid reliability/durability improvement during and after ART/ADT of systems, subsystems, or components.

This method is both time and cost-effective. Figure 3.15 illustrates the strategic process of this approach for accelerated reliability/durability/maintainability improvement.

If the correlation is not sufficient, the reasons for test subject degradation and the character of this degradation during ART usually do not correspond to the degradation in the field. Therefore, the conclusion from ART may be wrong and may increase the cost and time required for test subject improvement and development.

Often in this misleading circumstance, designers and reliability engineers thought that they understood the reason for failure (degradation). The design or manufacturing process was changed. However, after reviewing further test results for the "improvement" in the field, the improvement was shown to be invalid. It was then necessary to look for other reasons for degradation. This process does not allow for rapid improvement or the development of test subject reliability (quality). Moreover, it increases cost and time. Two practical examples follow to illustrate the advantages of the author's approach to ART/ADT for the rapid improvement of product quality, reliability, and durability.

Example 1. The designer did not eliminate the problems with the harvester's reliability and durability over several years of field testing and data collection. A special complex was developed for ART/ADT and to provide an accurate prediction of the harvester's reliability/durability/maintainability.

As a result, after 6 months of the above research [105],

- Two of the harvester's specimens were subjected to evaluation for the equivalent of 11 years.
- Three variations of one unit and two variations of another unit were tested. The resulting information was satisfactory based on an evaluation for their equivalent service life (8 years).

The results of eliminating the rapid degradation process were the following:

- The harvester's design was changed based on the conclusions and recommendations drawn from the testing results.
- The units with the incorporated design changes were then field tested.
- This reduced the cost for the harvester's development by a factor of 3.2 and the time by a factor of 2.4.

The reliability increase was validated in the field. Also, the design changes were validated for the harvester's basic components that limited their durability. Usually, this volume of work requires a minimum of 2 years for an accurate comparison. Thus, this improved process was four times faster.

Example 2. There was a problem with the durability of another type of wide belts (10 in.)—working heads for another type of harvester. For several years, beginning with the design stage and continuing through the manufacturing stage, the low durability of the belts limited the reliability of the whole machine.

Designers increased the strength of the belt design but could not increase durability of the belts more than 7% for several years.

During this period, the cost of the belts increased twice. The author created special test equipment for accurate simulation of field input influences on the harvester, including the belts. After several months of testing (studying), the reason for the low durability of the belts was discovered. The designers developed recommendations for improving the machinery design (the roller design was the root cause for the belt failures). The field-testing results showed that the average life of the belt in the updated harvester was increased by a factor of

2.2. The cost of the above work increased the harvester's cost by 1%. The cost increase included the cost of the testing equipment and all the work involved in finding the root cause of failure and thereby increasing the belts' durability.

If accurate initial information is not provided by the results of ART/ADT, the best methods of reliability prediction cannot be useful.

There is another approach for reliability/durability/maintainability prediction. Mathematical models can describe the relationship between the product reliability and the different stress factors in manufacturing or the relationship of reliability to the operating conditions (field). These factors (influences) should be evaluated using the results of the ART/ADT. Other factors such as special field testing (for the evaluation of the operator's reliability on the product reliability) can also be evaluated by using mathematical models from specific field testing. The coefficients must be changed to fit the desired goals. The authors of this approach recommended equations for the prediction of time to failures and time to maintenance. This approach was undeveloped until the current time; therefore, it was not useful for effective product improvement.

Another approach by the author and colleagues [106] is a multivariant Weibull model utilizing ART results of the components to predict system reliability with reduced test length and minimized cost.

The description of the results is especially important for a new product. The testing results should show no more than several failures and are important when one cannot approximate the product reliability function.

The use of applied statistics for reliability prediction is shown in International Electrotechnical Commission (IEC) standards (Section 8.1).

3.4 ART AND LCC

3.4.1 General

A life cycle is the time interval from a product's conception through its disposal. ART has a direct influence on the product's LCC.

A life cycle consists of three basic phases: research and development, production or construction, and operation and maintenance.

The phase research and development includes

- Concept development
- Product research
- Product and life cycle planning
- Engineering design and development
- System/product software design and development
- System/product technological test and evaluation
- System/product ART/ADT
- Prediction of dependability parameters

The phase production or construction includes

- Production/construction management
- Industrial engineering and operations analysis
- Manufacturing
- Construction
- Quality control
- Initial logistics support
- Recall and complaints after operation commences.

Operation and maintenance support consists of

- System/product life cycle management
- System/product operations
- System/product distribution
- System/product maintenance
- Inventory—spares and material support
- Operator and maintenance training
- System/product modifications
- Technical data

The design phase usually consists of 10–12% of the total LCC; the manufacturing phase consists of 23–32% of the total LCC; and the usage phase consists of 55–65% of the total LCC.

The distribution of LCC that corresponds to the data in Reference 107 is shown in Figure 3.16.

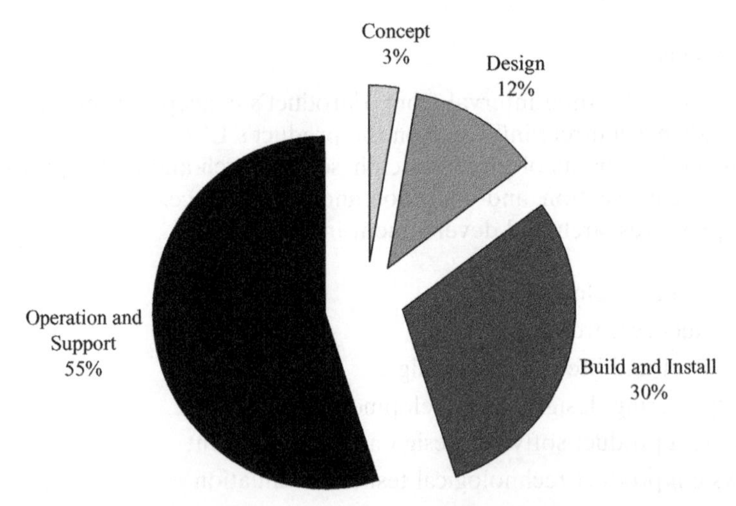

Figure 3.16. Percentage of the total life cycle cost components [107].

Desirable ART directly influences the product's LCC during the above-mentioned basic phases for the following reasons:

- A slight increase occurs in the design phase cost, for the equipment and testing using ART, but this is offset by providing more accurate predictions of quality, reliability, maintainability, durability, and supportability.
- Decreased costs in the manufacturing phase due to a dramatic decrease in product quality deficiencies, resulting in fewer complaints and recalls. This results in a significantly reduced cost for redesign and changes in manufacturing to eliminate failure modes and greater customer satisfaction.
- Decreased usage phase costs because there is a dramatic decrease in maintenance and quality costs, as well as increases in reliability and supportability.

For this reason, ART impacts the LCC of the product.

To minimize the LCC, use ART to decrease the usage cost that has the largest influence. For example, accurate prediction, which is one of the results of using ART, decreases maintenance frequency and costs.

It is essential to understand a product's life cycle during all phases to understand the relationship of these activities to the product's performance, reliability, maintainability, and other characteristics that contribute to LCCs.

There are six major life cycle phases of a product:

- Concept and definition
- Design and development
- Manufacturing
- Installation
- Operation and maintenance
- Disposal

From an owner's perspective, the total costs incurred during these phases can also be divided into acquisition cost, ownership cost, and disposal costs:

$$\text{LCC} = \text{acquisition costs} + \text{ownership costs} + \text{disposal costs}.$$

Acquisition costs are generally visible and can be readily evaluated before an acquisition decision is made. This may or may not include installation cost.

Ownership costs exceed acquisition costs and are not as readily visible. These costs are difficult to predict and may include the cost associated with installation.

Disposal costs may represent a significant proportion of the total LCC. Legislation may require activities during the disposal phase that involve a significant expenditure for major projects.

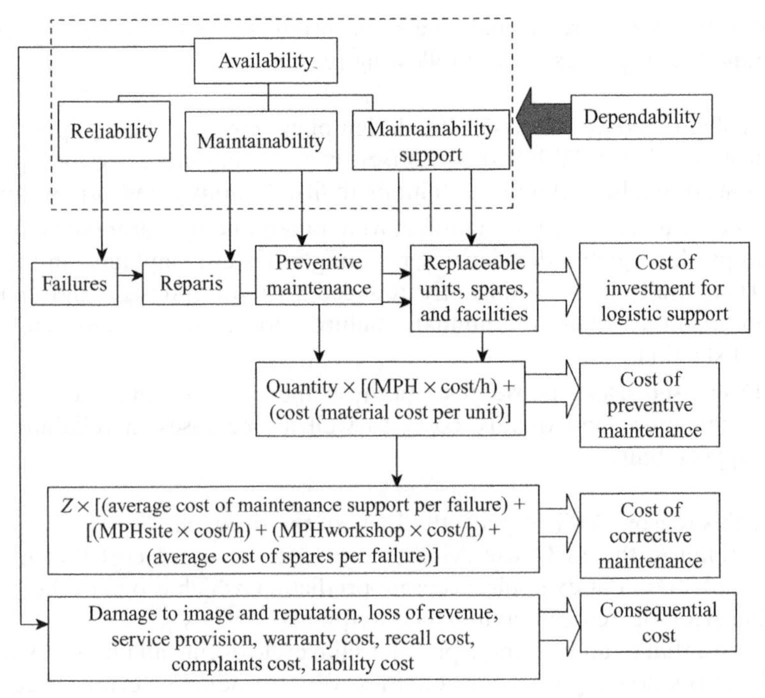

Figure 3.17. Typical relationship between dependability and LCC for the operation and maintenance phase.

Costs associated with dependability elements (reliability + maintainability + maintenance support) may include the following, as appropriate (Fig. 3.17):

- System recovery cost including corrective maintenance cost
- Preventive maintenance cost
- Consequential cost due to failure

3.4.2 Consequential Costs

When a product or service becomes unavailable, a series of consequential costs may be incurred. These costs may include [108]

- Warranty costs
- Costs due to recalls and complaints
- Liability cost
- Cost due to loss of revenue
- Costs for providing an alternative service
- Image

- Reputation
- Prestige

These consequential costs may result in loss of clients and market share.

3.4.3 Warranty Costs

Warranties provide protection to customers, insulating them from the cost of correcting product failures, particularly during the early stages of product operations. The cost of warranties is usually borne by the suppliers and may be impacted by reliability, maintainability, and the maintenance support characteristics of the product. Suppliers can exercise significant control over these characteristics during the design, development, and manufacturing phases, thereby influencing their warranty costs.

Warranties also limit vendor liability. Warranties usually last for a limited period of time, and a number of conditions and limitations generally apply. Rarely do warranties include protection against consequential costs incurred by the customer as a result of product unavailability.

Warranties may be supplemented or replaced by service contracts where the supplier, or a third party, performs all preventive and corrective maintenance for a fixed period of time. Service contracts can be renewed for any period of time up to the whole product lifetime. Suppliers who provide long service contracts are motivated to build an optimum level of reliability and maintainability into their product, usually by incurring higher acquisition costs.

3.4.4 LCC Breakdown into Cost Elements

In order to estimate the total LCC, it is necessary to break down the total LCC into its constituent cost elements. Identify these cost elements individually to be distinctly defined and estimated. The identification of the cost elements and their corresponding scope should be based on the purpose and scope of the LCC study.

The cost element links cost categories with the product/work breakdown structure. The selection of cost elements should be related to the complexity of the product and to the cost categories of interest in accordance with the required cost breakdown structure.

One approach often used to identify the required cost elements involves the work breakdown of the product to lower levels, cost categories, and life cycle phases. This approach can best be illustrated by the use of a three-dimensional matrix. This matrix involves the identification of the following aspects of the product:

- The work breakdown of the product to lower indenture levels (the product/work breakdown structure)
- The period in the life cycle when the work/activity is to be carried out (the life cycle phases)

- The applicable cost category of resources such as labor, materials, fuel/energy, overhead, and transportation/travel (the cost categories) to be applied.

This approach has the advantage of being systematic and orderly, thus giving a high level of confidence that all cost elements have been included.

3.4.5 Estimation of Cost

Examples of methods that may be used to estimate the parameters of a cost element include the following:

- Engineering cost method
- Analogous cost method
- Parametric cost method

An example of the application of the engineering cost method follows in Section 3.4.6.

In order to reduce different types of uncertainties involved in the analyses, perform sensitivity analyses by introducing minimum and maximum values for the parameters of the model into the cost estimation equations.

3.4.6 Engineering Cost Method

The cost attributes for the particular cost elements are directly estimated by examining the product component by component or part by part.

Often, standard established cost factors, for example, the current engineering and manufacturing estimates, are used to develop the cost of each element and its relationship to other elements. Older available cost estimates may be updated by the use of appropriate factors, for example, annual discounting and escalation factors.

The engineering cost method is illustrated by the following example, which examines the cost related to a recurring cost element.

The labor cost for the manufacturer of a power supply is to be estimated. The following information is given:

- *Product*: Power supply
- *Life Cycle Phase*: Manufacturing phase
- *Cost Category*: Labor cost

3.4.7 Review of Life Cycle Costing Results

A formal, possibly independent review of the analysis may be required to confirm the correctness and integrity of the results. The following elements should be addressed:

- Review of the objectives and scope of the analysis to ensure that they have been appropriately stated and interpreted

- Review of the model (including cost element definitions and assumptions) to ensure that it is adequate for the purpose of the analysis
- Review of the application to ensure that the inputs have been accurately established, that the model has been used correctly, that the results (including those of sensitivity analysis) have been adequately evaluated and discussed, and that the objectives of the analysis have been achieved
- Review of all assumptions made during the analysis to ensure that they are reasonable and that they have been adequately documented

3.4.8 Analysis Update

It is advantageous in many LCC studies to keep the LCC model current so that it can be exercised throughout the life cycle of a product. For example, it may be necessary to update the analysis results that were initially based on preliminary or estimated data with more detailed data as they become available later in the product life cycle. Maintaining and updating the LCC model may involve modifications to the LCC breakdown structure and changes to the cost estimating methods as additional information sources become available, and alterations in the assumptions embodied in the model change.

The updated analysis should be documented and reviewed to the same extent as the original.

3.4.9 Implementation Result

The approach included in this book in Chapter 7 has been implemented in industrial companies. The implementation of the ART/ADT effect on the LCC of an industrial product is included in Figure 3.18.

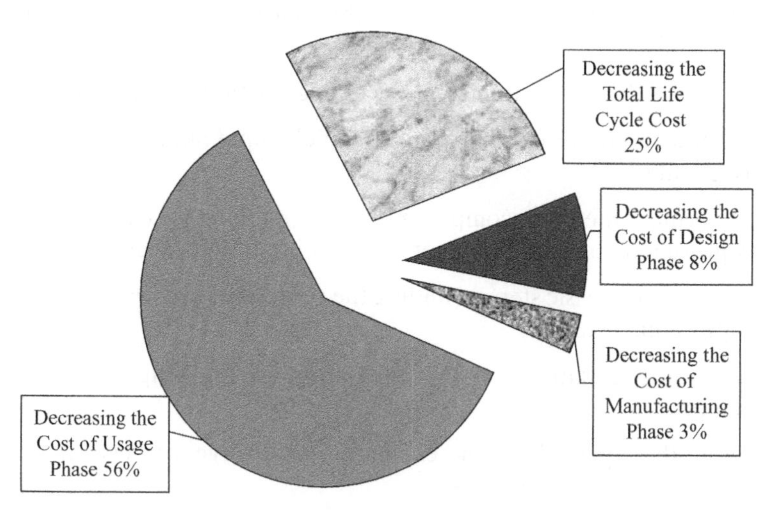

Figure 3.18. Distribution of the average life cycle cost after using the author's approach.

EXERCISES

3.1 ART/ADT may be successful or unsuccessful. What are the key elements of successful ART/ADT?

3.2 Identify and describe an acceptable set of input influences and output variables in the field for a product to be tested.

3.3 What is an accurate physical simulation of input influences? Show the equation for accurate physical simulation.

3.4 The simulation of field input influences in the laboratory can be accurate or inaccurate. Show and describe the basic concepts for providing accurate input influences and describe how they are different from those that are inaccurate.

3.5 Schematically show what a simultaneous combination of field input influences means for any type of product.

3.6 Write an equation that shows when the reliability distribution function cannot accurately evaluate reliability after accelerated testing. Describe this situation.

3.7 When does the reliability distribution function in accelerated testing conditions not correspond to the reliability distribution function in the field?

3.8 Show the equation and describe the role of the physical degradation mechanism (chemical and mechanical) during ART/ADT to provide an accurate simulation of the field input influences in the laboratory.

3.9 How can the degradation mechanism be practically evaluated during and after testing?

3.10 Describe the parameters of the basic types of degradation mechanisms.

3.11 Describe the connection of physical degradation with ART/ADT.

3.12 What does "representative region" of field input influences mean? What is its value?

3.13 Determine the basic components of the methodology for selecting the representative input region of the field input influences.

3.14 Describe the basic steps in finding the representative region for the field input influences.

3.15 Show the equation for the evaluation of the length of the input region.

3.16 Show the equation for the estimation of divergence between any two regions of input influences.

3.17 Describe the seven basic steps of estimation of the representative input region.

3.18 List 11 basic procedures of ART/ADT preparation and performance.

3.19 Describe the first procedure of ART/ADT preparation: collection of the initial information from the field.

3.20 Describe the second procedure of ART/ADT preparation and performance: analysis as a random process of the initial information from the field.

3.21 Describe and show schematically the full hierarchy of the complete product with the connections of its components

3.22 Describe the third procedure of ART/ADT preparation and performance: establishing the statistical criteria for the physical simulation of the input influences on the product.

3.23 Describe the meaning of the acceleration coefficient and the limit of this coefficient when conducting ART/ADT.

3.24 Describe the fourth procedure of ART/ADT preparation and performance: development and use of test equipment that simulates the field input influences on the actual product. What are the basic requirements of this equipment?

3.25 Describe the fifth procedure of ART/ADT preparation and performance: determining the numbers and types of test parameters to be analyzed during the testing.

3.26 Describe the steps in this book's approach to choosing the area of influences that introduces all the basic variations of the field.

3.27 Demonstrate the sixth procedure of ART/ADT preparation and performance: selecting a representative input region for providing ART.

3.28 Demonstrate the seventh procedure of ART/ADT preparation and performance: procedures for the ART/ADT preparation.

3.29 Show the eighth procedure of ART/ADT preparation and performance: use of statistical criteria for a comparison of ART/ADT results and field results.

3.30 Demonstrate the criterion for the correlation of the results of ART/ADT with the results of field testing.

3.31 Show the ninth procedure of ART/ADT preparation and performance: collection, calculation, and statistical analysis of ART/ADT data.

3.32 Describe the tenth procedure of ART/ADT preparation and performance: evaluation and prediction of the dynamics of the test subject reliability, durability, and maintainability during its service life.

3.33 Describe the eleventh procedure of ART/ADT preparation and performance: using ART/ADT results for the rapid cost-effective development and improvement of the test subject.

3.34 Show examples using ART/ADT results for rapid and cost-effective development and improvement.

3.35 Identify all the life cycle phases of the product.

3.36 What percentage of the total LCC is traditionally incurred during the design phase, the manufacturing phase, and the usage phase?

3.37 Why must one evaluate the cost of ART/ADT not only for the present time but also for the future?

3.38 What must one take into account when calculating the ART/ADT cost?

3.39 Show the basic components of the LCC.

3.40 Schematically show the typical relationship between dependability and the LCC for the operation and maintenance phases.

3.41 What kind of consequential costs may occur when the product or service becomes unavailable?

3.42 What does a warranty cost mean?

3.43 Describe the estimation cost and its elements.

3.44 Provide an outline of the LCC with examples from this chapter.

3.45 Show the resulting distribution of the LCC after the implementation of the author's approach to ART/ADT.

3.46 How does reliability influence the LCC?

Chapter 4

Accelerated Reliability and Durability Testing Methodology

4.1 ANALYSIS OF THE CURRENT SITUATION

Product and technology quality and reliability are two of the most critical elements for achieving success today and in the future for both industry and service areas. Accelerated reliability testing (ART)/accelerated durability testing (ADT) is one of the basic components needed to achieve success.

Let us analyze the current situation in ART/ADT methodology.

Many of the examples in Chapter 1 enable each user to understand the definition of ART/ADT. Now consider the methodological examples presented by other authors. In Reference 109, other authors describe ongoing research for an ART strategy called multiple environment stress analysis (MESA). The goal is to find potential reliability problems quickly and efficiently during the design phase of high-volume consumer products. This test method has shown promising results in its capacity to induce failures, especially those formerly identified as no-fault-found (NFF) failures. The methodology applies combinations of stressors in a statistically efficient test scheme. However, in fact, this is not reliability testing.

A description of Alion Science and Technology Corporation reported about an accelerated stress testing methodology that its authors called accelerated reliability tests. The definitions in Chapter 1 show that this is not ART.

A publication [110] reported, "... Accelerated testing, just like other *forms of reliability testing*, is a reliable test method supported by scientific logic." The

Accelerated Reliability and Durability Testing Technology, First Edition. Lev M. Klyatis.

authors were unable to see the difference between accelerated testing and reliability testing. One can see the difference from the definitions in this book for these types of testing.

In "Reliability Tests on Power Devices" [111], the authors wrote, "Environmental tests (temperature and humidity) were applied, to the devices in a climatic box for more than 1800 hours." The authors did not recognize the difference between temperature–humidity testing and reliability testing.

The publication "Accelerated Reliability Test Results: Importance of Input Vibration Spectrum and Mechanical Response of Test Article" [112] noted the installation of two types of vibration screening systems:

- Repetitive shock machine, which produces vibration
- A second vibration screening system

The use of two vibration systems for ART clearly contradicts the true ART definition.

In the publication "Accelerated Reliability Tests: Solder Defects Exposed" [113], it was reported: "This paper describes a methodology to perform risk assessment and establish the quality and reliability of solder joints utilizing accelerated reliability life testing techniques. Temperature cycling tests were conducted on an equal number of 100-pin Quad Flat Pack (QFP) package mounted on printed circuit board assemblies from a reworked and non-reworked soldering process. The data showed that the accelerated reliability tests exposed the latent defects inherent in reworked solder joints." Testing by temperature cycling alone is not equivalent to ART.

Truck body vibration testing with vertical sinusoidal loads reported [114] "body durability testing, in which a vehicle's body structure is evaluated for reliability." Vibration testing alone contradicts the basic principles of durability and reliability testing described herein, as well as other qualified engineering literature.

Reference 115 contains a description of the development of multiaxis accelerated durability tests for commercial vehicle suspension systems: "The procedure starts with a definition of the vehicle's duty cycle based on the expected operational parameters, namely: road profile, vehicle speed, and warranty life." The procedure is to

1. Determine the proving ground test schedule for durability such that the accumulated pseudodamage is representative of the vehicle's duty cycle
2. Develop a multiaxis laboratory rig test for the suspension system to replicate the proving ground accumulated damage in a compressed time
3. Design a single-axis, accelerated durability test for the suspension subsystem

The authors demonstrated the procedure by using a trailing arm in the front suspension of a class 8, tractor-semi-trailer combination.

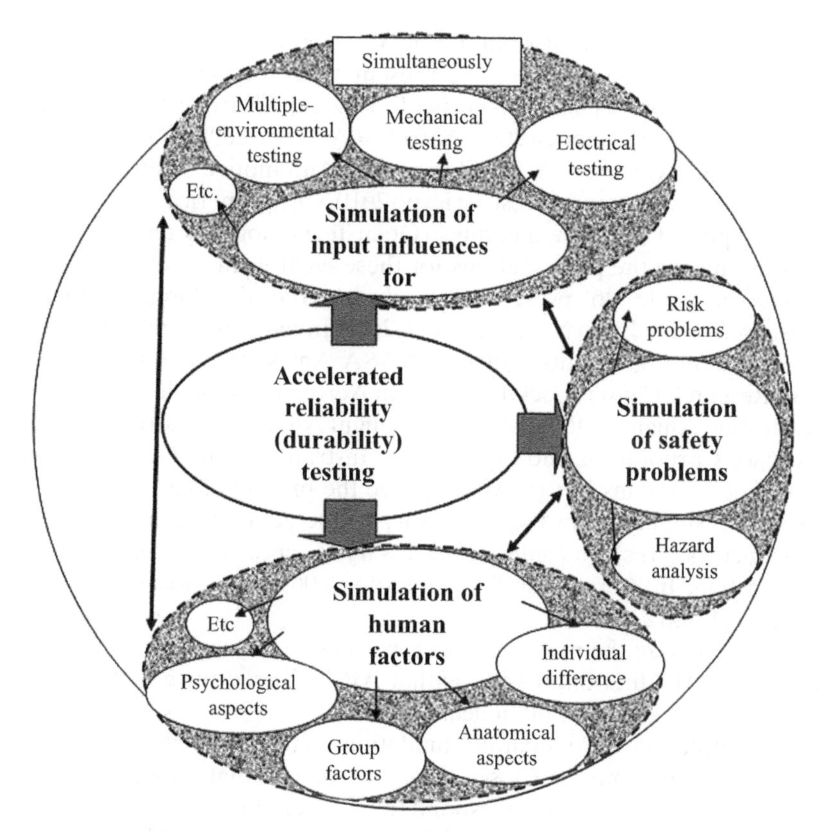

Figure 4.1. Consideration of accelerated reliability (durability) testing as an interconnected system of systems approach.

This "durability testing" does not correspond to the requirements shown in Figure 4.1. It includes the simulation of only part of the real-life input influences. Therefore, it cannot help obtain initial information for the accurate prediction (AP) of the product durability. Many other examples characterize the situation in the following literature: References 38, 63, 78, and 116–119.

Many practical engineers and managers who have read the above publications fail to understand the difference between accelerated testing, reliability testing, and ART methodologies. This lack of understanding results in problems continuing to occur in quality, reliability, and maintainability (QRM) in industrial and service areas.

Therefore, one needs a precise definition to substantiate and describe the term "ART." The definition must clearly state how ART differs from other types of testing like accelerated testing, reliability testing, vibration testing, and durability testing. If the difference is well-known and understood, then understanding true ART makes it possible to accurately evaluate and predict QRM. This will reduce many of the mistakes made in these areas.

Let us analyze a current situation in ART/ADT in the aerospace, aircraft, and automotive industries. Use the presentations for the Aerospace Testing Expo North America 2006, Automotive Testing Expo Europe 2008, Automotive Testing Expo North America 2008, Aero Test America 2008, SAE International 2009 AeroTech Congress & Exhibition, the Automotive Testing Expo 2009 North America, Automotive Testing Expo 2010 North America, and Automotive Testing Expo 2011 Europe meetings (Open Technology Forums).

Let us analyze the presentations for these eight testing expos.

Speakers made 46 presentations during the development forum at Aerospace Testing Expo North America 2006. Individuals from well-respected companies and organizations such as NASA Ames Research Center, NASA Glenn Research Center, Lockheed Martin Aeronautics, Sandia National Labs, Aero Testing Alliance ONERA/DNW, Vought Aircraft Industries, Gulfstream Aerospace Corporation, and the National Instruments and General Atomics-Aeronautical Systems Inc. (GA-ASI) gave the presentations.

Accurate simulation in the laboratory of the actual field input influences, human factors (HFs), and safety is necessary for accelerated reliability (durability) testing. In Aerospace Testing Expo 2006, simulation solutions were presented [120–122].

We next analyze the presentations according to the above-mentioned principles of AP. The first impression is that APs occurred for a high number of separate simulation input influences.

For example, there are separate simulations of temperature, humidity, vibration, and other parameters. Closer inspection reveals that sometimes, the titles of the presentations do not correspond to their contents. For example, the title is "Material and Components Models," but the contents of the presentation addressed only the airframe and turbine.

Another presentation stated that the vibration data are "measured on all three major axes." However, mobile machinery feels vibration along six major axes—vertical, transverse, longitudinal, and the three angular axes around them. The purported simulation of actual field vibration is insufficient for accurate quality and reliability prediction.

The next presentation shows that a new technique has been developed ". . . for ground-test simulation of flight environments in a combined acceleration, axial spin, and vibration test. Until recently, ground testing could only simulate each environment separately, and the synergistic effects of the combined environments experienced during flight were not accounted for." However, professionals know that a much wider combination of influence types than the three cited compose a real flight environment. Therefore, the simulation presented does not represent the real complex of the field conditions. It cannot provide ART sufficient for determining the initial information needed for accurate reliability and durability prediction during a given period of use, such as warranty period or service life.

The Development Forum of Aerospace Testing Expo North America 2006 did not include any presentations related to reliability or durability testing.

Now consider the automotive industry. First, we analyze the Technology Forum of Automotive Testing Expo 2008 Europe. The contents of the technology forum mentioned only one durability testing paper and no reliability testing paper in its nine sessions.

Next, analyze the Open Technology Forum of the Automotive Testing Expo 2008 North America, October 22–24, 2008, Novi, Michigan.

Over 300 of the world's leading automotive test equipment manufacturers and test service providers displayed their latest technologies for the North American automotive industry. Product sectors included

- Acoustical evaluation
- Quality assurance
- Alternatives to physical testing
- Quality management
- Durability testing
- Sensors and transducers
- Dynamometers
- Simulation software
- EMC testing
- Standard certification and legislation
- Emissions measurement
- Stress analysis
- Environmental equipment
- Subassembly testing
- Materials
- Suspension kinematics and compliance
- Track simulation and laboratory testing
- Occupant protection
- Vehicle aerodynamics
- Onboard diagnostics
- Wind tunnel technology

World-famous companies attended the exhibition and Open Technology Forum. The forum included 54 presentations in seven sessions. Only two testing-related papers applied to durability testing:

1. "Standards and approaches for ball joint durability testing" by Dr. Eric Little of MTS Systems Corporation
2. "The application of dynamometers and auxiliary equipment for durability testing of hybrid electric vehicle motors," presented by Keith McCormick of Horiba Automotive Test Systems.

It is not possible to analyze the contents of these two papers because the Open Technology Forum papers are unpublished.

The forum Vehicle Dynamics Expo North America occurred as a component of Automotive Expo North America 2008. This forum included 41 presentations in seven sessions that did not include any presentations related to durability testing or reliability testing.

The Aero Test America 2008 met at the Fort Worth Convention Center (Texas), from November 18 through 20. This event targeted the aerospace industry's testing, R&D, evaluation and inspection professionals. Attendees had a chance to hear more than 100 expert presentations.

It included the Society of Flight Test Engineers (SFTE) 2008 International Symposium and Global Wind Tunnel Symposium. The SFTE 2008 International Symposium included 18 presentations. It was evident that not one presentation in 3 days related to reliability or durability testing.

Only one presentation included the word "reliability." It was the "Strain Gage Installation of Quality, Reliability and Performance" presentation by Mike McFarland, Ron Rutledge, J.R. Norman, and Paul Davis (Lockheed Martin Aeronautics). They reported: "RESULTS: The Fort Worth Strain Gage lab has maintained that in order for their mechanics to be proficient they must have constant training and repetition. Instead of 100 part time mechanics with questionable skills, due to the importance of these data it is preferable to have 10 highly skilled mechanics. Unfortunately, we've had to justify the way the strain gage lab does business at the start of every program, and it seems to become more difficult each time."

The SAE 2009 AeroTech Congress & Exhibition took place on November 10–12, 2009 in Seattle, Washington. This event reported the newest technological developments in materials composites, designing for green, safety and cyber security, and avionics. The SAE AeroTech Congress and Exhibition is the forum for the aerospace community to address solutions to current and future challenges, to create opportunities, and to determine requirements for the next-generation R&D, products, and systems.

The program included presentations in aerospace operations, autofastening/assembly and tooling, aviation cyberphysical security, avionics, business/economics, environment, and flight sciences. Not one presentation mentioned reliability or especially reliability testing.

Once again, the Automotive Testing Expo 2009 North America was held on October 29–29 in Novi, Michigan. The Open Technology Forum included all 43 technical presentations.

No presentations were included for reliability testing or ART. Only one presentation related to durability testing: "Virtual Engine Dynamometer and Transmission Durability Testing; Complete ICE Torque Reproduction for Wear and Fatigue Effects on the Driveline."

The Global Wind Tunnel Symposium included 18 presentations with no mention of reliability or durability testing.

The Automotive Testing Expo 2010 North America was held on October 26–28 in Novi, Detroit, Michigan. The Open Technology Forum included all 40 technical presentations.

No presentations were included for reliability testing or ART. Only one presentation related to durability simulation: "Hybrid System Response Convergence (HSRC) an Alternative Method for Hybrid Durability Simulation."

The Automotive Testing Expo 2011 Europe was held on May 17–19 in Stuttgart Messe, Germany. The program of Open Technology Forum at Automotive Testing Expo Europe 2011 included 43 technical presentations.

No presentations were included for reliability testing or durability testing, or ART, or ADT.

The SAE 2009, SAE 2010, and SAE 2011 World Congresses in Detroit included special exhibition booths and presentations with titles that included ART or ADT, or durability testing. Most of their contents do not correspond to the author's definitions of ART/ADT (Section 1.4).

But the above-mentioned SAE World Congresses included more presentations in reliability and accelerated testing than other forums in the world. These congresses included a special session, "Reliability and Robust Design in Automotive Engineering: Reliability and Accelerated Testing."

4.2 PHILOSOPHY OF ART/ADT

ART/ADT is one of the elements needed to predict field values for quality, reliability, durability, and maintainability extended over the expected life of each product/process with confidence. Another element is the development of input data for ART/ADT that is representative of actual influencing conditions experienced during field operations. Tracking of the unit begins in the design stage and continues throughout production, packaging, transfer, use, and service of each product/process.

Most current methods of accelerated stress testing, sometimes mistakenly called reliability testing or ART or ADT, do not determine the appropriate initial information to accurately solve many of the problems encountered during continuing design and the corresponding effect on and input to quality, reliability, durability, maintainability, and other design parameters. ART/ADT provides the means to enhance

- The AP of product or technology serviceability, availability, and the dynamics of reliability/durability parameters during a predicted time, such as service time or a warranty period
- The AP of the product/technology dynamics of maintainability during a predicted time such as time between maintenance, the cost of maintenance, and maintenance time

- The AP of product/technology dynamics of the quality parameters during a predicted time of interest
- The AP of the life cycle cost for the product/technology
- The successful development of product quality, reliability, maintainability, and durability during the design and manufacturing phases

ART/ADT, as a specific type of accelerated testing, uses only random stresses.

Without ART/ADT, there are technical and economic problems that lead to an increase in recalls and customer complaints due to design and manufacturing deficiencies. Additionally, designing the product/technology utilizes more time and resources than necessary, thus increasing time to market, redesign, and production costs. Increasing redesign time for improvement to eliminate emerging deficiencies further increases costs and causes greater delay in providing a satisfactory product.

ART/ADT is a new and better way of providing accelerated product/technology development consisting of accurate physical simulation (APS) of the field situation, ART/ADT, reduced design time and results in more AP of product quality, reliability, maintainability, and durability during a given time. Thus, the philosophy of the ART/ADT processes includes the following elements:

1. Reliability testing and durability testing under artificial conditions (laboratory, proving grounds) need to provide an accurate simulation of the field situation. Otherwise, it would be impossible to obtain the initial information for the AP of a product's quality, reliability, durability, maintainability, supportability, and other factors. This approach can also quickly find the reasons for failures and degradation and changes in the characteristics of the product's quality for accelerated product development and improvement at a minimal expense. Figures 2.5 and 3.2 provide examples of field situations for accurate simulation. The field situation will vary for different types of products.
2. The general requirements of the author's philosophy consist of two basic groups of components. The first group of components includes the following:
   - To achieve high levels of effectiveness in development and improvement, it is necessary to provide a globally integrated complex of
     - Accurate simulation
     - ART
     - AP of reliability/quality/maintainability/durability during a given time interval
     - Successful development of accelerated quality and reliability solutions

  - When providing ART separately from all the other components of the above-mentioned complex, it is necessary to first use an accurate simulation of the field input influences on the product/technology followed by conducting ART again.
- Simultaneously simulate the whole complex of input influences such as temperature, humidity, air fluctuations, air, chemical and mechanical air pollution, solar radiation, vibration, corrosion, and input voltage (Fig. 3.2) on the product/process in combination with each element.
- Accurately simulate the whole complex of field input influences simultaneously, as well as each influence using given criteria.
- The physics or chemistry of the degradation process of the test subject must be similar to this process in an actual service environment while using the given criteria.
- ART consisting of laboratory testing and special field testing (in Chapter 3)

The second group of components of the author's philosophy includes the description of accelerated reliability and durability testing as interconnected systems using a system of systems (SoS) approach (Fig. 4.1), which consists of the following:

- Accurate simulation of the field input while providing the simultaneous integrated combination of multi-environmental testing, mechanical testing, electrical testing, and other testing
- Accurate simulation of safety issues including risk problems and hazard analysis
- Accurate simulation of HFs including psychological aspects, anatomical aspects, individual differences, and group factors

The philosophy relates to strategic thinking that requires the different steps of ART/ADT to be conducted and directed by engineers and managers who think in terms of capabilities. Achieving a capability requires many systems to work in unison. We can model the entire SoS, treating it as a "system" and the individual systems of the SoS as "subsystems." It is possible to simulate the operation of the entire system when each step and component of ART/ADT are known; QRM metrics are known; and the characteristics of each of these subsystems are known.

The assessment can be continually refined as more ART/ADT data become available and are analyzed and then applied at appropriate levels of integration. Depending upon the step of the testing and amount of data, confidence intervals are established for the measures of QRM indices. These assessments will be much more robust than the point estimates made earlier on the basis of analysis alone. Using operational tests and predictions, a more "accurate" assessment of QRM is possible prior to fielding a system.

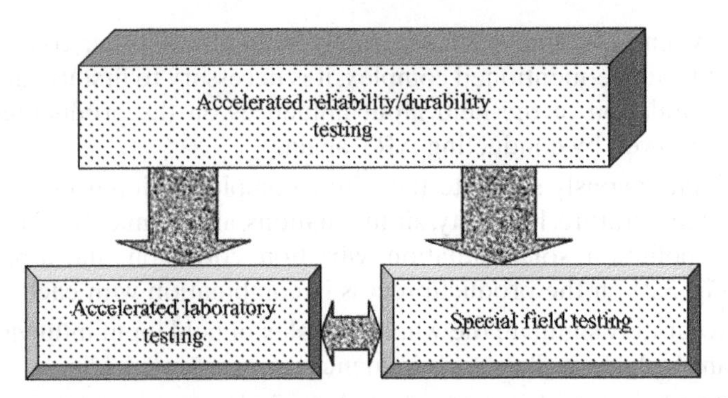

Figure 4.2. Combination of accelerated laboratory testing and special field testing as basic interconnected components of accelerated reliability/durability testing.

Accelerated reliability and durability testing technology are initially more expensive than testing each influence separately (Fig. 1.2). Therefore, one may choose to provide a step-by-step implementation of the ART/ADT technology. For example, as a first step of technology implementation, one could build a test chamber with a necessary line of communications with only the equipment for vibration testing along six axes (three linear and three angular). For testing wheeled equipment, one can use vibration equipment along three linear axes. Later on, as a second step, one can build/add temperature–humidity equipment to the test chamber and continue this process until a fully equipped test chamber is completed.

4.3 ART/ADT METHODOLOGY AS A COMBINATION OF DIFFERENT TYPES OF TESTING

As mentioned in the definition of the ART/ADT of a test subject (Chapter 1), two specifics of this type of testing one must consider are the following:

- The physics (or chemistry) of the degradation mechanism compared with the degradation mechanism in actual use
- Does a high correlation exist between the reliability parameters from testing and the same parameters in actual use?

To achieve this goal, one must provide a combination of different and interconnected test types. As mentioned earlier, the condition of the test subject (its reliability, quality, and maintainability) is the result of these interconnections and influences and their interaction.

This is one of the basic methodological requirements of ART/ADT. Let us consider the methodology of the above-mentioned combination.

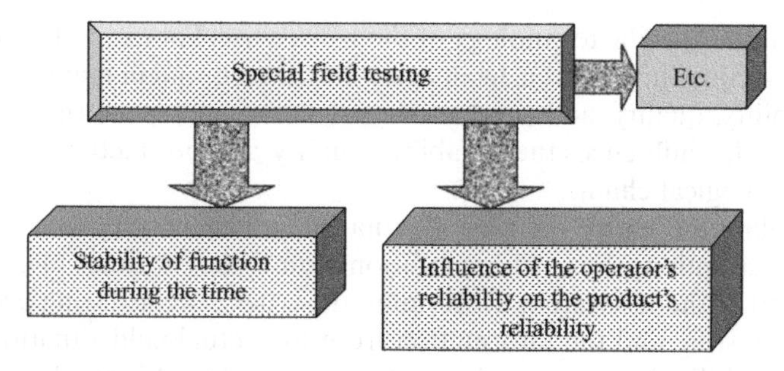

Figure 4.3. Scheme of special field testing for the automobiles.

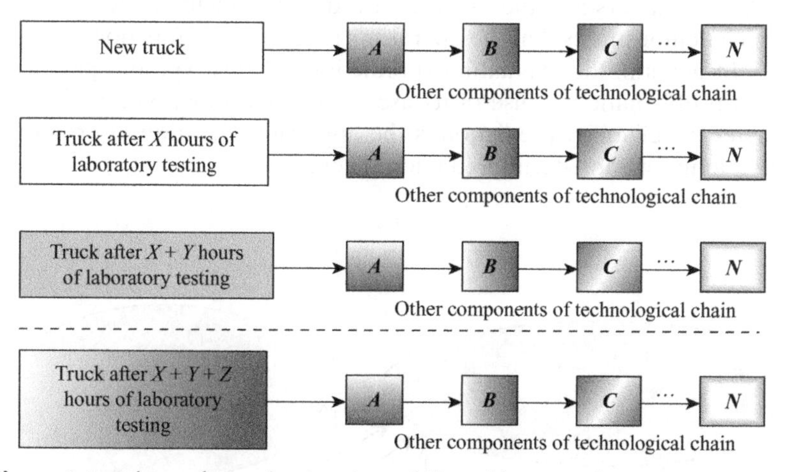

Figure 4.4. Scheme for evaluating the stability of function during a given time.

4.3.1 The First Component of This Methodology

The ART/ADT methodology has to provide a combination of accelerated laboratory testing and special field testing (Fig. 4.2).

Use special field tests for the evaluation of field influences that are impossible or very expensive to simulate in the laboratory. The selection of appropriate methodologies of accelerated laboratory testing and special field testing depends on the specifics of the test subject and its use. For example, it is necessary to provide special field testing for the evaluation of the stability of a product/technology function for an automotive product and how the operator's (driver's) reliability influences the product's reliability (Fig. 4.3).

There are different types of automotive products including cars, trucks, recreational vehicles, vans, and buses. Each of the products has a specific purpose. For example, trucks often transport different products and connect with other operations (Fig. 4.4) in technology chains. A technology chain depends on the specific technology of the truck's use. For example, a truck can be used in a construction area, in an agricultural area, and in many other areas. The reliability, quality, and productivity of the truck's performance change during use. This influences the reliability, quality, and productivity of the complete technological chain.

To simulate the above-mentioned situation, it is necessary to compare the actual field situation (the truck's condition after different times of use) to that experienced in the laboratory. Determine the truck's wear during accelerated laboratory testing and the information from the actual field situation for calculating reliability, quality, productivity, and maintainability. The operator's reliability also influences the product's reliability.

Consider the methodology for the evaluation (prediction) of the driver's reliability. To understand the essence of a driver's reliability, refer to Figure 4.5, which demonstrates that the operator's reliability is a function of different factors. In general, the evaluation (prediction) of this type of operator reliability is problematic because of its use for specific situations. An important parameter of reliability prediction is the applicable duration of operation. It can be a long- or short-term prediction. Therefore, the

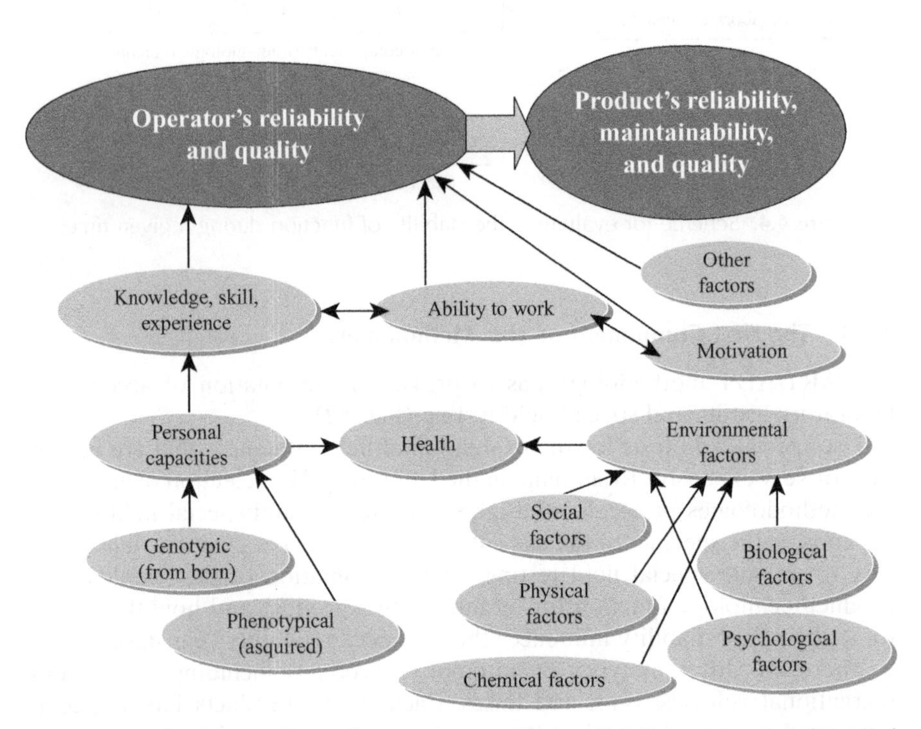

Figure 4.5. The factors that influence the product's reliability, maintainability, and quality through the operator's reliability and quality.

- Short-term prediction assesses the reliability for executing specific tasks during the workday, for example, preflight tests for pilots or prerun tests for vehicle drivers.
- Long-term prediction utilizes longer time segments (several days, weeks, months, years) to predict when applicants are capable of working the required period. For example, use this period to assess and resolve the problems of psychophysiology (PP) orientation and selection as one of the basic components of HFs.

Setting limits on the levels of the abilities for specific tasks consisting of functional conditions is also a concern. An example is setting limits on the complex characteristics of the operator's functions and personal qualities that impact the quality of the operator's actions controlling the vehicle over a given time period.

The influences of complex and personally disturbing factors on the homeostatic level of a person's basic tolerance for psychological or emotional distraction must be included. There are circumstances that prevent a person from performing favorably under unusual and stressful conditions. Nevertheless, these mechanisms are limited and lead to a decline in adaptability and to a decrease of the operator's reliability. Sometimes in situations resulting in a high level of stress and responsibility (life threatening, inadmissible psychological or economic consequences of mistakes), functional mobilization necessarily affects an operator's situation management and reliability for a limited time, which eventually may change to a dramatic failure.

Figure 4.5 describes factors from the "environment of movement" that relate to the basic characterization of an operator's functional condition with respect to the reliability of the system "operator-mobile vehicle." The circumstances in other industries, for instance, aerospace or any stationary equipment, are similar.

In the environment of movement (Fig. 4.6), one can note the following:

- Adequate reaction to functional conditions and to disturbing influences
- Duration of the conservation of the homeostatic level for basic functions with the use of minimal energy expenditures during the period of action
- The time required to execute a given action at a functional level and its reinstatement from its initial start point to its completion.

To provide a short-term prediction, the definition of the current functional condition offers additional information that makes it possible to eliminate alcohol, drugs, medical, and stress influences.

The results of this research are available to solve the problem of psychophysiological professional selection of drivers for city buses as an example of long-term prediction.

The methodological complex of driver selection includes

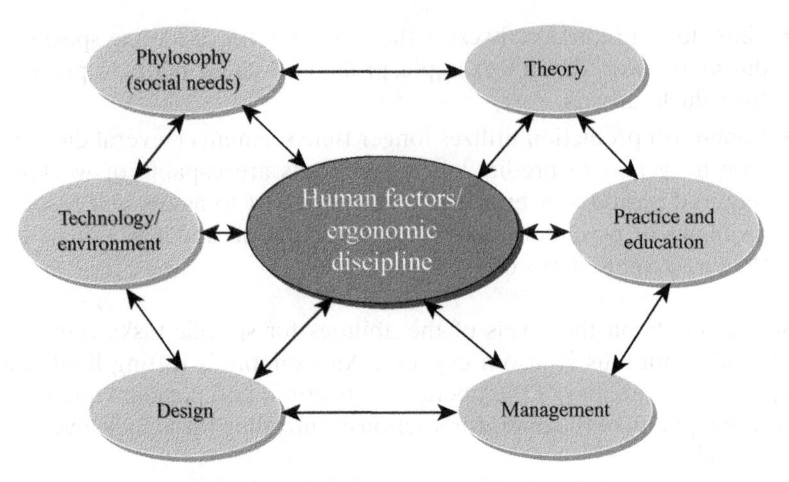

Figure 4.6. General dimensions of human factors.

- Research into reactions to a moving subject (capacity for perception and definition of special-time relations)
- Different sensorimotor reactions
- Patience for enduring monotony
- Long duration and operational memory
- Emotional stability
- Capacity for adequate conduct in stress situations
- Volume and concentration of attention
- Capacity for probabilistic prediction
- Inclination of risk
- Extra (intra) vision

The following example studied 2000 people who finished driving school during 1 year of driving city buses. The study divided them into two groups based on the number of traffic accidents: successful (no accidents) and unsuccessful (one or more accidents).

As a practical example, compare the average levels that are important professionally for PP indices exhibited by both successful and unsuccessful drivers (Table 4.1). The research results show the higher average levels registered in the successful group for functional conditions. These functional conditions include the capacity for reliability prediction, adequate reaction to a moving object, and the ability to withstand monotony. The group of unsuccessful drivers did better in the characteristic of attention. This means that an operator's reliability is characterized not only by separate PP functions or the sum of these functions but also by their more complicated dynamic interconnections.

TABLE 4.1 Comparison of Average Levels of Psychophysiological Indexes by Successful and Unsuccessful Drivers [146]

Number	Indexes	Successful $M + -m$	Unsuccessful $M + -m$	P (Less Than)
1	Capacity to probabilistic prediction	26.8 + –6.6	9.8 + –2.5	0.01
2	Research of reaction on moving subject (capacity to perception and definition of spatial-time relations	30.7 + –3.5	45.7 + –3.3	0.01
3	To withstand monotony	1.4 + –0.39	3.2 + –0.67	0.025
4	Volume and concentration of attention	7.5 + –0.53	6.1 + –0.52	0.05

The generalization of the results of the PP research assists using the method's factor and cluster analysis in the development of a decision rule for an operator's selection. This selection based on the operator's reliability in the system "driver–bus–city's environment of movement."

The vectors of training selection determined allowable limits for the suitability of a person for a given activity based on psychophysiological parameters. The study reviewed 450 auto school entrants to evaluate the effectiveness of the proposed method. A comparative analysis studied an entire year of predicted and actual reliability of their work as bus drivers. The predicted outcome confirmed the real parameters of reliability for the people examined at least 80% of the time.

The data show that the proposed methodology is useful based on the dynamics of interconnected psychophysiological parameters for the prediction of an operator's reliability under given conditions of use.

4.3.2 The Second Component of This Methodology

A simultaneous simulation combining many input influences in one complex is required to conduct effective ART/ADT. Separate simulation of input from field influence types or from only a portion of the many field influences contradicts the basic principles of ART. Accelerated laboratory testing must include a simultaneous combination of several groups of interacting testing influences in one integrated complex to account for the synergistic effect. The number of these groups needs to equal the number of groups that act on the test subject in the field.

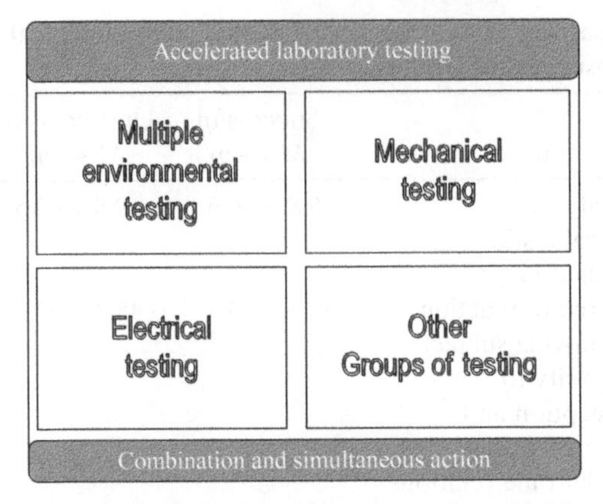

Figure 4.7. Interconnected groups of combined accelerated laboratory testing as a component of accelerated reliability testing.

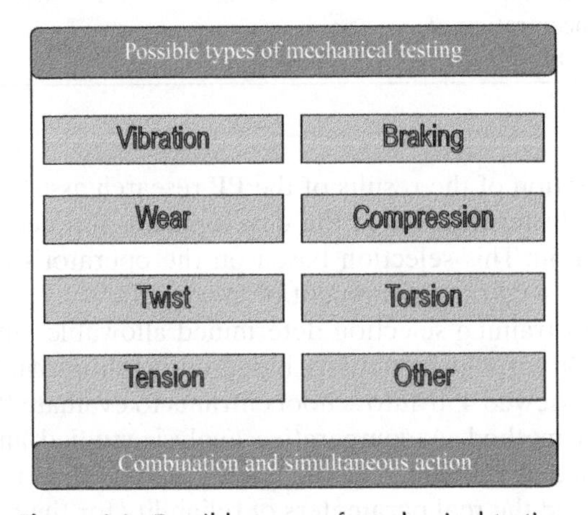

Figure 4.8. Possible types of mechanical testing.

Figure 4.7 shows an example with the following groups:

- Multiple environmental testing
- Mechanical testing
- Electrical testing
- Other necessary testings

One must use each of these testing groups as a complex for specific types of testing. For example, mechanical testing consists of different types of testing that mutually interact simultaneously. These types of testing depend upon the specifics of the test subject and the types of input influences that act on this subject in the field. Figure 4.8 shows the possible types of mechanical testing with their interconnections.

A similar situation exists with multi-environmental, electrical, and other groups during laboratory testing.

As an example of the above-mentioned approach in aerospace, the experience of Arnold Engineering Development Center's (AEDC) in simulation and the propulsion wind tunnel ground test complex for durability testing and prediction of test object (system) performance can be demonstrated. AEDC operates five active wind tunnels in two primary facilities, the Propulsion Wind Tunnel Facility (PWT) and the Von Karman Gas Dynamics Facility (VKF). The above-mentioned complex includes two wind tunnels at remote operating locations—the Hypervelocity Wind Tunnel 9 in Maryland and the National Full-Scale Aerodynamics Complex in California.

AEDC wind tunnels are used for testing in areas including vehicle aerodynamic performance evaluation and validation, weapons integration, inlet/airframe integration, exhaust jet effects and reaction control systems, code validation, proof of concept, large- and full-scale component research and development, system integration, acoustics, thermal protection system evaluation, hypersonic flow physics, space launch vehicles, operational propulsion systems, and captive flight.

Propulsion Wind Tunnel 16T provides flight vehicle developers with the aerodynamic propulsion integration and weapons integration test capabilities needed for an AP of system performance. Pressure in the test section can be varied to simulate unit Reynolds numbers from approximately 0.03–7.2 million feet or altitude conditions from sea level to 86,000 ft. Air-breathing engine and rocket testing can also be performed in Tunnel 16T using a scavenging system to remove exhaust from the flow stream.

Sea-Level Test Cells SL-2 and SL-3 provide the capability to economically conduct durability testing on large augmented turbine engines at near-sea-level conditions by eliminating the cost of running inlet and exhaust plant machinery. In addition, they also provide the capability of using the engine testing facility plant to run ram conditions.

4.4 ACCELERATED MULTIPLE ENVIRONMENTAL TESTING

Figure 4.7 shows that multiple environmental testing is a necessary subcomponent of accelerated laboratory testing in ART/ADT. Many mistakes occur with this type of testing. Some forget that the word "multi" means many, not one to three. For example, some companies use testing chambers that simulate only temperature and humidity.

4.4.1 Principles of Accelerated Multiple Environmental Testing

The first principle of the establishment of accelerated multiple environmental testing (AMET) is an integrated simultaneous combination within the

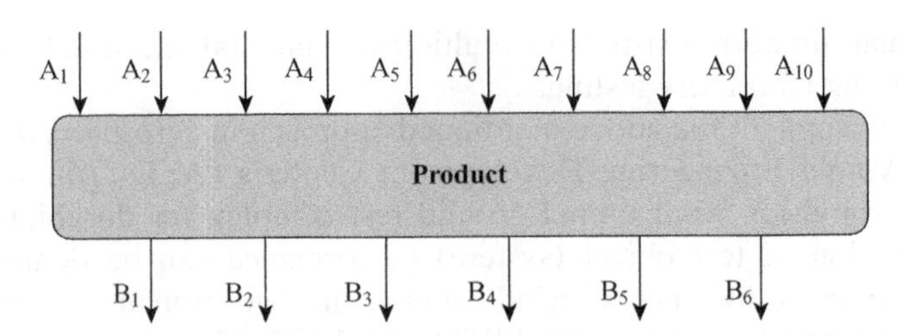

Figure 4.9. The scheme of basic environmental influences on the actual product and the results of their action.

laboratory simulation of multiple environmental factors that act on the test subject in actual service.

Advanced AMET applies the principles of laboratory simulation of the field environmental influences on the actual product. These influences may include temperature (A_1), humidity (A_2), pollution (A_3), radiation (A_4), water (A_5), gas pressure (A_6), air fluctuations (A_7), rain (A_8), snow (A_9), and wind (A_{10}) (see Fig. 4.9). The combination of these simulated influences depends upon the types of specific field influences acting on the test subject.

The basic principle of AMET is the simulation of the simultaneous combination of field input influences. The results of the simultaneous actions on the product are metal corrosion (B_1); destruction of polymers (B_2), rubber (B_3), and wood (B_4); and a decrease in the protective effects of grease (B_5) and paint (B_6). When combined with other subcomponents, the result is a decrease in the product's quality and reliability through different types of degradation and failure processes.

An accurate simulation of the input influences is essential to conduct AMET successfully.

The common principles applied to obtain accurate information for field input influences on the product for the simulation are the following:

1. The simultaneous combination of input influences that in real life influence the product
2. These influences are interconnected and simultaneously interact with each other.
3. The interaction of the input influences results in an action on the product.
4. Their effect on the product (degradation and failure) is a cumulative reaction to the combination of input influences.

The following information is necessary for an accurate evaluation of the environmental conditions for accelerated testing. The ideal ratio of simulated conditions in the test chambers to the number of naturally occurring conditions equals one. Nevertheless, this is difficult to achieve in actual practice due to the presence of their conditions and constraints. If the influences on the product under operating conditions differ from influences in the laboratory by no more than a fixed limit, then the results in the test chamber correspond to the results in the field.

Failures in the field should differ from failures in test chambers because normally, the conditions of the test chambers only partly simulate field conditions. More details about these issues are included in the author's other publications [18, 101, 105, 123].

The most important aspect of this strategy is finding the optimum conditions for AMET (regimens in the test chambers).

The basic steps of the physical simulation of environmental conditions are the following:

- The analysis of the real-life input environmental influences on the product provide guidance for testing influences in accordance with the above-mentioned principles.
- The analysis of the kinds of influences that are important depends on a result of their action on the degradation and failures of the product.
- The development and use of the laboratory conditions for the simulation of the necessary combination of influences (important for failures, reliability, and serviceability of the test subject) during testing

An analysis of the work and the storage of the test subject in the field guide the choice of influences for AMET. Therefore, it is very important to evaluate the effect of probabilistic characteristics of the regional field influences to include when building physical models of the environmental field conditions in the test chamber. Different regions have different influences on the product.

4.4.2 The Mechanism of the Influence of Environmental Factors on Product Reliability

The mechanism of this influence is the basis for the methodology of accelerated multi-environmental testing (see Fig. 4.10).

Many types of environmental influences act on a product in real life. Some influences are studied, but many other influences are not studied.

A description of solar radiation action is included in Section 4.4.4. There are fewer studies of solar radiation than for temperature, humidity, or some components of pollution. People who want to simulate the field situation accurately must take into account that the environmental factors are integrated with mechanical, electrical, and other factors. The environment acts on a product through input influences like temperature, humidity, air pollution, solar radiation, and many others. A result of these actions and interconnections is output variables such as loading, tension, voltage, amplitude, and frequency of vibration (see Fig. 3.1). The result of these actions is a degradation of the product that one can measure in terms of metal corrosion, plastic and

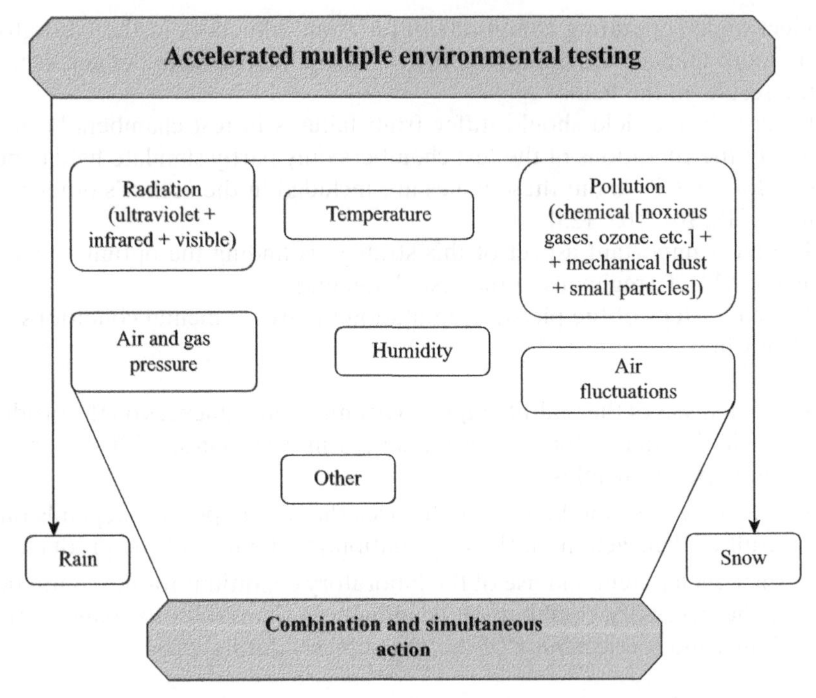

Figure 4.10. Scheme of accelerated multiple environmental testing.

rubber aging, the decrease of grease and lubricant properties, and failures. The final result affects the product reliability, durability, and maintainability, and other measures.

The mechanism for the influence of environmental factors in the field is very complicated. One of the most complicated problems is the integrated cause-and-effect relationship of different factor steps including input influences, output variables, and degradation (see Fig. 2.7).

The integrated cause-and-effect relationship of input influences such as pollution with degradation from corrosion is an example. The corrosion of materials and systems is often partially responsible for accidents that cause pollution.

A description follows for the mechanisms of corrosion and pollution combining to become one of the most complicated environmental factors.

The environmental damage caused by pollution is large and is rapidly becoming worse because of the increase in the industrialization of many countries throughout the world. Air is 99.9% nitrogen, oxygen, water vapor, and inert gases. There are several types of pollution. The well-known effects of pollution commonly discussed include smog, acid, rain, the greenhouse effect, and "holes" in the ozone layer.

For an accelerated reliability test goal, one type of air pollution is the release of particles into the air from burning fuel. A good example of this particular matter is diesel smoke. The particles comprising this type of smoke are very small, measuring about 2.5 microns. This type of pollution is sometimes labeled "black carbon" pollution.

Another type of pollution is the release of noxious gases including sulfur dioxide (SO_2), hydrogen sulfide (H_2S), nitrogen dioxide (NO_2), chloride (Cl_2), and ozone (O_3). Once released into the atmosphere, these noxious gases can enter into further chemical reactions producing smog formation, acid rain, and formation of other pollutions. Most of these gases damage materials, electronic components, modules, devices, and structures.

Environmental corrosion is also a by-product of pollution, particularly when dealing with atmospheric corrosion and the corrosion of historical landmarks or artifacts.

4.4.3 Accelerated Environmental Testing of Composite Structures

The use of composites for structural components has increased because composites have a number of advantages compared to metals. The main advantage is that composites reduce the weight of a structure because their specific strength and stiffness is greater than metal. Another advantage is that composites enable production of complex three-dimensional structures in one piece.

Composites reduce the number of parts and the assembly costs significantly, although part of the cost reduction goes to compensate for more expensive tools and materials.

The use of composite structures is a recent development compared to the use of metallic, wood, and plastic structures. Accelerated testing for composite structures has experienced less development, including accelerated environmental testing. The current practice for accelerated environmental testing of composite structures follows. The ability to predict stresses and failures caused by out-of-plane loads has gained more and more importance as airframe manufacturers begin to use integral (co-cured or bounded) composite structures to minimize weight and maximize performance. Development of rapid and accurate analysis methods reduces the amount of testing required and ensures confidence in integral composite structure performance.

Simple two-dimensional analysis methods exist to predict the out-of-plane failure strength of composite airframe structures [124]. These analyses were primarily used to address induced stresses in laminate corner radii, direct stresses due to fuel pressure loads, induced stresses from panel buckling, and induced stresses from stiffener runouts or other load path changes.

The development and verification of analyses for out-of-plane failures have demonstrated the need for out-of-plane material properties and strength that are not traditionally determined for composite laminates. Accurate stress and failure predictions depend heavily on these experimentally determined values. The test methods and experience gained in this program [124] indicates the need for new techniques to determine the interlaminar properties required for accurate stress and failure predictions. This program suggested test techniques for determining some of the values necessary for these predictions.

In Reference 125, Vodicka reported that composite materials lose mechanical properties during exposure to aircraft operating environments. This is mainly due to the absorption of moisture from humid air by the matrix material. The RAAF uses composite materials extensively for both major structural components on the F/A-18 and for bonded repairs and doublers. The performance of these materials under long-term environmental exposure is an important aspect for both aircraft certification and in the understanding of how the components will perform as they age. The report [125] provides an overview of environmental effects on composite materials and methods that predict long-term behavior. Reducing the testing duration makes use of accelerated testing environments in the laboratory an attractive proposition. A number of accelerated testing methodologies and their implications are in Reference 125. If the exposure conditions are representative and the failure modes of the material during mechanical testing reflect those seen in service, then one can conduct accelerated testing with confidence.

In Reference 126, Higgins et al. selected the composite grid-stiffened structure concept for the payload fairing of the Minotaur launch vehicle. Compared to sandwich structures, this concept has an advantage of lower manufacturing costs and lighter weight. The skin pockets may buckle visibly up to about 0.5-cm peak displacement to reduce weight.

The above-mentioned paper examines various failure modes for the composite grid-stiffened structure. The controlling criterion for this design is a joint failure in the tension between the ribs and skin of the structure. The identification of this failure mechanism and the assessment strains necessary for control required an extensive testing and analysis effort. Increasing skin thickness to control skin buckling resulted in reduced strains between the skin and ribs.

A final fairing design followed the identification of the relevant failure criteria. Testing results show that the applied load envelope exceeded worst-case dynamic flight conditions with an added safety factor of 25%.

Data collected during the testing came from a variety of sensors including traditional displacement transducers and strain gages. Additionally, monitoring the full field displacement occurred at the critically loaded fairing sections a digital photogrammetric workstation. The results of testing document the overall performance of the fairing under the test loads, correlate test response, and analyze and identify lessons learned. Work continues at the Air Force Research Laboratory and Boeing to identify means to further control tensile failure of the unreinforced polymer-bonded joint between the ribs and skin.

4.4.4 Solar Radiation: Global Dimming

Scientific publications in climatology show the current situation and the global changes in solar radiation (in References 127, 128, and other publications). These changes are influencing the testing regimes as well as the accuracy of reliability and durability predictions. Researchers at the Zurich-based World Radiation Monitoring Center (ZWRMC), an organization that measures the amount of solar radiation reaching the ground around the globe, discovered an interesting result.

Global dimming is the observed effect of an increase in particulate matter in the atmosphere. This particulate matter produced a substantial drop in the solar radiation reaching the planet's surface over the past 60 years, approximately 22% globally and up to 30% in some locations. Pollution from burning coal, oil, and wood appears to cause global dimming. Because it reduces the amount of sunlight reaching the ground, it cuts the amount of heat trapped by greenhouse gases. This has troubling implications.

One of the most alarming implications is that global dimming may have led scientists to underestimate the true power of the greenhouse effect. Scientists have a measurement of the extra energy trapped in the earth's atmosphere by the extra carbon dioxide produced on the planet. Surprisingly, this extra energy has resulted in a temperature rise of only 0.6°C so far. This has led many scientists to conclude that the present-day climate is less sensitive to the effects of carbon dioxide than it was during the ice age when a similar rise in CO_2 led to a 10 times higher temperature rise of 6°C.

This result was a very surprising finding for the researchers at the ZWRMC [129].

As part of these studies into climate and atmospheric radiation, the researchers were checking levels of sunlight recorded around Europe when they made an astonishing discovery. Compared to similar measurements recorded by their predecessors in the 1960s, the results suggested that the levels of solar radiation reaching the earth's surface had declined by more than 10% in three decades. Solar radiation was declining [128].

A function of solar radiation is its influence on the sweetness of white wine grapes. The more sun a grape plant's leaves absorb, the more sugar the plant produces and the more sweetness is infused into the fruit. By paying close attention to the global meteorological records, particularly the geographic distribution of solar radiation, it should be possible to infer a wine's origin by sensing its sweetness.

The ZWRMC reviewed national meteorological records and current publications, copying and pasting solar radiation measurements into a database until they had observations from 1600 locations around the world, including the oldest continuous measurement recorded by the Swedish professor Anders Angström in 1922.

Prof. Angström is famous for crawling up onto the roof of the University of Stockholm to set up his novel nyranometer that was the first device to measure accurately direct and indirect solar radiation. This initiated the science that would lead Prof. Ohmura from ZWRMC to the missing sunlight. Never before had anyone drawn such a thorough picture of worldwide solar radiation levels. This discovery was hard to believe at first. However, the data were extensive. Prof. Ohmura presented this information about the "missing chunk of energy" at the 1988 International Radiation Symposium in Lille, France.

Israeli researchers Shabtai Cohen and Gerald Stanhill confirmed these findings in 2001. They reported a 10% radiation loss between 1958 and 1992 that is virtually the same conclusion that Ohmura had reached in 1988. In 2003, Graham Farquhar and Michael Roderick, climatologists at the Australian National University in Canberra, discovered corroborating evidence in the global evaporation record. The ongoing explanation for the loss of sunlight is that particulate pollution-like soot plugs up the clouds. The result follows that when it is cloudy, the atmosphere is darker than before.

Clouds are made of water droplets that form by latching onto tiny particles called condensation nuclei. Condensation nuclei occur naturally in the atmosphere and humans produce even more by emitting more particulate pollution into the atmosphere.

The result is that many smaller water droplets form, instead of fewer larger water droplets. In effect, this is like the difference between two sieves when one is coarse and the other fine. Like a coarse sieve, the cloud with fewer, larger particles lets more solar radiation through to the ground, whereas like the fine sieve, the cloud with lots of very small particles lets less sunlight pass through, resulting in less solar radiation reaching the ground.

For some time, scientists have wondered why the temperature remained relatively stable or even became colder during the 1960s and 1970s. At the same time, the greenhouse gases were sufficient to increase the temperature. To illustrate the point, the ZWRMC scientists cite the average annual melt rates for each of the last four decades for 40 glaciers around the world:

- In the 1960s, the glaciers retreated by 200 mm/year.
- In the 1970s, the glaciers retreated by 180 mm/year.
- In the 1980s, the glaciers retreated by 260 mm/year.
- In the 1990s, the glaciers retreated by 480 mm/year.

The results published in Reference 128 show that during the 1960s and 1970s, global dimming buffered the climate against global warming caused by greenhouse gases. As increasing amounts of greenhouse gases warmed the earth, growing amounts of particulate pollution reduced the amount of sunlight that reached its surface and cooled the planet. In other words, one form of pollution counteracted the other, hence the lower melt rates and stable temperatures of the 1970s.

Scientists then realized that particulate pollution was almost entirely responsible for deaths related to air pollution—pollution that still causes a staggering

135,000 early deaths in the United States every year (6% of all deaths from any cause). While this number may seem large, the U.S. Environmental Protection Agency (EPA) found that if Congress had not adopted the Clean Air Act in 1970 and its amendments to the act in 1977, particulate matter would have caused the early deaths of 184,000 more Americans per year by 1990. Clearly, this type of pollution was, and still is, a big problem. Wisely, the industrialized world began curtailing emissions of soot and smoke.

What happened to the particulate matter buffer?

As the emissions of deadly particulate matter decreased, so did their cooling power. Clouds let the solar radiation through and—behold—the greenhouse effect's disguise diminished. Because of this double punch of more solar radiation and more greenhouse gases, the global temperature increased enormously.

Preliminary results based on radiation records from 1992 to the present support this theory. Key monitoring stations show a resurgence of radiation levels during the 1990s, not to pre-1958 levels, but enough to expose the true warming potential of greenhouse gases.

The implication is the impact of both particulate matter and greenhouse gases on the climates has been underestimated. Traditional global climate models will have to be revised, and although the interactions between climatological components are complex and uncertain, revised models and the above-mentioned results will undoubtedly raise challenging questions for the public and for policy makers. For instance, will further reductions in particulate pollution, which are necessary to alleviate serious, deadly, and widespread respiratory illnesses, mean even more global warming? If so, do even progressive policies including the Kyoto Protocol underestimate the potential damage of climate change?

Until Prof. Ohmura and his colleagues examined the radiation records, no one had noticed that between 1958 and 1988, a whopping 10% of the expected solar radiation had disappeared.

Critics charged that the numbers had to be wrong. They are too big to be accurate. Moreover, the idea of a much darker planet did not fit into conventional climate models that predicted a much brighter planet. The theory went largely unnoticed for an entire decade while faced with this criticism.

4.5 ACCELERATED CORROSION TESTING

The components that relate to accelerated corrosion testing are in Sections 4.5.1–4.5.4.

4.5.1 Cost of Corrosion

Corrosion is one of the most serious problems in modern societies. The losses from corrosion each year are hundreds of billions of dollars. Several countries including the United States, the United Kingdom, Japan, Australia, Kuwait, Germany, Finland, Sweden, India, and China have studied the cost of corrosion. The studies have ranged from formal and extensive efforts to informal and modest efforts. The common finding of is that the annual corrosion costs ranged from 1 to 5% of the gross national product (GNP) of each nation.

Two annual reports, both numbered FHWA-RD-010-156 [130, 131], describe the annual direct and indirect costs of metallic corrosion in the United States and preventive strategies for optimum corrosion management. The estimated total direct cost of corrosion is $276 billion per year, which is 3.1% of the 1998 U.S. gross domestic product (GDP). The direct cost of corrosion consists of the following [130]:

- Utilities (47.9 B—34.7%)
- Transportation (29.7 B—21.5%)
- Infrastructure (22.6 B—16.4%)
- Government (20.1 B—14.6%)
- Production and manufacturing (17.6 B—12.8%)

Analyzing 26 industrial sectors and extrapolating the results for a nationwide estimate determined this cost. The sectors divide into five major categories.

The infrastructure category consists of

- Highway bridges
- Railroads
- Airports
- Hazardous material storage sites
- Waterways and ports
- Gas and liquid transportation pipelines

The utilities category consists of

- Telecommunications
- Electrical utilities
- Gas distribution

The transportation category consists of

- Ships
- Aircraft
- Motor vehicles

- Hazardous material transport
- Railroad cars

The government category consists of

- Nuclear waste storage
- Defense

The production and manufacturing category consists of

- Agricultural
- Food processing
- Electronics
- Chemical, petrochemical, and pharmaceutical
- Petroleum refining
- Mining
- Home appliances
- Pulp and paper
- Oil and gas exploration–production

The estimated indirect cost of corrosion is equal to the direct cost. Thus, total direct cost plus indirect cost is approximately 6% of the GDP. Components of the large indirect corrosion costs are lost time and damaged facilities leading to lost productivity due to outages, delays, failures, and litigation.

4.5.1.1 Indirect Costs.

Indirect costs include the

- Cost of labor attributed to corrosion management activities
- Cost of the equipment required because of corrosion-related activities
- Loss of revenue due to disruption in the supply of the product
- Loss of reliability
- Cost of corrosion abatement

The indirect costs total $275.5 billion per year. The drinking water and sewer systems ($36 billion), motor vehicles ($23.4 billion), and defense ($20 billion) sectors have the largest cost impact from direct corrosion.

Spending on corrosion control methods and services totals $121 billion per year. The current study [130] shows that technology changes provide many new ways to prevent corrosion and improve the use of corrosion management techniques.

However, using preventive strategies in nontechnical and technical areas can achieve better corrosion management. These preventive strategies include the following:

- Increase awareness of large corrosion costs and potential savings.
- Change the common industry misconception that nothing can be done about corrosion.
- Change policies, regulations, standards, and management practices to increase corrosion savings through sound corrosion management and best practices.
- Improve staff education and training in the recognition of corrosion control.
- Advance design practices for better corrosion management.
- Advance life prediction and performance assessment methods.
- Advance corrosion technology through research and development.

The direct costs associated with the corrosion of a valve, an oil tanker, and a gas pipeline can be tremendously amplified when the subsequent events result in increased pollution. Releases of pollutants can contaminate drinking water and crops, cause expensive property damage, cause fish kills, and create explosions and fires. For example, the lack of proper plant maintenance in Bhopal, India, resulted in the worst industrial catastrophe in history including loss of life.

4.5.2 Mechanism of Corrosion

The literature study of corrosion processes and corrosion testing primarily addresses the corrosion of metallic materials. The corrosion process of a material usually differs from the corrosion process of a product created from this material. The concentration of the loading, friction, and the interconnection between different units and details of the product may strongly influence the corrosion of the product. The corrosion of metals and corrosive degradation of paint and coatings are the result of a combined action from multi-environmental factors in combination with mechanical factors, plus the interconnection of these factors (see Fig. 4.11).

Finally, the corrosion of metals is the degradation of metals. The degradation of our quality of life due to the presence of unwanted metals in our body is a consequence of corrosion. Additionally, it is the degradation of the planet due to pollution, global warming, and global dimming. The author's publications [42, 103, 104], among others, describe in detail the modern approach to the corrosion process.

Additional information follows about the mechanism of corrosion. Publications, including Reference 132, and virtual information from the Corrosion Technology Laboratory, Kennedy Space Center (KSC) in Florida, are used.

4.5.2.1 Why Metals Corrode. Metals corrode upon exposure to environments where metals are chemically unstable. The only metals found in their

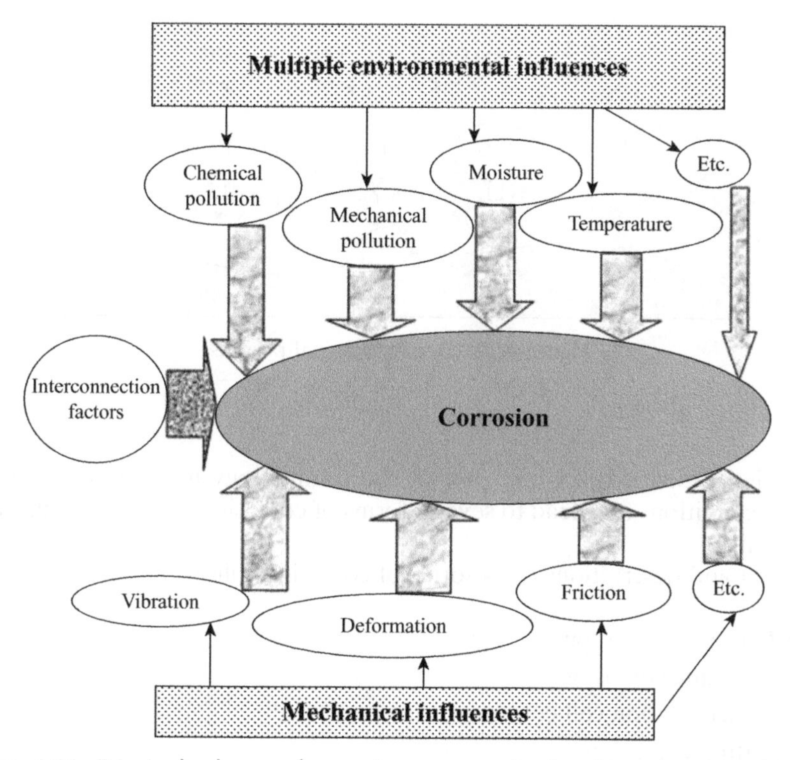

Figure 4.11. Principal scheme of corrosion as a result of multi-environmental influences, mechanical influences, and interconnection factors.

pure metallic state in nature are copper and the precious metals gold, silver, and platinum. Processing minerals or ores that are inherently unstable in their environments produces all other metals including the most commonly used metal, iron.

4.5.2.2 The Electrochemical Process. Corrosion is an oxidation reaction and a reduction reaction of the surface of the corroding material. The oxidation reaction generates metal ions and electrons. The reduction reaction consumes electrons. The conversion of oxidants and water to hydroxide ions consumes electrons in environments with water present including moisture in the air. In iron and in many iron alloys, these hydroxide ions in turn combine with iron ions to form a hydrated oxide [$Fe(OH)_2$]. Subsequent reactions form a mix of magnetite (Fe_3O_4) and hematite (Fe_2O_3). This red-brown mixture of iron oxides is rust. Figure 4.12 illustrates the basic oxidation/reduction reaction behind corrosion.

4.5.2.3 Forms of Corrosion. There are several methods of classifying corrosion, and no universally accepted terminology is currently used. The forms of

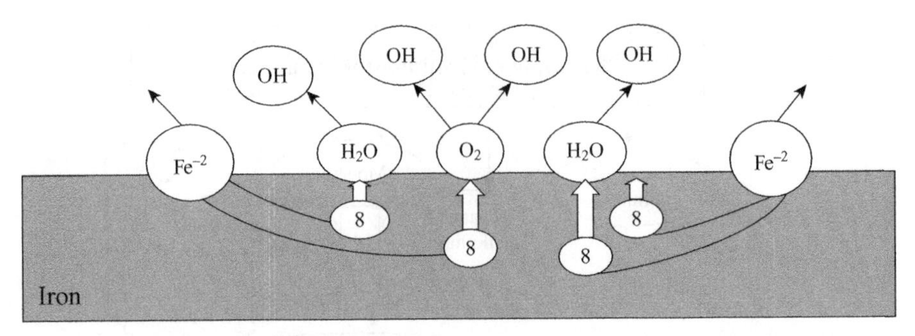

Figure 4.12. Electrochemical process.

corrosion described next are part of the terminology used at NASA KSC. Given conditions may lead to several forms of corrosion on the same piece of material.

A list and descriptions of 14 forms of corrosion follow:

- Uniform corrosion
- Galvanic corrosion
- Concentration cell corrosion
- Pitting corrosion
- Crevice corrosion
- Filiform corrosion
- Intergranular corrosion
- SCC
- Corrosion fatigue
- Fretting corrosion
- Erosion corrosion
- Dealloying
- Hydrogen damage
- Microbial corrosion

4.5.2.4 Uniform Corrosion. General corrosion is another name for uniform corrosion. The surface effect produced by the direct chemical attack by an acid is a uniform etching of the metal. On a polished surface, this type of corrosion appears as a general dulling of the surface and, if corrosion continues, the surface becomes rough and possibly frosted in appearance. Discoloration or general dulling of metal created by exposure to elevated temperatures is not uniform etching corrosion. The use of chemical-resistant protective coatings or other resistant materials will control these problems.

While this is the most common form of corrosion, it is generally of little engineering significance. Structures will normally become unsightly and will attract maintenance long before they become structurally degraded.

4.5.2.5 Galvanic Corrosion. Galvanic corrosion is an electrochemical action of two dissimilar metals in the presence of an electrolyte and an electron conductive path. It occurs when dissimilar metals are in contact with one another.

It is recognizable by the presence of a corrosion buildup at the joint between the dissimilar metals. Galvanic corrosion may occur on and accelerate the corrosion of the aluminum or magnesium when aluminum alloys or magnesium alloys are in contact with carbon steel or stainless steel.

4.5.2.6 Concentration Cell Corrosion. Concentration cell corrosion occurs w hen two or more areas of a metal surface are in contact with different concentrations of the same solution. Three general types of concentration cell corrosion are

- Metal ion concentration cells
- Oxygen concentration cells
- Active-passive cells

Metal Ion Concentration Cells. In the presence of water, a high concentration of metal ions will exist under faying surfaces. A low concentration of metal ions will exist adjacent to the crevice created by the faying surfaces. An electrical potential will exist between the two points. The area of the metal in contact with the low concentration of metal ions will be cathodic and protected. The area of metal in contact with the high concentration of metal ions will be anodic and corroded.

Sealing the faying surfaces to exclude moisture will eliminate this condition. Applying a protective coating of inorganic zinc primers is effective in reducing faying surface corrosion.

Oxygen Concentration Cells. A water solution in contact with a metal surface will normally contain dissolved oxygen. An oxygen cell can develop at any point where the oxygen in the air does not diffuse uniformly into the solution, thereby creating a difference in oxygen concentration between two points. Typical locations for oxygen concentration cells are under either metallic or nonmetallic deposits (dirt) on the metal surface and under faying surfaces such as riveted lap joints.

Oxygen cells can also develop under gaskets, wood, rubber, plastic tape, and other materials in contact with a metal surface. Corrosion will occur at the area of low-oxygen concentration (anode). Sealing the surfaces, maintaining clean surfaces, and avoiding the use of material that permits the wicking of moisture between faying surfaces will minimize the severity of this type of corrosion.

Active-Passive Cells. Active-passive cells can corrode metals that depend on a tightly adhering passive film (usually an oxide) for corrosion protection (e.g., austenitic corrosion-resistant steel).

The corrosive action usually starts as an oxygen concentration cell. Salt deposits on the metal surface in the presence of water containing oxygen can create the oxygen cell. If the passive film is broken beneath the salt deposit, the active metal beneath the film is susceptible to corrosive attack. An electrical potential will develop between the large area of the cathode (passive film) and the small area of the anode (active metal). Rapid pitting of the active metal will result. Frequent cleaning and the application of protective coatings will prevent or delay active-passive cell corrosion.

4.5.2.7 Pitting Corrosion. Passive metals, such as stainless steel, resist corrosive media and can perform well over long periods. However, when corrosion does occur, it randomly forms in pits. Pitting is most likely to occur in the presence of chloride ions combined with depolarizers like oxygen or oxidizing salts. Pitting can be controlled by maintaining clean surfaces, applying a protective coating, and using inhibitors or cathodic protection for immersion service. Molybdenum additions to stainless steel (e.g., in 316 stainless) are intended to reduce pitting corrosion.

Rust bubbles or tuberculosis on cast iron indicate that pitting is occurring. Researchers have found that the environment inside the rust bubbles is usually higher in chlorides and lower in pH than the overall external environment. This leads to a concentrated attack inside the pits.

Similar changes in the environment occur inside crevices, stress corrosion cracks, and corrosion fatigue cracks. These forms of corrosion are sometimes included in the term "occluded cell corrosion."

Pitting corrosion can lead to unexpected catastrophic system failure. Sometimes pitting corrosion can be quite small on the surface but very large below the surface. Pitting can lead to stress corrosion fracture. A complete discussion of this corrosion is contained in Reference 133.

4.5.2.8 Crevice Corrosion. Crevice or contact corrosion is the corrosion produced at the region where metals contact with metals or nonmetals. It may occur at washers, under barnacles, at sand grains, under applied protective films, and at pockets formed by threaded joints. Whether or not stainless steels are free of pit nuclei, they are always susceptible to this type of corrosion because a nucleus is not necessary.

Cleanliness, the proper use of sealants, and protective coatings are effective means of controlling this problem. Molybdenum-containing grades of stainless steel (e.g., 316 and 316L) have increased crevice corrosion resistance.

Crevice corrosion happens when an aerospace alloy (titanium–6 aluminum–4 vanadium) is used instead of a more corrosion-resistant grade of titanium. Special alloying additions added to titanium make alloys that are crevice corrosion resistant even at elevated temperatures. Screws and fasteners are common sources of crevice corrosion problems.

4.5.2.9 Filiform Corrosion. Filiform corrosion normally starts at small, sometimes microscopic defects in the coating. It occurs under painted or plated surfaces when moisture permeates the coating. Lacquers and "quick-dry" paints are most susceptible to this problem. Avoid these paints unless field experience shows the absence of an adverse effect.

Where a coating is required, it should exhibit low water vapor transmission characteristics and excellent adhesion. Also, consider zinc-rich coatings for coating carbon steel because of their cathodic protection quality.

Minimize filiform corrosion by careful surface preparation prior to coating, by the use of coatings that are resistant to this form of corrosion, and by careful inspection of the coatings for developing holes.

4.5.2.10 Intergranular Corrosion. Intergranular corrosion is an attack on or adjacent to the grain boundaries of a metal or alloy. A highly magnified cross section of most commercial alloys will show its granular structure. This structure consists of quantities of individual grains, and each of these tiny grains has a clearly defined boundary that chemically differs from the metal within the grain center. Heat treatment of stainless steels and aluminum alloys accentuates this problem.

4.5.2.11 Exfoliation Corrosion. A form of intergranular corrosion, exfoliation manifests itself by lifting up the surface grains of a metal by the force of expanding corrosion products occurring at the grain boundaries just below the surface. It is visible evidence of intergranular corrosion and most often occurs on extruded sections where the grain thickness is smaller than in rolled forms.

This form of corrosion is common on aluminum, and it may occur on carbon steel.

The expansion of the metal caused by exfoliation corrosion can create stresses that bend or break connections and lead to structural failure.

4.5.2.12 Stress Corrosion Cracking (SCC). The simultaneous effects of tensile stress and a specific corrosive environment cause SCC. Stresses may be due to applied loads, residual stresses from the manufacturing process, or a combination of both.

Cross sections of SCC frequently show branched cracks. This river branching pattern is unique to SCC and is used in failure analysis to identify this form of corrosion.

This type of failure occurs on an aluminum alloy subjected to residual stresses and salt water. Changes in alloy heat treatment as recommended by McDaniel [132] eliminate this problem.

Several years ago, the widespread use of plastic tubing began in a new house construction and the repair of old systems. The faucets connect to the fixed metal piping using flexible tubing. The KSC Corrosion Technology Laboratory identified SCC after 8 years in service. The tubing bent and stress cracks started at the outer tensile side of the tube. Now, services seldom use flexible plastic piping for this application—especially for hot water service.

4.5.2.13 Corrosion Fatigue. Corrosion fatigue is a special type of stress corrosion caused by the combined effects of cyclic stress and corrosion. No metal is immune to some reduction of its resistance to cyclic stressing if the metal is in a corrosive environment. Damage from corrosion fatigue is greater than the sum of the damage of both cyclic stresses and corrosion. Lowering the cyclic stresses or controlling corrosion limits corrosion fatigue.

An infamous example of corrosion fatigue occurred in 1988 on an Aloha Airlines Boeing 737-200 aircraft flying between the Hawaiian Islands. This disaster, which took one life, prompted the airlines to inspect their airplanes for corrosion fatigue.

4.5.2.14 Fretting Corrosion. The rapid corrosion that occurs at the interface between contacting, highly loaded metal surfaces when subjected to slight vibratory motions is fretting corrosion. This type of corrosion is most common on bearing surfaces in machinery, such as connecting rods, splined shafts, and bearing supports; it often causes a fatigue failure. It can occur in structural members including trusses where highly loaded bolts are used and some relative motion occurs between the bolted members.

Delay fretting corrosion by lubricating the contacting surfaces to exclude direct contact with air. For example, lubricate machinery-bearing surfaces to prevent direct air contact.

The bearing race is a classic location for fretting corrosion to occur. Reduce fretting corrosion by keeping the contacting surfaces well lubricated.

4.5.2.15 Erosion Corrosion. Erosion corrosion results from of a combination of an aggressive chemical environment and high fluid surface velocities. This can be the result of fast fluid flow past a stationary object, which is the case for an oilfield check valve, or it can result from the quick motion of an object in a stationary fluid, which happens when a ship's propeller erodes from cavitation.

The surfaces experiencing erosion corrosion are generally clean, unlike the surfaces from many other forms of corrosion.

Control erosion corrosion by using harder alloys including flame-sprayed or welded hard facings and by using a more corrosion-resistant alloy.

Alterations in fluid velocity and changes in flow patterns can also reduce the effects of erosion corrosion.

Erosion corrosion is often the result of the wearing away of a protective scale or coating on the metal surface. The oilfield production tubing shown in Reference 130 corroded when the pressure on the well became low enough to cause multiphase fluid flow. The impact of collapsing gas bubbles caused the damage at the joints where the tubing was connected and the turbulence was greater.

Many people assume that erosion corrosion is associated with turbulent flow. This is true because all practical piping systems require turbulent flow (the fluid would not flow fast enough with lamellar [nonturbulent] flow). Most, if not all, erosion corrosion is attributable to multiphase fluid flow. The check valve [130] failed due to sand and other particles in an otherwise noncorrosive fluid. The tubing failed due to the pressure differences caused by collapsed gas bubbles against the pipe wall that destroyed the protective mineral scale that was limiting corrosion.

4.5.2.16 Dealloying. Dealloying is a rare form of corrosion found in copper alloys, gray cast iron, and some other alloys. Dealloying occurs when the alloy loses the active component of the metal and retains the more corrosion-resistant component in a porous "sponge" on the metal surface. It can also occur by depositing the noble component losses of the alloy on the metal surface.

Control is by the use of alloys that are more resistant, inhibited brasses and malleable or nodular cast iron.

4.5.2.17 Hydrogen Damage. Hydrogen can cause a number of corrosion problems. Hydrogen embitterment is a problem with high-strength steels, titanium, and some other metals. To control hydrogen embitterment, it is necessary to eliminate hydrogen from the environment or to use resistant alloys.

Hydrogen blistering can occur when hydrogen enters steel via a reduction reaction on a metal cathode. Single-atom nascent hydrogen atoms then diffuse through the metal until they meet with another atom, usually at inclusions or defects in the metal. The resultant diatomic hydrogen molecules are too big to migrate elsewhere and remain trapped. Eventually, a gas blister builds up and may split the metal.

Control hydrogen blistering by minimizing corrosion in acidic environments. Hydrogen blistering is not a problem in neutral or caustic environments or with high-quality steels that have low impurity and inclusion levels.

4.5.2.18 Microbial Corrosion. Microbial corrosion is due to the presence and activities of microbes. This corrosion can take many forms. Control microbial corrosion with biocides or by conventional corrosion control methods.

There are a number of mechanisms associated with this form of corrosion, and detailed explanations of the mechanisms are available in the literature.

Most microbial corrosion takes the form of pits that develop underneath colonies of living organic matter, minerals, and biodeposits. This biofilm creates a protective environment where conditions can become quite corrosive and accelerate corrosion.

Microbial corrosion can be a serious problem in stagnant water systems, including fire protection systems that produce pitting. The use of biocides and mechanical cleaning methods can reduce microbial corrosion, but microbial corrosion can occur anywhere that stagnant water is likely to collect. Corrosion (oxidation of metal) can only occur if some other chemical is present for reduction.

In most environments, the reduced chemical is either dissolved oxygen or hydrogen ions in acids. Some anaerobic bacteria can thrive in anaerobic conditions (no oxygen or air present). These bacteria provide the reducible chemicals that allow corrosion to occur. That is what created the limited corrosion found on the hull of the Titanic.

Most microbial corrosion involves anaerobic or stagnant conditions, but it can also occur on structures exposed to air. In Reference 18 the author describes this phenomenon in more detail.

In addition, using corrosion-resistant alloys for limiting microbial corrosion involves the use of biocides and cleaning methods that remove deposits from metal surfaces. Bacteria are very small, and it is often very difficult to get a metal system smooth enough and clean enough to prevent microbial corrosion.

4.5.3 Pollution by Oil Pipeline Releases and Corrosion

According to an Environmental Defense Fund (EDF) analysis, pipelines release an average of more than 6.3 million gallons of oil and other hazardous liquids each year. This is more than half the amount released from the *Exxon Valdez* disaster. Since 1995, the amount of oil released into the environment increased each year. On average, tens of thousands of gallons of oil escaped from pipelines approximately every other day throughout the 1990s. Corrosion is the next most common cause of pipeline spills following excavation accidents. The pipeline industry has developed a range of technologies to eliminate or reduce corrosion.

Cathodic protection, which uses a constant low-voltage electrical current running through the pipeline to counteract corrosion, is required on all interstate pipelines and has been required for decades. Recent improvements in pipeline coating materials also help reduce the risk of corrosion-related failure. The U.S. Department of Transportation is revising the pipeline safety regulations to incorporate more stringent corrosion prevention rules—a change supported by the oil pipeline industry.

Lois Epstein, an EDF engineer stated, “The two upward trends in aggregate oil pipeline releases and release size clearly need to be reversed.” He also stated, “Numerous oil pipeline companies are not preventing pollution from their pipelines, and the Office of Pipeline Safety (OPS) is not forcing them to do so. The majority of these releases are from corrosion, operational incidents, and material defects." The EDF's analysis also examined property damage from oil and other hazardous liquid pipeline releases, which averaged over $39 million annually in the 1990s. The average property damage cost per incident is more than $194,000 with a median cost of $20,000. The OPS has failed to issue any environmental protection regulations, despite Congress mandating the office to do so in the Pipeline Safety Act of 1992.

Bhopal, India, is most likely the site of the greatest industrial disaster in history [134, 135]. Between 1977 and 1984, Union Carbide India Limited (UCIL), located within a crowded working class neighborhood in Bhopal, was licensed by the Madhya Pradesh Government to manufacture phosgene, monomethylamine (MMA), methylisocyanate (MIC), and the pesticide carbaryl, which is also known as Sevin.

Severe corrosion was blamed for the environmental catastrophe that struck the coast of France in the last days of 1999.

There are different specifics in the corrosion process in different industrial areas. Now examine specific corrosion occurring in marine and aircraft environments.

4.5.4 Corrosion in Marine Environments

Many industries like shipping, offshore oil and gas production, power plants, and coastal industrial plants use seawater systems. The primary use of seawater is in cooling systems, but it also has applications for fighting fires, in oilfield water injection and desalination plants. The corrosion problems in seawater systems have been studied over many years, but failures still occur despite published information on the behavior of materials in seawater [128].

Most of the elements found on earth are present in seawater, at least in trace amounts. However, 11 of the components alone account for 99.9% of the total solutes with chloride ions being the largest component by far. The concentration of dissolved materials in the sea varies greatly with location and time because rivers, rain, and melting ice dilute seawater or evaporation concentrates seawater. The most important properties of seawater are

- Remarkably constant ratios of the concentrations of the major constituents worldwide
- High *salt concentrations*, mainly sodium chloride
- High electrical conductivity
- Relatively high and constant *pH*
- Buffering capacity
- Solubility for gases including oxygen and carbon dioxide that are of particular importance in the context of corrosion
- The presence of a myriad of *organic compounds*

- The existence of biological life either as microfouling (e.g., bacteria or slime) or macrofouling (e.g., seaweed, mussels, barnacles, and many kinds of animals or fish)

The U.S. flag fleet [131] can be divided into several categories: the Great Lakes with 737 vessels at 62 billion ton-miles, inland with 33,668 vessels at 294 billion ton-miles, oceans with 7014 vessels at 350 billion ton-miles, recreational with 12.3 million boats, and cruise ships with 122 boats serving North American ports (5.4 million passengers). The estimated total annual direct cost of corrosion to the U.S. shipping industry is approximately $2.7 billion. This cost includes costs associated with new construction ($1.1 billion), maintenance and repairs ($0.8 billion), and corrosion-related downtime ($0.8 billion) (http://corrosion-doctors).

4.5.5 Aircraft Corrosion

Corrosion damage of aircraft fuselages is an example of atmospheric corrosion, a topic described in detail in a separate section. Airports located in marine environments deserve special mention. The risk and cost of corrosion damage are particularly high in aging aircraft. In the United States alone, aircraft corrosion is a multibillion dollar problem. For example, corrosion maintenance hours outstrip flight hours on some military-type aircraft.

The current approach for dealing with corrosion is to remove it as soon as it is found and to repair the corroded structure or replace the component. This is costly in terms of increased maintenance time and decreased aircraft availability. Treating the corrosion with corrosion prevention compounds (CPCs) and leaving it in place until there is easier access to the affected areas during a scheduled service would increase aircraft availability. However, this approach requires detailed knowledge of the propagation rates for the specific corrosion type after treatment with CPCs, and this information is not currently available.

Corrosion manifests itself in many different forms. Concentration cell corrosion is the most common type found on airplanes. It occurs whenever water is trapped between two surfaces, such as under loose paint, within a delaminated bond-line or in an unsealed joint. This corrosion can quickly develop into pitting or exfoliation corrosion, depending on the alloy, form, and temper of the material be attacked.

Crevice corrosion damage in the lap joints of aircraft skins has become a major safety concern, particularly after the Aloha Airlines incident. On April 28, 1988, a 19-year-old Boeing 737-200 aircraft, operated by Aloha Airlines, lost a major portion of the upper fuselage near the front of the plane, in full flight at 24,000 ft. The Aloha Airlines incident marked a turning point in the history of aircraft corrosion recognition and control.

In 1998, the combined commercial aircraft fleet operated by U.S. airlines was more than 7000 airplanes [131]. At the start of the jet age (1950s–1960s), corrosion and corrosion control received little or no attention. The continued aging of the airplanes beyond the 20-year design life is a corrosion concern. Only the most recent designs (e.g., the Boeing 777 and late-version 737) have incorporated significant improvements in corrosion prevention and control into the design and manufacturing process. The estimated total annual direct cost of corrosion to the U.S. aircraft industry is $2.2 billion including the cost of design and manufacturing ($0.2 billion), corrosion maintenance ($1.7 billion) and downtime ($0.3 billion) in an online report [131].

4.5.6 Accelerated Corrosion Testing of the Product

Use accelerated corrosion methods for accelerating corrosion damage and for obtaining test results in a relatively short time. Most include the following actions:

- Increasing the temperature to boost corrosion reaction kinetics and applying cyclic temperature ranges
- Acidifying the corrosive medium
- Increasing the medium's corrosiveness by other chemical changes (e.g., salt concentration, degree of aeration, and addition of oxidizing species)
- Alternate wetting and drying
- Increasing the relative humidity
- Cycling through different humidity ranges/different degrees of moisture exposure
- Application of stress, particularly relevant to SCC testing
- Combining mechanical degradation including erosion, cavitation, and impingement with corrosion damage

To apply the acceleration factors listed earlier, testing conditions must move away from the actual service environment conditions.

Typically, alternate wet and dry cycles in the life of a system can create serious corrosion problems as a result of the formation of partially dry corrosion products that can further the accumulation of corrosive agents by the absorption of moisture and the creation of underdeposit corrosion. The corrosion attack using alternating wetting and drying cycles accelerates corrosion and provides a relatively fast estimate of corrosion prevention measures.

The rapidness of the damage growth due to these cycles increases when operating at higher temperatures. This is typically the case with corrosion under insulation, a very rapid corrosion attack resulting from crevice corrosion. The above-mentioned processes apply to providing accelerated corrosion testing. The International Organization for Standardization (ISO) 11474 standard, for example, describes an accelerated outdoor corrosion test performed by the intermittent spraying of a salt solution onto the exposed pieces of equipment. This severe corrosion test, also called the "Volvo" outdoor for

Simulated Corrosion Atmospheric Breakdown (SCAB) test, is typically run for long periods (e.g., 12 months) at a 45° angle of exposure with a twice-daily salt spray application. Also, one can use the SCAB test to rank materials, processes, and variables of interest with regard to their impact on outdoor corrosion performance.

This is one of the simplest methods of accelerated corrosion testing. This method is inappropriate for the AP of equipment and systems corrosion degradation because it does not take into account that corrosion in the field is a result of complex input influences. Moreover, the corrosion does not act separately in actual service. This is only one of the interconnected components of environmental action.

Therefore, corrosion testing is a component of multiple environmental testing, which is a component of ART. As a result, *separate* corrosion testing cannot offer accurate initial information for reliability (maintainability, durability) evaluation and the prediction of details, units, and the whole equipment. The usefulness of separate corrosion testing relates to materials (metals, polymers, and composite materials).

Accelerated corrosion testing is a component of AMET and ART.

Begin with simple, but insufficiently accurate, types of accelerated corrosion testing. Accelerated atmospheric corrosion testing (ACT) is based on cyclic exposure conditions rather than the continually wet ("constant exposure") conditions for a more simplistic result. Traditional salt fog testing is based on unrealistic test conditions that lead to poor correlation with real-life performance.

Electrochemical corrosion testing relies on electrochemical theory and electrochemical corrosion measurements to characterize corrosion damage and, where possible, estimate corrosion rates. Fundamentally, there are two types of electrochemical measurements:

- Apply an external current to generate electrochemical data away from the free corrosion potential. These tests generally explore the relationship between electrochemical potential and current.
- Conduct electrochemical measurements at the free corrosion potential without applying an external current.

The usefulness of electrochemical corrosion testing for the short term is based on the following fundamental considerations:

- The application of external current allows corrosion damage, such as pitting corrosion, to be induced and studied in very short time frames, typically within minutes.
- Electrochemical corrosion rate measurements are available almost instantly, long before a conventional weight change would be measurable.

Immersion corrosion testing can take the following forms [136]:

- Conduct relatively simple testing involving constant immersion. The acceleration of corrosion damage can be achieved from the length of immersion in the corrosive medium and other accelerating factors.
- Testing involving alternate immersions/drying in the form of cyclical testing
- Instrumentation of test specimens during immersion (connections to electrochemical instrumentation) to facilitate measurements other than simple weight loss measurements
- From simple beaker-type testing to immersion of flush-mounted flow loops to simulate real-life conditions more closely

Historically, immersion testing occurs extensively to generate uniform corrosion data from alloys used in the process industries under immersion conditions.

Generally, actual corrosive environments are more complex and less carefully controlled than accelerated laboratory tests in salt spray chambers. As a result, one can use salt spray chambers to measure the relative field performance in terms of a particular corrosion mechanism. Some factors that make the correlation of accelerated laboratory testing results to actual in-service performance difficult include those cited in References 136 and 137:

- Accelerated corrosion testing designed to deviate from actual service exposure to produce results in significantly shortened time frames
- Laboratory testing is generally more simplistic and standardized, with fewer variables than actual service conditions.
- Multiple failure mechanisms/modes and their interaction in the field are not easily reproduced in the laboratory. Artificially accelerating one corrosion mode may delay another mode.
- The "acceleration" of corrosion by adjusting a few selected variables in laboratory testing usually does not represent the complex interplay of multiple variables under actual service conditions.

4.5.6.1 Corrosion Testing in the Automotive Industry. Consider an example of corrosion testing in the automotive industry. The Auto/Steel Partnership Corrosion Project Team completed a test program for perforation corrosion consisting of on-vehicle field exposures and various accelerated corrosion tests [138].

Steel sheet products with eight combinations of metallic and organic coatings were tested utilizing a simple crevice coupon design. On-vehicle exposures were conducted in St. John's and Detroit for up to 7 years in an effort to establish a real-world performance standard.

TABLE 4.2 Description of Test Materials [138]

ID	Material Description	Average Sheet Thickness (mm)*	Actual Metallic Coating Mass (g/m^2)†
1	Cold-rolled steel (CRS)	0.841	None
2	Electroplated Zn, 60 g/m^2 (EG60)	0.779	75
3	Hot-dipped ZnFe, 50 g/m^2 (A50)	0.874	63
4	Electroplated ZnNi, 30 g/m^2 + organic	0.783	34
5	Electroplated Zn, 40 g/m^2 (EG40)	0.786	48
6	Electroplated Zn, 60 g/m^2 + 3-μm e-coat (EG60) + EC)	0.782	73
7	Hot-dipped ZnFe, 50 g/m^2 + 3-μm e-coat (A50 + EC)	0.876	59
8	Electroplated Zn, 100 g/m^2 (EG100)	0.755	112

*Average of all test specimens.
†Coating mass calculated from metallic thickness measurements, assuming a density of gram per cubic centimeter.

Identical test specimens were exposed to the various accelerated tests and the results compared to the standard.

The accelerated test results (including SAE J2334, GM9540P, Ford APGE, CCT-1, ASTM B117, South Florida Modified Volvo, and Kure Beach [25-m exposures]) were compared to the on-vehicle tests. Five criteria compared the results: the extent of corrosion, rank order or material performance, degree of correlation, accelerated factor, and control of test environment. Three of the laboratory accelerated tests—SAE J2334, GM9540P, and Ford APGE—have results that were reasonably consistent with the on-vehicle behavior.

Table 4.2 shows the eight materials selected to represent the currently available sheet products and provide a range of responses.

4.5.6.2 Test Performed. Table 4.3 lists the tests included in this study and information about test duration, the total minimum time required to perform the test, the number of replicates included in this study, and the organization responsible for performing the test.

The total minimum time calculated assumed tests ran 7 days a week. However, laboratory tests SAE J2334, GM9540P, and Ford APGE often run only 5 days a week, and this would affect the test by increasing the total minimum test time by a factor of 7/5.

4.5.6.3 Results of On-Vehicle Testing. The two sites selected for this study—Detroit, Michigan, and St. John's, Newfoundland—are severe and extremely severe corrosion regions.

For each location, replicate test specimens of each material were placed on test vehicles for times of up to 7 years. One of the St. John's vehicle test setups

TABLE 4.3 Tests Included in the Perforation Corrosion Study [138]

Test	Description	Duration	Total Minimum Test Time (Days)	Replicates	Conducted by
Detroit	On-vehicle test	2 years	730	5 (11 for matl. 2)	Ford
		7 years	2555	12 (11 for matl. 5)	
St. John's	On-vehicle test	2 years	730	5	GM
		4 years	1460	5	
		7 years	2555	6	
SAE J2334*	Cyclic corrosion test	40 cycles	40	3	Bethlehem Steel
		80 cycles	80	3	
		120 cycles	120	3	
		160 cycles	460	3	
GM9540P*	Cyclic corrosion test	40 cycles	40	3 2 (matl. 3)	GM
		80 cycles	80	3 (2 for matl. 3, 4 for matl. 2)	
		160 cycles	160	3	
Ford APGE*	Cyclic corrosion test	60 cycles	60	5	Ford
ASTM B117	Neutral salt spray test	20 weeks	140	2	Bethlehem Steel
CCT-1	Cyclic corrosion test	100 cycles	33	3	LTV
		200 cycles	67	3	
		400 cycles	133	3	
South Florida	Static outdoor exposure with salt application	18 weeks	126	2 (matl. 1–4) 3 (matl. 5–8)	BASF
Modified Volvo		9 months	270	3	
		18 months	540	2 (3 for matl. 1)	
25-m Kure Beach	Static outdoor exposure	7 months	210	2	LaQue Center
		14 months	420	Unknown	
		28 months	840	2	

*Total minimum test time is based on a 7-day week. If a 5-day week is used, the total minimum test time increases by a factor of 7/5.

is in Table 4.3. For each material, four or more specimens were removed for evaluation after 2, 4 (St. John's only), and 7 years.

4.5.6.4 Laboratory Tests. Laboratory tests provide a quick and relatively inexpensive means of evaluating corrosion resistance. A summary of the laboratory tests included in this study is in Table 4.4.

TABLE 4.4 Test Parameters for Salt Solutions and Environmental Conditions in Laboratory Tests

Cyclic Test	Salt Application	Drying	Wet Exposure
SAE J2334*	0.25-hour dip 0.5% NaCl + 0.1% $CaCl_2$ + 0.075% $NaHCO_3$ Ambient	17.75 hours 50% RH 60°C	6 hours 100% RH 50°C
GM9540P*	Several intermittent spray applications with 0.9% NaCl + 0.1% $CaCl_2$ + 0.25% $NaHCO_2$ during an 8-hour hold at 40–50% RH 25°C	8 hours <30%RH 60°C	8 hours 95–100% RH 49°C
Ford APGE*	0.25-hour dip 5% NaCl Ambient	1.25 hours <50% RH 25% C	22.5 hours 85% RH 50°C
ASTM B117	Continuous fog 5% NaCl 35°C	None	Continuous 100% RH 35°C
CCT-1	4-hour fog 5% NaCl 35°C	2 hours <30% RH 60°C	2 hours >95% RH 50°C

The sequence for these cyclic tests is salt application/wet exposure, except for GM9540P, which is salt application/wet exposure drying [138].
*SAE J2334, GM9540P, and Ford APGE are typically conducted manually with one cycle per day for 5 days/week, with weekend holds in the drying stage. These tests can also be run in automated cabinets with one cycle per day for 7 days/week, without a weekend hold.

Davidson et al. [138] wrote that the results from using the above-mentioned methodologies of corrosion testing demonstrated the following:

1. Three of the laboratory accelerated corrosion testing methodologies, namely, SAE J2334, GM9540P, and Ford APGE, gave results that are reasonably consistent with the range of observed on-vehicle behavior and can be used with reasonable confidence to predict on-vehicle behavior.
2. The other laboratory tests (ASTM B 117 and CCT-1) were less than satisfactory.

4.5.7 Advanced Testing Methods for Accelerated Corrosion

Advanced testing for accelerated corrosion is a modernized type of testing. Its basic advantage is usefulness as a component of multi-environmental tests and, as a result, as a component of ART/ADT. There are several advanced testing methods for accelerated corrosion with published and approved results.

This information, including a detailed description, is in the paper [139] that discusses the development of the theoretical and experimental basis of ACT for metals, complete machines, and/or their components.

The current techniques for ACT need to be improved. One approach for establishing the necessary environmental conditions for ACT on various products is to determine the range of metallic corrosion based upon the periodical wettings with KCl solution and the drying process of a film of moisture, changing temperature, and humidity in the test chamber.

As an example of the methodology, an equation was developed to describe one range of test chamber parameters for calculating the quantity of metallic corrosion as a function of these parameters. Checking the related statistics confirms the experimental results.

The author's research results and the results of other researchers demonstrate that the corrosive destruction of mobile machinery is dependent upon a complex of interconnected factors, such as temperature, humidity, chemical and mechanical pollution, vibration, mechanical wear, and other factors. This is very important for professionals working in the field of corrosion testing.

The proposed methodology for accurate simulation of the field input influences as an example, on a sample of 1020 steel, with polyurea-thickened grease, and/or paint DBU-4273W protection, is demonstrated in test chambers [139] showing

- The limit of environmental influences
- The required temperature and humidity
- Determination of the dependence of steel corrosion on the number of wettings
- Measured accelerated corrosion versus the field

The corrosion degradation mechanisms and the fatigue failure mechanisms confirm that the proposed method of simulation using accelerated test conditions is both practical and useful.

The proposed test methodology for complex mobile products takes into account the mechanical damage to all protective films, especially during exposure to surface drying, winter road salt, the impact of road sand and dust, combined with any mechanical wear, exposure of solar radiation, and/or electrical actions. These mobile products include automotive, aircraft, off-highway vehicles, farm machinery, and construction machinery (whole machinery and their components).

4.5.7.1 Development of the Technique of Accelerated Corrosion Testing. Now observe the development and substantiation of accelerated corrosion testing conditions for components and complete machinery by reviewing the examples for various steels, automobiles, and machinery. It is useful to work with material specimens for the substantiation of the corrosion test regimen of the product.

For steels, it is necessary to solve the following problems to establish the conditions of accelerated corrosion testing [139]:

- The impact of environmental conditions on various products
- Determine the physical-chemical processes causing product deterioration under the action of environmental factors.
- Identify the parameters that characterize the above-mentioned processes.
- Establish the intensity of possible environmental influences and determine the evaluation methods.
- Find the limit of intensity of the physical-chemical mechanism for the field deterioration processes.
- Determine the optimal conditions for accelerated corrosion testing of a particular product with similar degradation mechanism as in the field.
- Check the correlation of the established conditions of ACT results with the field results.

The practical establishment and application of ACT conditions for metals and metallic components leads to the following brief description. The magnitude of metal corrosion is usually a function of the wetting of the metallic surface. The maximum rate of corrosion should occur when the moisture film is drying.

4.5.7.2 Establishing the Limits of Environmental Influences on Steel Corrosion and Protection [43, 103, 104]. Examine the approach for a sample of 1020 steel. The basic environmental factors are temperature, humidity, and chemical pollution using wettings with a KCl solution. Figure 4.13 demon-

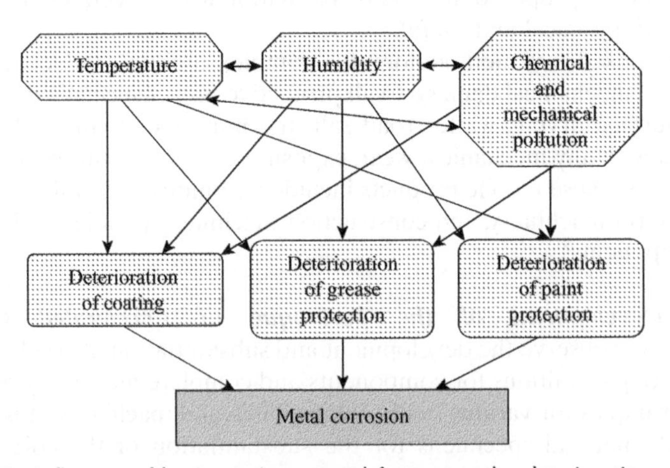

Figure 4.13. Influence of basic environmental factors on the deterioration of coating, greases, paints, and, as a result, corrosion of the steel.

strates the interaction of environmental influences in the deterioration of coatings, greases, paints, and metals.

The limit of the basic environmental factors determines the result of their action on the 1020 steel, polyurea-thickened grease, and a standard paint DBU-4273W. The following problems have to be solved [139]:

- Establish the limits of environmental stress influences on the paint DBU-4273W.
- Determine the necessary humidity conditions in the test chamber to evaluate or estimate the speed of the atmospheric corrosion of the 1020 steel under a film of moisture.
- Determine the necessary temperatures for the test chamber to evaluate or estimate the speed of atmospheric corrosion of the 1020 steel under a film of moisture.
- Determine the time interval necessary for the evaluation of experimental dependence of "corrosion of 1020 steel on the duration of wettings by 0.01 normal solution of KCl" applied during selected conditions of temperature and humidity.

4.5.7.3 The Establishment of Temperature in a Test Chamber [139]. The protective quality of grease is limited by the temperature that causes a flow down from the surface [140]. The grease melts first and then flows down, and the corrosion process develops afterward. Our experimental research was conducted at a relative humidity of 50% with temperatures of 30°C (85°F), 35°C (95°F), 40°C (104°F), and 45°C (113°F). The results are in the graph in Figure 4.14. After 6 days and nights at temperatures of 30 and 35°C, the electrical resistance of the metal had not changed. The surface properties during these temperature changes have not changed. The indications of oxidation shown by an increase in the resistance of the metal appear first at 40°C after only 72 hours and at 45°C after 48 hours. A trace of grease appears at the top of the electrolyte coats, resulting in a slower process of evaporation of the solution from the surface of the specimens.

As shown in Figure 4.14, two opposing forces are acting on the grease element that is under the coat of the water solution of the electrolyte. The liquid exerts a force from one side and the force of cohesion with neighboring elements attracts the grease element from another side. If the temperature increases, the fluidity of the grease increases too, but the force of cohesion is weakened. After the grease element tears away from the common mass and floats up to the surface of the liquid, the thickness of the protective coat continuously decreases, and the corrosion process begins. Testing specimens with a 0.005-mm thickness of a greased coat confirms the results of the test for this factor.

Therefore, to avoid destructive corrosion of the metal under the coat of grease during accelerated "wetting–drying" cycles of the accelerated corrosion test, the temperature in the test chamber must be higher than 35°C (95°F) for the particular grease studied.

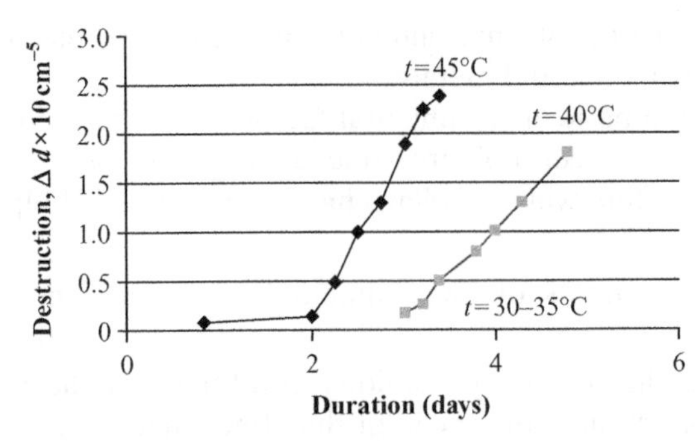

Figure 4.14. Dependence of the iron corrosion with protective grease as a function of different temperatures.

4.5.7.4 The Establishment of Necessary Humidity [139]. This work, based on representing atmospheric corrosion by a model, employs many short-circuited microelements of the system on the metal surface [141] in contact with a corrosive film. Therefore, one can estimate one measure of the steel deterioration from the electrical current generated by the microcorrosion elements on the surface of the metal.

In our experiments, the test objective was to create a battery with metal plates and to exploit the different potentials.

The microcathode consisted of 40 copper plates and the microanodes consisted of 40 plates of 1020 steel. The thickness of the plates was 0.5 mm. These batteries were studied at relative humidity values of 50, 60, 70, 80, and 90% across a temperature range of 5–35°C with intervals of 5°C. The thickness of electrolyte solution film was 100 μk. The availability of current in the battery determines the duration of the corrosion process. The area under the curve of the graph of current versus time measured the anode destruction.

The common corrosive effect depends entirely on the time duration of the moisture on the wetted surface. The mean rate of corrosion is constant while the surface is wet because the speed of oxygen diffusion to the cathode does not depend upon the relative humidity when under a thick, wet film.

The access of oxygen depends on the thickness of the electrolyte coat. This confirms the observation that the corrosion current increases with high humidity.

The test plans have to have long periods between the uses of the sprinkling system for the stability of the environment in the test chamber.

The dependence of "corrosion versus temperature" shows that the increasing corrosion is not proportional to the increase in temperature. The curves are at the maximum in the interval 10–15°C. The effect of increasing the speed of the reaction by 15°C is smaller than the effect of slowing the reaction rate. The results of these experiments are in Table 4.5.

This research shows that the corrosion current time duration is dependent on drying duration and this is shown in Figures 4.15–4.17.

TABLE 4.5 The Influence of Humidity and Temperature on the Corrosion Speed under Dried and Wet Films [43]

Relative Humidity (%)	Temperature (°C)	Value of Corrosion, $C \cdot 10^{-1}$	Time (Hour)	Speed of Corrosion, $V_k \cdot 10^{-1}$ mg/cm^2 /h
50	5	0.14	0.81	0.17
	8	0.21	0.67	0.31
	12	0.25	0.55	0.45
	15	0.26	0.46	0.56
	20	0.21	0.32	0.66
	25	0.17	0.24	0.71
	30	0.12	0.17	0.70
	35	0.07	0.11	0.64
60	5	0.16	1.01	0.16
	8	0.25	0.85	0.29
	12	0.29	0.67	0.43
	15	0.30	0.55	0.55
	20	0.25	0.40	0.64
	25	0.21	0.31	0.69
	30	0.15	0.21	0.70
	35	0.12	0.17	0.70
70	5	0.22	1.39	0.16
	8	0.32	1.16	0.28
	12	0.37	0.88	0.42
	15	0.38	0.71	0.54
	20	0.33	0.54	0.62
	25	0.27	0.41	0.66
	30	0.20	0.30	0.67
	35	0.14	0.21	0.66
80	5	0.35	2.21	0.16
	8	0.50	1.80	0.28
	12	0.60	1.37	0.44
	15	0.61	1.11	0.55
	20	0.52	0.82	0.63
	25	0.41	0.60	0.68
	30	0.31	0.45	0.68
	35	0.23	0.35	0.67
90	5	0.80	4.73	0.17
	8	1.17	4.03	0.29
	12	1.36	3.01	0.45
	15	1.30	2.21	0.58
	20	1.01	1.58	0.64
	25	0.81	1.19	0.68
	30	0.58	0.86	0.67
	35	0.41	0.63	0.65

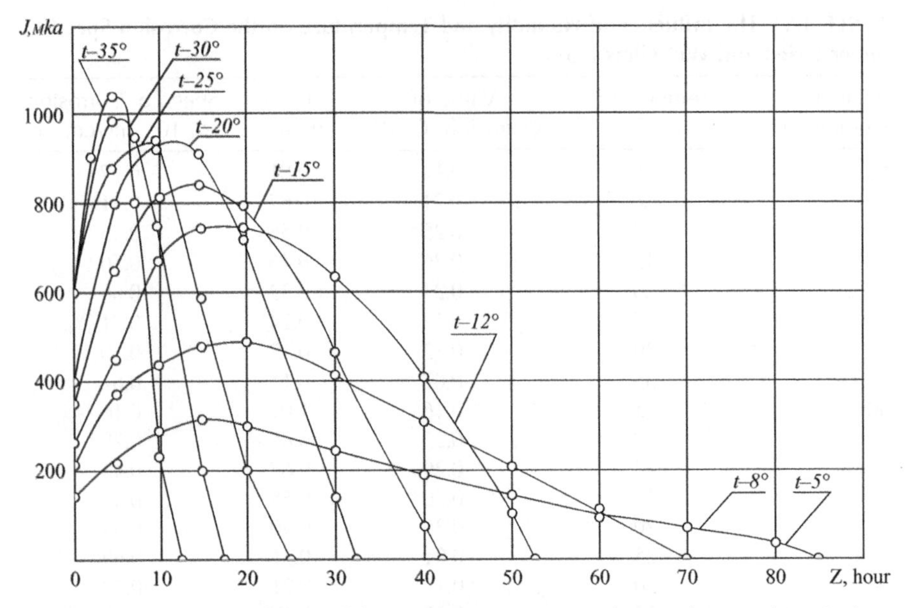

Figure 4.15. The dependence of the corrosion current on the conditions of Fe–Cu batteries drying at a humidity of 70%.

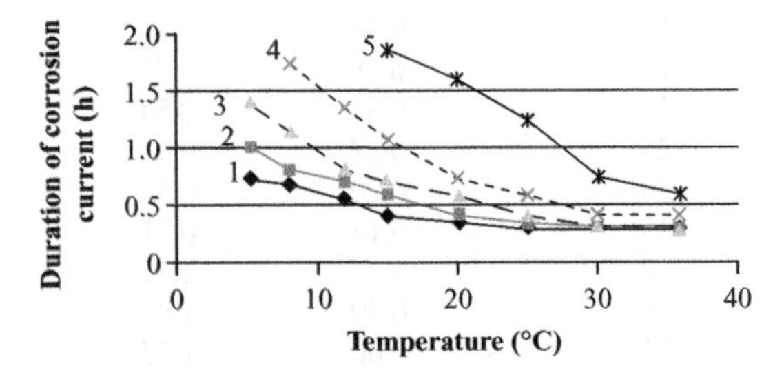

Figure 4.16. The dependence of the corrosion time versus different surface drying conditions of the batteries (humidities are 1—50%, 2—60%, 3—70%, 4—80%, 5—90%) [43].

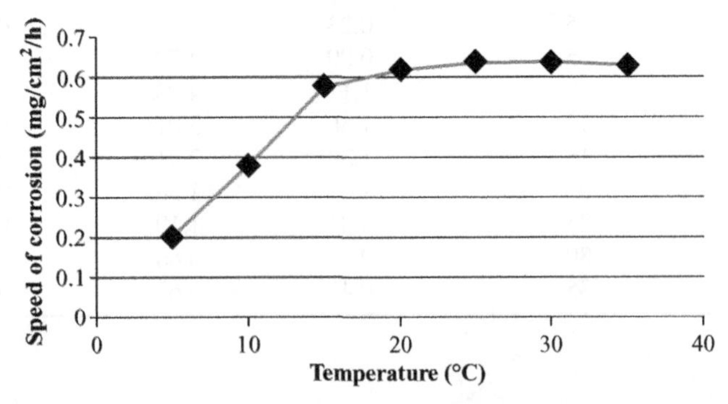

Figure 4.17. Dependence of the duration of corrosion speed versus the temperature at 90% humidity.

The character of the curve $v_k = f_2(t, W = 90\%)$ (Table 4.5), based on experimental results, shows that the value of v_k changes very slowly across the intervals of the temperature increase from 20 to 35°C. This low dependence is enough to determine the optimal temperature, which is $t = 30$°C. Therefore, the optimal parameters of a test chamber regimen for this material are $t = 30$°C, W (humidity) = 90%.

4.5.7.5 The Determination of the Dependence of "Steel Corrosion versus the Number of Wettings" [139]. Compare the field and laboratory results for the evaluation of the acceleration coefficient during the time interval. This experiment used specimens of 1020 steel of size 50 × 80 mm with a thickness of 1.6 mm. The dependence of the amount of the corrosion versus the number of wettings with 0.01 normal KCl solutions is the result of this experiment.

The drying time of the surface determined the microcorrosion. The time between wettings was 55 minutes. The specimens changed places after 10 wettings. The loss of the mass determined the level of corrosion. An agreed-upon method eliminated the product of corrosion. The results of the approximation for these experiments are included in the curve in Figure 4.18.

The dependence of steel corrosion on the number of wettings for $N_w > 100$ is approximated by the equation

$$C = 0.546\, N_w^{\,0.97}, \tag{4.1}$$

where

C is the quantity of the metal's corrosion in gram per square centimeter,
N_w is the number of wettings of the corroded surface, and
0.97 is the experimental coefficient.

Fill in Equation 4.1 with the accumulated steel corrosion in any region. Calculate the required number of wettings needed to simulate this accumulated corrosion in the test chamber for the particular region. If one knows the interval between two wettings and how many years are required for accurate

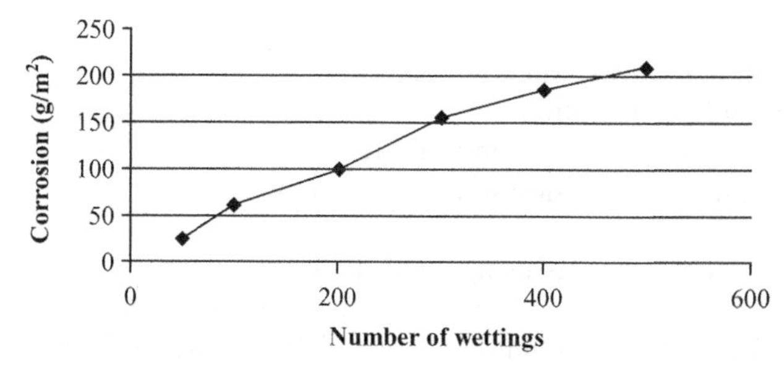

Figure 4.18. Dependence of 1020 steel corrosion values on the number of wettings.

TABLE 4.6 Example of Calculating the Acceleration Coefficient (The Time between Wettings Is 1 Hour)

How Many Years (Days) Are Required to Simulate	How Long the Test Time in the Chamber Should Be (Day)	Acceleration Coefficient
1 year (365 days)	13.77	26.50
2 years (730 days)	30.42	24.00
3 years (1095 days)	43.80	25.05
4 years (1460 days)	61.34	23.80
5 years (1825 days)	72.85	25.05
6 years (2190 days)	87,25	25.10
7 years (2555 days)	106.41	24.01
8 years (2920 days)	154.50	18.90

simulation, then one can determine how long the test time in the chamber should be. The ratio of this time to the time of 1 year (365 days) will show the testing acceleration coefficient. For an example, see Table 4.6.

4.5.7.6 ***Checking the Laboratory Corrosion versus the Field Corrosion*** [18]. Consider an example. The sample plates are 1020 steel of size $50.0 \times 80.0 \times 1.6$ mm and are coated with grease to a thickness of 0.5 mm. Other samples are

- The same plates of 1020 steel with paint DBU-4273W protection
- Glass plates with iron coated with grease to a thickness of 0.2 mm
- Specimens of 5140 steel and 1045 steel of the same size

The test conditions are $t = 30°C$, $W = 90\%$. From Equation 4.1, there are 456 wettings with a normal solution of KCl. The wetting is for 0.9 minutes with intervals of 1 hour. The results of the laboratory testing (a degradation mechanism) and real life will correspond only for sample "1" (Table 4.7). The variants are $(\sigma_1^2 - \sigma_c^2) \leq \Delta_{1\sigma}$ and $(\mu_1^2 - \mu_c^2) \leq \Delta_{1\mu}$; $\Delta_{1\sigma}$ and $\Delta_{1\mu}$ are permissible squares of standard deviations and mathematical expectation (mean) between the expected value for real-life and test chamber conditions

where

μ is the mathematical expectation,
σ is the standard deviation,
σ_1 is the standard deviation in real life, and
σ_c is the standard deviation in the test chamber.

The experimental results using 1020 steel for 1 year under real-life conditions yield a mean corrosion μ of 198.4 g/m^2 with a standard deviation σ_1^2 of 203.0 g^2/m^4. The hypothesis of normality, which was checked using the Pearson

TABLE 4.7 Possible Variants of Standard Deviation and Mathematical Expectation

Example	Standard Deviation	Mathematical Expectation	Variants
1	$\sigma_1^2 = \sigma_c^2$	$\mu_1 = \mu_c$	1–1
		$\mu_1 \neq \mu_c$	1–2
2	$\sigma_1^2 \neq \sigma_c^2$	$\mu_1 = \mu_c$	2–1
		$\mu_1 \neq \mu_c$	2–2

TABLE 4.8 The Corrosion in Real Life and in Test Chamber (Calculated from Eq. (4.1)) for 1 Year (Number of Wettings is 8760)

Where	Mean of Corrosion (g/m^2)	Standard Deviation of Corrosion (g^2/m^4)
In real life	198.4	203.0
In test chamber	179.1	250.0

criterion, does not contradict the experimental data ($P = 0.62$ for $\chi^2 = 3.56$). The necessary number of wettings for the test chamber is calculated from Equation 4.1 as $N_w = 436$. Table 4.8 shows that the value of the calculated mathematical expectation is 179.1 g/m^2, and the variance is 250.0 g^2/m^4.

The degradation corrosion mechanism in our example is measured through the comparison of the mathematical expectations and standard deviations of both distributions, under the real field conditions and laboratory conditions. The result of this comparison evaluates the level of similarity of the test chamber conditions to the field conditions.

Hypothesis checking provides

$$H_0 \left| \frac{\sigma_M^2}{\sigma_N^2} = 1 \right|.$$

The alternative hypothesis is

$$\sigma_m^2 > \sigma_n^2.$$

In this case, the sampling ratio is

$$\frac{S_m^2}{S_n^2},$$

where S_m^2 and S_n^2 are estimations of the standard deviations for the corrosion model and the natural results. Use the percentage points of the F distribution (tables of mathematical statistics) as a limiting value of this ratio.

The level of importance $\alpha = 1\%$ is found from the above-mentioned tables with 1%, ∞, 49 yielding a value of 1.523. These tables are in different handbooks on the theory of probability.

The actual ratio of standard deviations derived from the data was 1.232.

$$\frac{S_m^2}{S_n^2} = 1.232.$$

Since 1.232 < 1.523, the hypothesis that $\sigma_m^2 = \sigma_n^2$ does not contradict the experimental data.

Typical engineering estimates usually suggest that the deviations from the model should not be more than 8–10%. Using this measure, $\delta = |m_n - m_m| =$ 15 g/m^2 with a level of importance equal to 1%, the critical value of acceptance statistic is equal to 2.33.

From the data, we obtain a value of $V = 1.84$, which is less than 2.33. Therefore, the suggested hypothesis does not contradict the experimental data.

Using the known ratio for standard deviation to check the null hypothesis $H_0/m_n^* - m_m^* = \delta/$. With the help of the limiting values of the statistic, the following relationship shows:

$$V = \frac{\eta - \delta}{\sqrt{\lambda_m S_m^2 + \lambda_n S_n^2}}, C = \frac{\lambda_1 S_1^2}{\lambda_1 S_1^2 + \lambda_2 S_2^2},$$

where

$$\lambda_n = \frac{1}{N_n}; \lambda_m = \frac{1}{N_m}; \eta = m_n^* - m_m^*; \delta = |m_n - m_m|.$$

n_n and m_m are values for the field results and test chamber results.

The standard fatigue testing results for the metallic specimens establish the final correspondence of the recommended conditions in the test chamber with real-life conditions. The experimental results (Fig. 4.19) show that corrosion destruction of 5140 steel specimens due to fatigue decreases from 18.0 to 14.8 kg/mm^2 after testing in test chambers. The "Shenck" test machine performed 5×10^6 cycles of fatigue testing. After field testing (Fig. 4.17), the fatigue of the same specimens decreases from 18.6 to 13.8 kg/mm^2.

Similar results occurred after testing 1045 steel specimens (see Fig. 4.17).

It provided micro- and macroscopic research for the destroyed specimens. One example is included in Figure 4.20.

4.5.7.7 *Accelerated Corrosion Testing of the Complete Product.* For accelerated corrosion testing of a complete product like whole (especially mobile) machinery and their major components, it is necessary to take into account other influences, in addition to those factors demonstrated earlier. One type of influence is mechanical damage of protective films during surface drying. Other influences causing corrosion include road salt, impact of sand dust thrown up by wheels, any mechanical wear, and solar radiation. To obtain

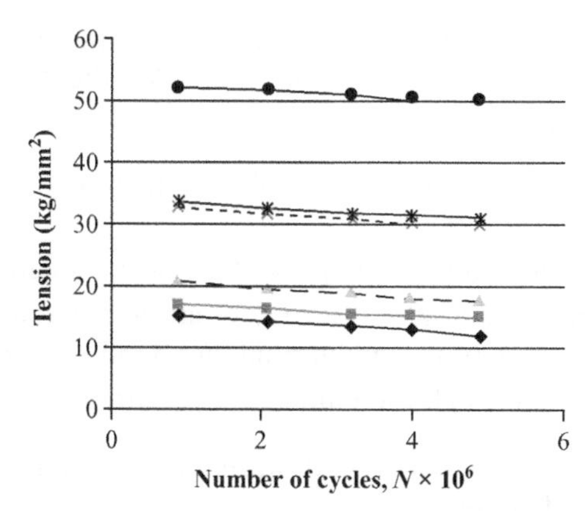

Figure 4.19. Results of the fatigue testing of steel specimens [43]: ● is 1045 steel before exposure; * is 5140 steel after testing in test chambers; **x** is 1045 steel after field testing; ▲ is 5140 steel before exposure; ■ is 5140 steel after testing in test chamber; and ◆ is 5140 steel after field testing.

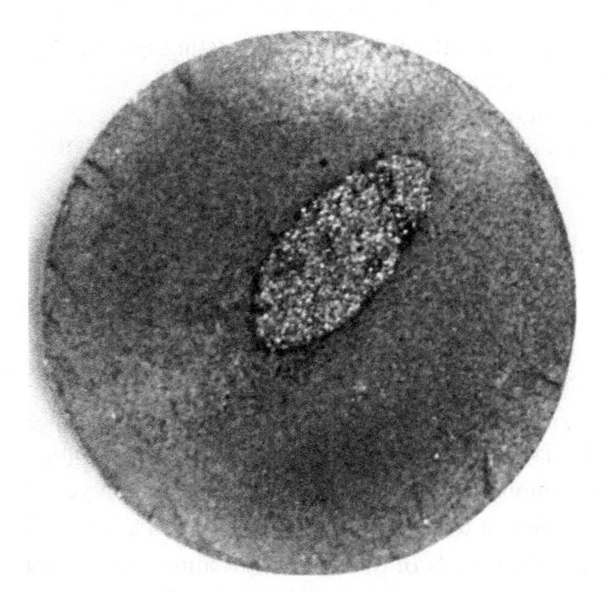

Figure 4.20. The surface of fatigue fracture of the steel 1020 after the influence of climate conditions (the size increased by six times).

accurate information for the real corrosive processes of machinery in the field, it is necessary to simulate these factors correctly.

It is becoming more common for mobile product life testing to include a combination of motion from special road or off-road simulation with a bath or spray of a saline-type solution with the corrosion test solutions.

Different testing programs correspond to this methodological approach.

For example, one of these programs includes the following conditions during a single day [142]:

- Six hours in the corrosive test chamber (salt spray or fog, with humidity and high temperature)
- Movement through a bath or spray of a 5% salt solution
- Motion to simulate a 40-mile drive along a dirt road, combined with simulating 14 miles of macadam road and 3 miles of cobblestone road
- Multiple passes through the bath or spray with a 5% salt solution

The equivalency of the corrosive destruction of an automobile by accelerated testing and by field use is

$$D_f = D_a + D_s(N),$$

where

D_f is the maximum depth of design corrosion for life in the field in mμ;
D_a is the maximum depth of corrosion life testing, ignoring the time of stress influences in mμ; and
D_s is the maximum depth of corrosion for time of testing in a regime of stress influences of moving cycles with the use of the corrosive test chamber, in mμ.

Corrosion depth has a constant speed; thus,

$$\lambda_f T_f = \lambda_{s.r.}(T_{s.r.} - T_s) + D_s(N),$$

where

T_f is the life of automotive components under given conditions, such as the normative index of corrosive endurance of the test subject in years;
$T_{s.r}$ is the common duration of special roads testing in years;
T_s is the duration of testing stress regimes in years;
λ_f and $\lambda_{s.r.}$ are the speeds of distribution of atmospheric corrosion in given field conditions and accelerated testing conditions in mμ per year;
N is the number of cycles of accelerated stress corrosive influences; and
$D_s(N)$ is the dependence of corrosion depth on the number of accelerated stress testing cycles.

For new components, use T_f and $T_{s.r.}$ as values of time for the basic types of studied components from actual use experience. These components can be new technical conditions, prototypes, and estimates of accelerated testing duration. Evaluate the function $D_s(N)$ to measure the value of corrosion influences on accelerated testing to obtain the corrosive results identical to field testing. $D_s(N)$ measures the correlation of corrosion depth with the number of cycles of the accelerated influences.

The following research result measures the dependency of automobile component corrosion depth due to the number of accelerated testing cycles for the above-mentioned methodology. It is

$$D_s^0(N) = 11.2 \cdot N^{0.7}.$$

The dispersion of the evaluated function of deviation is

$$D_d = 536 \text{ m}\mu^2.$$

The confidence interval of function $D_\sigma^0(N)$ for N_I can be evaluated as the confidence interval of its mathematical expectation in given experiments for $n = 8$ random observations and the common variance of their deviations.

If one will use the Student's criterion, this interval is available in the following equation:

$$T_\beta = D_s^0(N_I) \pm t_\beta \sqrt{D_d / n},$$

where

Index β is determined by the confidence probability of the calculated interval of function D_s^0 values and

t_β is evaluated by Student's distribution density tables for the accepted value β and $n - 1$ DOF.

Figure 4.21 shows the common approach for determining this 99% interval.

Use the following formula to calculate the number of influence cycles necessary for the corrosion testing of automobiles, their components, and other types of mobile products:

$$11.2 \cdot N^{0.7} - k_1 N - k_0,$$

where $k_1 = T_c / 365 \cdot v_{s.r.}$; $k_0 = v_f T_f - v_{s.r.} T_s$, and where

1 = approximated mathematical expectation of function $D_D^0(N)$;

2 = the limit of its 99% confidence probability intervals;

T_c is the duration of one cycle of stress influences by testing in days; and

v_f and $v_{s.r.}$ are speeds of corrosion development under given climatic conditions in mμ per year.

The above-mentioned approach for accelerated corrosion testing evaluates different types of components in the automotive industry, farm machinery, off-highway equipment, food technology, and other industrial products.

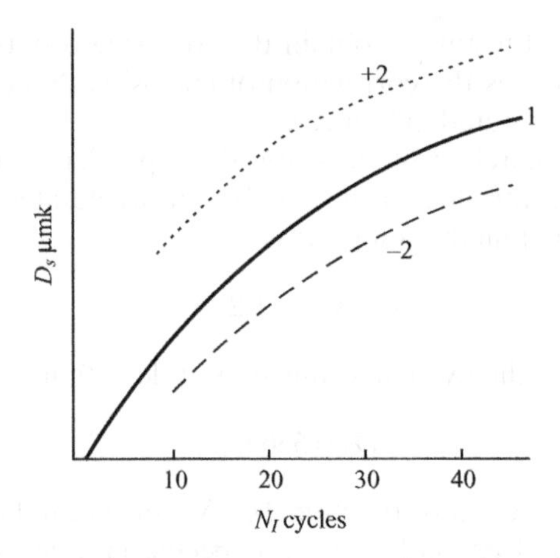

Figure 4.21. Statistical probability generalization of the data of corrosion depth by dependence on the cycle number of stress influences combined with the used corrosion chamber and normalized road motion [43].

Another approach to accelerated corrosion testing of complete vehicles [143] is used for military vehicles. Corrosion testing combined with "durability" (Baziari's term) testing improves the ability to evaluate the interaction between corrosion and the physical stresses that act upon a vehicle.

The army service environment is different from the commercial automotive vehicle environment. Driving will include more off-road applications and the hauling of heavy loads. Vehicles also undergo long periods of static exposure at training locations or at prepositioning locations on land and at sea. The maintenance performed by military personnel will differ from a commercial automotive owner. The impact of these variations is unknown, but the army service environment is (expected to be) at least as severe as the worst commercial environment. Table 4.9 provides the weight loss due to corrosion targets.

Corrosion test events by design are to

- Apply different corrosive contaminants that are likely to be present in different service environments
- Work those contaminants onto/into vehicle surfaces where they may occur during the expected use of the vehicle
- Create an environment where the corrosion caused by these contaminants will occur at an accelerated rate

Applying contaminants and working them onto/into all appropriate surfaces of the vehicle occurs simultaneously throughout the CDT procedure test events. Accelerate the corrosion mechanisms with an environmental

TABLE 4.9 Test Year Cumulative Control Coupon Corrosion Weight Loss Targets [143]

Test Year	Cumulative Weight Loss (g)	Test Year	Cumulative Weight Loss (g)
1	0.489	12	8.30
2	1.28	13	8.94
3	2.13	14	9.57
4	2.98	15	10.2
5	3.83	16	10.9
6	4.47	17	11.5
7	5.11	18	12.1
8	5.74	19	12.8
9	6.39	20	13.4
10	7.02	21	14.0
11	7.66	22	14.7

Note: Weight loss values are for a 1008 steel coupon measuring 1 × 2 × 1/8 in., having a total surface area of 30.65 cm^2.

chamber that maintains a high-temperature and high-humidity environment. Baziari [143] wrote that the amount of time spent within the chamber controls the severity of the corrosion on the test vehicles throughout the test. "Event" refers to the lowest level of vehicle driving activity. "Cycle" refers to the group of events completed in a driving period. "Phase" refers to a test year. The hierarchy of corrosion test events includes the following:

- Each individual corrosion or durability input is considered an event.
- The sequence of events is considered a cycle.
- One phase is composed of 10 cycles.

The corrosion/durability test utilizes an expected mission profile that includes driving on four different terrains

Surface	**Percentage of Travel**
Primary roads (paved road)	10%
Secondary roads (gravel road)	20%
Trails ("Belgian" blocks)	30%
Cross country (rolling hills)	40%

The corrosion/durability test [143] closely approximates the anticipated fielding of this vehicle. The test phases closely match the type of driving terrain

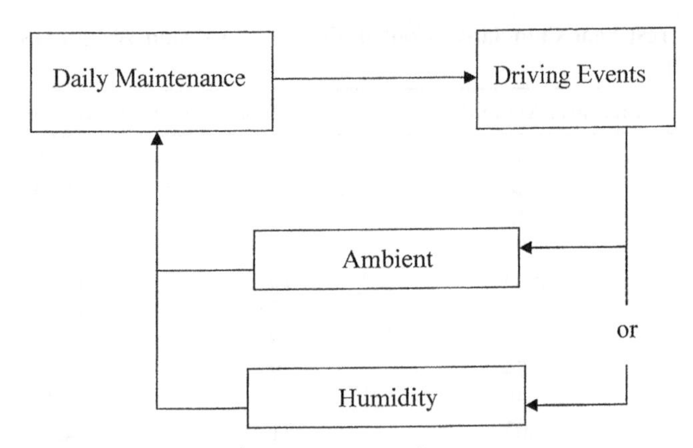

Figure 4.22. Test cycle events.

and accelerate the environmental influences in which the vehicle will be operated. The conceptual design of the daily test events is in Figure 4.22.

Several different test events are included in the accelerated corrosion testing. Each event is a corrosion input. The establishment of these test event requirements occurred during the development of the CDT in the 1980s. Recently, the requirements were modified to reflect the current state-of-the-art conditions in vehicle corrosion testing.

The salt solutions used for this test contain contaminants representative of elements found in the various natural environments where the vehicle will operate. They may include airborne chlorides and deicing salts among others. This corrosion test uses two solutions. The salt splash portion of the corrosion course uses the first solution exclusively: 5% by weight sodium chloride solution. The grit trough and salt mist portions of the corrosion course use the second solution. Its chemical makeup is listed with the following mixture percentages:

0.9% Sodium chloride ($NaCl$)
0.1% Calcium chloride ($CaCl$)
0.075% Sodium bicarbonate ($NaHCO_3$)

Both tests illustrate the modern essence of accelerated corrosion testing, as a component, in combination with other types of testing, for ART/ADT.

4.5.7.8 Accelerated Corrosion Testing by Corrosion Technology Laboratory at KSC. The Corrosion Technology Laboratory at KSC has the facilities, equipment, and experience to conduct indoor accelerated laboratory testing. The available equipment includes a salt fog chamber, a Q-Fog programmable

CCT-1100 cyclic corrosion salt fog chamber, a QUV/spray accelerated weathering tester, and a Q-Sun Xenon test chamber Xe-HDS with dual spray capability cabinets to perform the typical accelerated corrosion testing [132].

While the use of accelerated corrosion testing for quality control and preliminary investigations is routine, it is important to conduct real-time outdoor exposure tests in conjunction with accelerated laboratory tests to obtain meaningful service life data. The Corrosion Technology Laboratory at the KSC has a beachside atmospheric exposure site to conduct long-term corrosion performance testing. All facilities and expertise are available to customers outside NASA.

4.5.7.9 Development of Accelerated Corrosion Testing Mode. Oh SL [144] published that the simulation of accelerated corrosion for a vehicle in a controlled environment not only involves large chambers for actual vehicle testing but also requires careful consideration of the interactions between various parameters given the short time period that bounds the test. New corrosion testing modes now exist to reproduce various field conditions using salt spray, climatic sunlight simulation, and cold chambers. Verification of the test mode is verified using four actual vehicle corrosion tests that correlate to the used cars of North America and northern Europe.

4.6 TECHNOLOGY OF ADVANCED VIBRATION TESTING

Accelerated vibration testing may be one of the components of accelerated reliability (durability) testing, or an independent type of testing. It depends on the goal of vibration testing. Of greater value is the use of vibration testing as a component of ART. There are different types of vibration: random, sinusoidal, classical shock, and others.

The vibration as an output variable for all mobile products, as well as most stationary products, has a random character. Therefore, it is necessary to use random vibration for ART/ADT. Accordingly, this book considers the methodology and equipment for random vibration.

A basic criterion for the use of vibration testing technology as a component of ART/ADT technology is that the technology for vibration testing must be accurate in order to simulate real-life input influences of the vibration on the actual product. As shown earlier, the vibration of a product is an output parameter. This is a result of the input influences for a mobile product including (Fig. 4.23) road features (type of road, profile, density, hardness, and moisture), design and quality of the whole product, design and quality of wheels, coupling of wheels, surface of road, surface stability due to weather changes, speed of test subject movement, and wind speed and direction. The types of road surfaces include concrete, asphalt, cobblestone, brick, or dirt road. Most of the surface influences on the product have a random character.

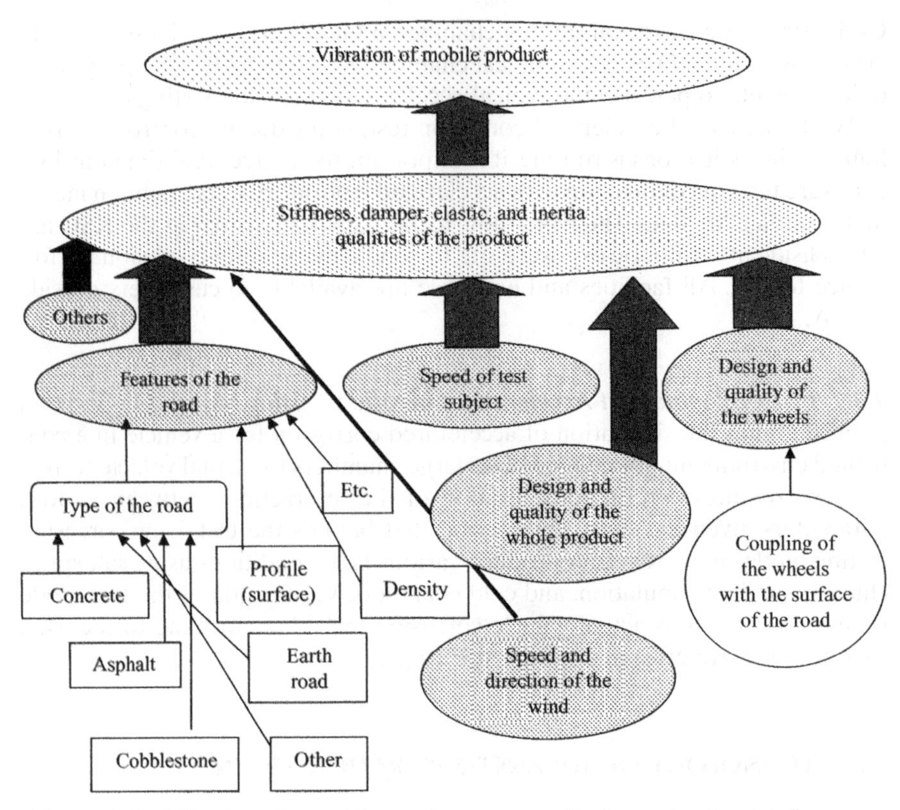

Figure 4.23. Vibration of a mobile product as a result of complex input influences.

The characteristics of the influences determine the stiffness, inertia, elasticity, and dampening qualities of the product. The result of the interconnection for these factors is the vibration of a mobile product on a particular surface under specified conditions. Therefore, treat the mobile product as a dynamic system that vibrates due to the complex input influences and interconnections.

Together with his colleagues, the author created a new technology of vibration testing, including methodology and equipment. The basis of this methodology is the theoretical and experimental research results with electronic products, car trailers, trucks, and farm machinery. The basic components of this methodology are the following:

- Accurate simulation of real field influences as shown in Figure 4.23
- Vibration in three linear axes for a wheeled mobile product and six axes (three linear and three angular) for another product
- Execution of a random process of input influences as it is in real life

- This random process of values for the statistical characteristics (normalized correlation, power spectrum, variance, and mean) of the amplitudes and frequencies of vibration for the test must be similar to the random process of corresponding values of statistical characteristics in the field, within given divergences (3%, 5% or other limit)

The advanced vibration testing technology requires unique features like working heads that provide accurate assigned random values for the acceleration amplitude and frequency. The system of control consists of a computer with a peripheral network system and sensors for vibroacceleration and tension. Vibration is available for both mobile and stationary products. Vibration is delivered by universal and special-purpose vibration equipment and equipment provided for special products. It covers wheeled and stationary applications acting in one to three linear and in one to three angular axes for a total of up to 6 degrees of freedom (DOF). This technology design is simpler than modern technologies that use electrohydraulic and electrodynamic equipment with no deterioration of performance. It also expands the functional applications to a wider range of markets.

Figure 4.24 shows the principal scheme for mounting sensors on a car's trailer. To execute this technology, the parameters of the vibration equipment are provided by a different combination of the following factors: high level of simulation, frequency of the drum rotation, and the number of lugs that act on the test subject during each moment.

When the large automotive company KAMAZ Inc. (Russia, 200,000 employees, about 100 plants) wanted to change its current multiaxis vibration equipment for new equipment to test completed trucks (30 tons of weight and

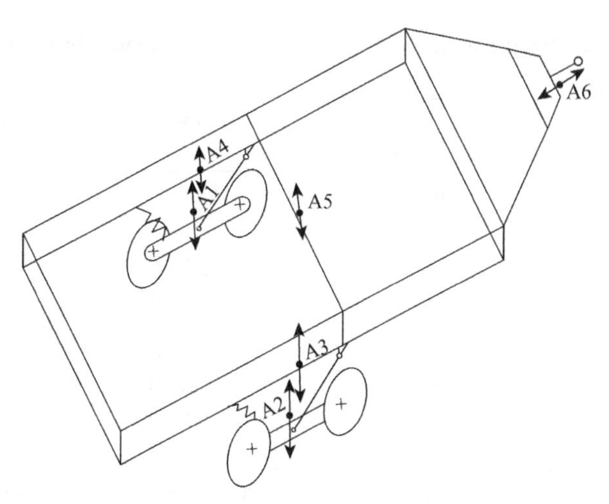

Figure 4.24. Example of sensor locations on a car's trailer.

with three axles) in one to three linear axes at the end of the 1980s, they ordered this equipment from Carl Schenk (Germany). Carl Schenk asked $9 million for this equipment.

The Russian company TESTMASH, where the author was once chairman and research leader, agreed to solve this problem for $1.5 million. Then TESTMASH implemented this type of equipment for KAMAZ Inc. The TESTMASH equipment is simpler to produce, has greater functional possibilities, and more accurately simulates the field situation. The basic advantage of this equipment is the accurate simulation of real vibration. It is still modern, up-to-date, and other companies continue to use it.

The components of this equipment with working heads are in Chapter 5 (Figures 5.38 and 5.39).

This technology is for electronics, car trailers, trucks, farm machinery, and other items. Tables 4.10 and 4.11 show some of the car's trailer testing results. The difference of the tension in the laboratory (new vibration technology) compared to the field was no more than 8%.

TABLE 4.10 The Frequency of Vibroaccelerations of the Car's Trailer Axle [43]

	Field			Under Laboratory Conditions		
Type of Road	Linear Speed (mph)	Minimum Frequency (Hz)	Maximum Frequency (Hz)	Linear Speed of Drum (mph)	Minimum Frequency (Hz)	Maximum Frequency (Hz)
Town road	37.50	26	248	5.60	16	324
Highway	50.00	36	224	5.60	24	357
Field	12.50	28	221	4.05	26	362
Cobblestone	12.50	30	271	3.08	23	314

TABLE 4.11 Results of Comparative Testing of the Car Trailer in the Field and in the Laboratory* [43]

	Linear Speed	Tension at the Most Loaded Point of Trailer Frame	
Type of Testing	mph	kg/cm^2	%
Empty trailer			
Town road	37.50	272.00	100.00
Laboratory	5.60	251.00	92.00
Full trailer			
Highway	37.50	442.00	100.00
Laboratory	5.60	424.00	96.00
Field	12.50	318.00	100.00
Laboratory	4.00	312.00	98.10

*The mathematical expectation is shown. The normalized correlation and power spectrum would have a difference similar to that of frequency and amplitude.

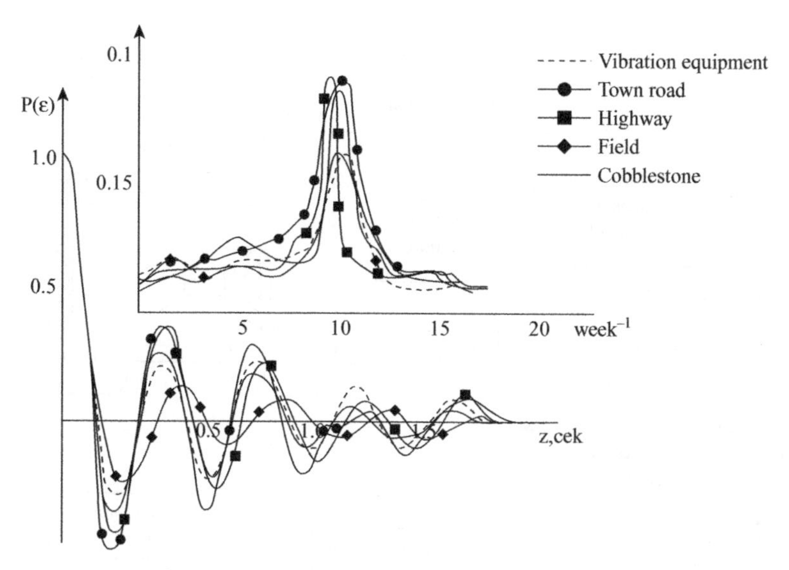

Figure 4.25. Normalized correlation and power spectrum of frame tension data for a car's trailer in field conditions and with the new vibration equipment [43].

Figure 4.25 shows that the characteristics of the random process with new vibration equipment are also similar to these characteristics in field conditions.

The test results show that this technology makes it possible to eliminate some of the negative aspects of the current vibration equipment.

As Chapter 5 shows, the equipment developed for this methodology has the following capabilities:

1. The equipment has the ability to simulate different physical-mechanical qualities of roads. This is not possible with the other current vibration testing equipment.
2. The equipment has the capability to simulate random vibration in three axes for equipment (cars and others) with wheels.
3. It can simulate torsion swing in the simultaneous combination of linear and angular vibrations. Other equipment cannot do this simulation.
4. This equipment can integrate vibration and dynamometer testing.

This equipment is easy to combine with other components for ART/ADT. It combines with a test chamber, transmission, engine, clutch, fans, or a power system to achieve their simultaneous vibration requirements. The design of this equipment is less complex and expensive than current electrodynamic and servohydraulic equipment.

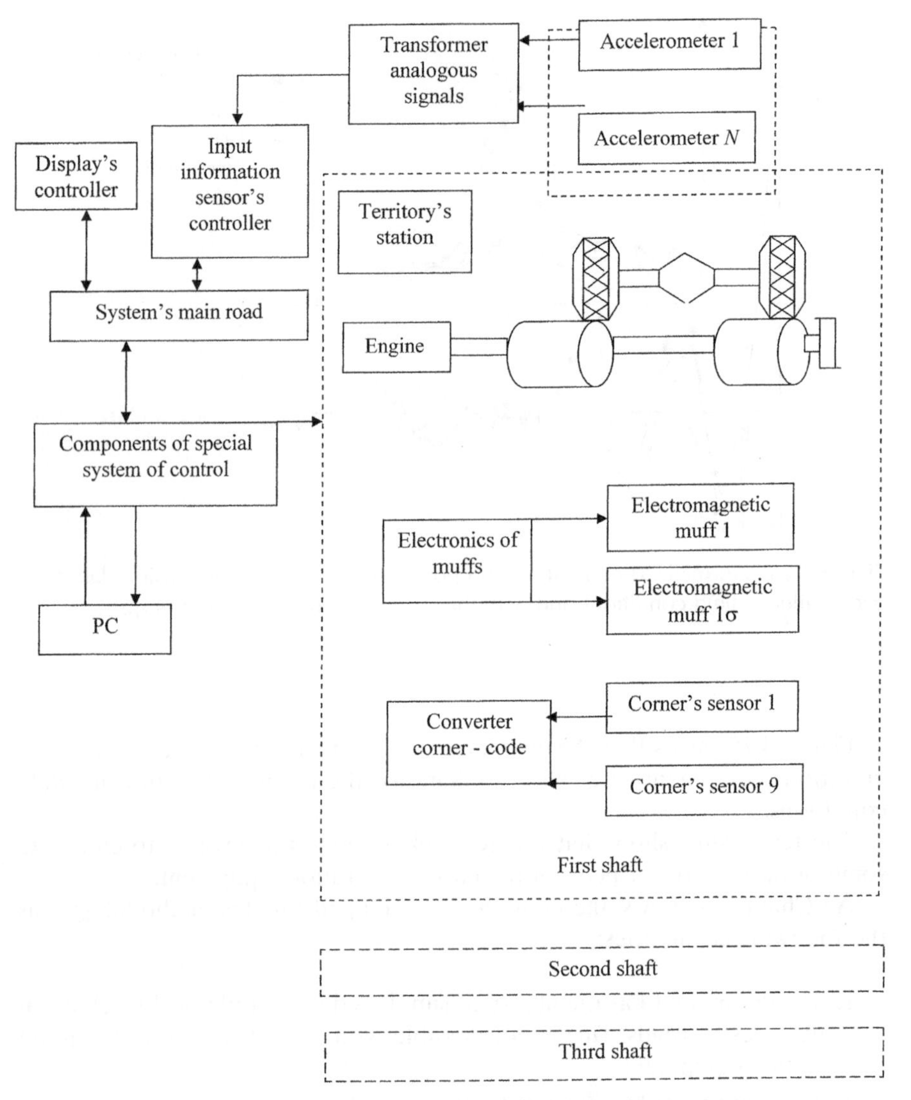

Figure 4.26. Scheme of system of control for truck vibration equipment.

The plan to achieve system control for truck vibration equipment is in Figure 4.26.

The second use for advanced vibration testing technology (simulation 6 DOF) is similar, but it is less complex and less expensive. The basic advantages of this type of vibration testing is that it's applicable to many types of stationary and mobile products, such as mechanical, electrical, and electronic products plus to more complicated products.

Specifically, the basic advance by this technology is vibration testing in 6 DOF: three linear and three angular. In the 21st century, the author developed the above technology for vibration testing concurrently with the development of ARTT and ADTT.

4.7 FIELD RELIABILITY TESTING

Usually when one conducts reliability field testing, one wants to evaluate reliability during a short time, usually one season of work. This applies to products that work for a short time during the year, such as some farm machinery including different types of harvesters and fertilizer spreaders.

These machines work in the field during one season, and the results of their work provide initial information for evaluation of their reliability and maintainability for this season. If one performs this field testing more intensively than under normal usage to obtain the volume of work for more than one season of normal usage, then the results of this test are not accurate for reliability and maintainability evaluation for this volume of work. For example, the normal volume of a harvester's work for one season is 150 hours. If the harvester operates for 300 hours (two normal seasons' volume of work) during one season, then one cannot accurately evaluate the reliability and maintainability of this harvester for 2 years (seasons) because this does not take into account the results of interacting environmental influences (corrosion, crack, and deformation) on the harvester's reliability and maintainability during the storage time. This is important because storage time is much longer than work time.

Field test results for reliability, quality, and performance and the evaluation of the ease of operation and servicing are one of the most important decision criteria for or against a farm machine. One cannot underestimate criteria like postsale service or machinery reliability and durability. Thus, in a survey of 760 combine owners, 66% reported service as a decisive reason for purchase [145].

Both customer service and spare parts supply through farm machinery dealers and manufacturers and the working lifetime and reliability cannot be precisely determined. To interpret the test results and decisions regarding the investment, the results from the field tests and on-farm surveys, analysis of technical data, and the results of scientific tests are all essential.

4.8 TRENDS IN THE DEVELOPMENT OF ART/ADT TECHNOLOGY

How can one improve the current situation and realize that ART/ADT technology gives initial information for the AP of field reliability, provides rapid and cost-effective solutions, and provides initial estimates of other quality (Q), durability (D), and maintainability (M) problems?

It is necessary to understand trends in the development of ART/ADT technology: the general trends and the trends for a specific product.

The basis of this technology is APS of the field input influences (II) on the actual product, as well as, safety (S) and HFs. The trends in the development of accelerated reliability testing technology (ARTT) and accelerated durability testing technology (ADTT) (Fig. 4.27) follow:

- Trend #1 includes
  - The basic concepts of development
  - Determination of how to obtain accurate initial information from the field for a product under consideration
  - Determination of how to accurately simulate the field situation for a considered product
  - Use the degradation (failure) mechanisms of the product as a criterion for an accurate simulation of field input influences
  - Considering the simulation of field input influences, safety, and human factors as interconnected components of one integrated complex

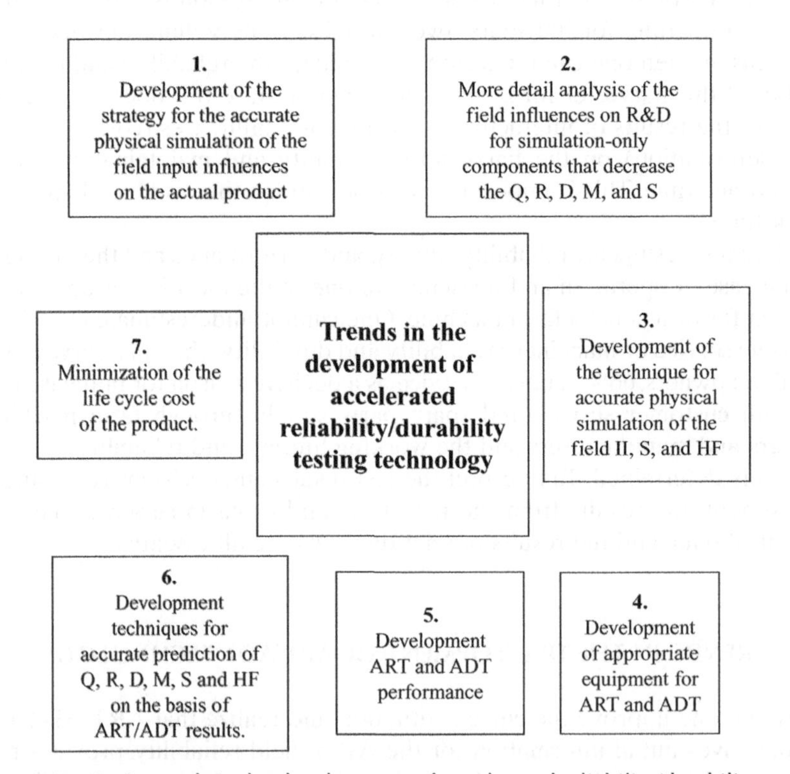

Figure 4.27. Basic trends in the development of accelerated reliability/durability testing technology (ARTT/ADTT).

  - Development of the methodology to select the representative input region for the APS of the field conditions
- Trend #2 includes
  - Use the field characteristics as the external conditions during the machinery use
  - Development of classifications and characteristics of the world's climate for their effect on technical goals
  - Performance of detailed analyses of the influences that quantitative field characteristics have on the product reliability and durability. These influences include thermal regimes, air fluctuations, air humidity and rain, wind speed, fog and dew, atmospheric phenomena, characteristics of their combined factors, biological factors, influences of solar radiation, fluctuations and rapid changes, and the characteristics of the combination of influences on basic climatic factors
  - Separation of all field factors by two groups
    1. Decrease in quality, reliability, durability, and maintainability (Q, R, D, and M)
    2. No decrease in quality, reliability, durability, and maintainability
- Trend #3 includes
  - Providing APS in the laboratory of each real-life input influence that decreases a product's reliability, quality, maintainability, and durability
  - Development of a technique to perform the necessary simultaneous combination of the numerous field influences for each product
  - Development of a system of control for the physical simulation of random input influences
  - Development of a technique to provide special field testing (in general, as well as for the specific product) in addition to the accelerated laboratory testing
  - Development of a testing technique that takes into account the interconnection and interactions of the product components as well as the interconnection of different input influences with safety and human factors
  - Creation of a new artificial media substitute for natural technological media when necessary
  - Development of an accurate simulation for safety problems and human factors
- Trend #4 Testing equipment on the current market is not suitable for ART and ADT. It does not provide accurate simulation of the field influences. Therefore, it is important to develop new equipment to meet this goal. For example, it is expensive to mechanically combine the various test equipment for accelerated multi-environmental testing/humidity,

temperature, chemical and mechanical pollution, all three parts of radiation (ultraviolet, infrared, and visible), air and gas pressure, wind, snow, and rain/.

One also needs to consider the design and overall cost problems. The particular system applies only to multi-environmental accelerated testing that is a component of ART and ADT. Complete ART is more complicated and needs more attention.

Develop new, less expensive equipment with more functional possibilities for the simultaneous combination of different types of field input influences.

Sections 5.2, 5.3, and 5.5 demonstrated how this trend developed. These sections describe the practical essence of the trends with new combined equipment for ART. The trends include the use of electronic devices, combined equipment for multi-environmental testing, equipment for combined vibration testing, and other useful test apparatuses. This was not apparent 10–20 years ago.

The process of testing equipment development is moving very slowly, mostly through the development of electronic systems of control, but not through the development of the working heads for the equipment. The fault lies with both the companies that produce test equipment and the companies that use the test equipment. Combined test equipment is more expensive, and users want to save money by purchasing simpler and less expensive test equipment. However, this approach is wrong. It does not take into account lower testing quality that requires more time for product development and raises the cost of the product's development, improvement, and use.

ART/ADT provides an effective means to reduce the time and cost of product development and to save money, especially in the manufacture and usage stages.

One needs to increase the speed of ART/ADT standards development and implementation to increase the speed of the ART/ADT equipment development.

Chapter 5 addresses this problem.

- Trend #5 consists of
  - Development of the entire technology for step-by-step ART/ADT
  - Implementation of advanced reliability/durability testing as a step-by-step technology, that is, implementation of a test chamber with one type of equipment, for example, vibration. Then, add to this test chamber a second type of equipment, for example, for temperature and humidity simulation. Then, add to this test chamber a third type of equipment, for example, for solar radiation simulation. The more effective method of implementation begins with the types of testing

that have the greatest weight for the reliability and quality of the test subject. This is usually an expert's estimate.

- ○ Providing laboratory accelerated testing technology as a simultaneous combination of multiple environmental, mechanical, electrical, and other components of ART/ADT.
- ○ Providing the previously mentioned accelerated laboratory and special field testing in combination with safety and human factors.

- Trend #6 The development of methods and equipment using ART/ADT results as a source of initial information for AP and accelerated development of reliability, quality, maintainability, availability, and safety.
- Trend #7 The life cycle of the product consists of three stages: design, manufacturing, and usage. ART/ADT can help to decrease the cost of all three stages and, as a result, to minimize the life cycle cost of the product. Usually, the usage stage incurs more than 50% of the life cycle cost. Therefore, decreasing the cost at this stage is the most important factor of minimizing the life cycle cost.

  How can ART/ADT help to do this? AP helps to minimize the time between maintenance or repairs and the duration of these procedures. As a result, a decrease in maintenance and repair costs reduce the life cycle cost. ART/ADT makes it possible to reduce the time and cost of product development and improvement. This saves money, especially in the manufacturing and usage stages.

  Life cycle cost is all costs associated with design, manufacture, and usage. To show that ART (ADT) is cost effective, it is necessary to develop the methods that take into account the cost of recalls and complaints, loss of business, the increase in the warranty period, the adjustment in the time and cost of maintenance, and to maintain the product reputation.

EXERCISES

4.1 What kind of testing do some other authors identify as ART/ADT? Why are they wrong?

4.2 Describe the basic components of ART/ADT philosophy.

4.3 Why base ART/ADT on the simulation of three complexes?

4.4 Why is ART/ADT a combination of different types of testing?

4.5 Describe the methodological requirements of ART/ADT.

4.6 What kind of human factors need consideration for general dimensions?

4.7 Why does one not need to simulate separate input influences or their small group when conducting ART/ADT?

4.8 Describe the basic principles of AMET.

4.9 Describe what AMET is as a component of ART/ADT. Show this schematically.

4.10 Describe the specifics of accelerated environmental testing of composite structures.

4.11 What is the essence of global dimming?

4.12 How does the direct cost of corrosion differ from the indirect cost of corrosion?

4.13 Describe and show corrosion resulting from different complexes and types of different groups of influences. What is the composition of each group?

4.14 Describe the forms of corrosion you know. What are the specifics of each of them?

4.15 Describe corrosion in the marine environments.

4.16 Describe the specifics of aircraft corrosion.

4.17 Briefly describe accelerated corrosion methods and their specifics. Why is it not possible to use all of them for the AP of the system's corrosion degradation in the field?

4.18 Why does separate corrosion testing not provide accurate information for reliability and durability prediction?

4.19 Describe an example demonstrating corrosion testing in the automotive industry.

4.20 What is the basic advantage of modernized accelerated corrosion testing?

4.21 What kind of problems must one solve to establish the accelerated corrosion testing conditions of steel?

4.22 How does one proceed to establish the necessary temperature in the test chamber if the temperature is one of the components of accelerated corrosion testing?

4.23 How does one proceed to establish the necessary humidity in the test chamber if humidity is one of the components of accelerated corrosion testing?

4.24 Describe a short methodology for determining the dependence of "steel corrosion versus the number of wettings."

4.25 Show and describe the equation for the approximation of steel corrosion based on the number of wettings when the number of wettings exceeds 100.

4.26 Draw the table that demonstrates the possible variants of standard deviation and mathematical expectation between the expected value for real-life and test chamber conditions.

4.27 After conducting corrosion testing in test chambers, how can one evaluate whether the testing conditions correspond to the field conditions? Show how the testing results correspond to the field results.

4.28 Accelerated corrosion testing of components differs from the accelerated testing of complete equipment. What are the basic differences?

4.29 Show schematically the complex of input influences that have an influence on the mobile product's vibration.

4.30 There are many types of vibration equipment. Which types of vibration equipment constitute advanced vibration equipment? What are the specifics of advanced vibration equipment?

4.31 The book includes seven trends in the development of ART/ADT technology. Describe one of these trends.

Chapter 5

Equipment for Accelerated Reliability (Durability) Testing Technology

5.1 ANALYSIS OF THE CURRENT SITUATION WITH EQUIPMENT FOR ACCELERATED RELIABILITY (DURABILITY) TESTING

One can use these recent exhibitions to analyze testing equipment: Aerospace Testing Expo 2006, Automotive Testing Expo 2008 Europe, Automotive Testing Expo 2008 North America, Aero Test America 2008, SAE International 2009 Aero Tech Congress and Exhibition, the Automotive Testing Expo 2009 North America, and Automotive Testing Expo 2010 North America.

The Aerospace Test Expo 2006 occurred on November 15–17 in the Anaheim Convention Center, California. David Pegler, Managing Director—Exhibitions Division, UkiP (United Kingdom) Media & Events Aerospace Testing Expo 2006, wrote that more than 250 of the world's leading suppliers of testing technologies, solutions, equipment, and special services attended. They addressed the full life cycle of airframe, power plant, subassembly, and component "test" applications within the critical areas of

- Research and development testing and evaluation
- Flight testing and certification
- Production, assembly testing, and quality management
- Maintenance engineering test and inspection plus much more

This expo consisted of three basic components: exhibition, seminars, and forums (Chapter 4 analyzed the forums). Participating worldwide companies

Accelerated Reliability and Durability Testing Technology, First Edition. Lev M. Klyatis.

and organizations included the Aero Testing Alliance (France), Boeing Technology Services (United States), Lockheed Martin (United States), MTS Systems Corporation (United States), NASA Glenn Research and Center Plum Brook Station (United States), Brüel & Kjær (United States), CEL Aerospace Test Equipment Ltd. (Canada), Fabreeka International (United States), Honeywell (United States), Kyowa Electronic Instruments Co Ltd. (Japan), LDS Test and Measurement (United States), National Instruments (United States), RENK LABECO Test Systems Corporation (United States), TechSAT GmbH (Germany), and Tyco Electronics Gath (United States).

The following analysis of the exhibition and development forum contents provides information on how physical simulation and testing technology develop, especially accelerated reliability testing (ART) technology. The first impression from the exhibition was that the booths did not demonstrate techniques and equipment for reliability testing.

Let us analyze the booth contents for Aerospace Testing Expo 2006.

East-West Technology Corporation demonstrated "Environmental Simulation Climatic."

It showed equipment for the simulation of

- *Corrosive Atmosphere*: Salt fog/sulfur dioxide (SO_2)/copper acidified salt solution
- *Blowing Sand and Dust*: Requirements of MIL-STD-810, Method 510, and procedures with wind velocities to 6000 fpm
- *Explosive Atmosphere Environments*: Hexane, aviation fuels, hydrogen, propane, and butane to altitudes of 100,000 ft and temperature of 475°F
- *Fungus Resistance*: Certified spores and inspections 28- to 90-day exposures
- *Temperature Extremes*: –320 to 2000°F with transition rates up to 1000°F/min
- *Humidity*: Temperature humidity cycling from 2 to 100% condensing and noncondensing

The above-mentioned simulation equipment is sufficient for the separate simulation of climate parameters, but it is not sufficient for the simulation of the complete environmental complex. Therefore, it is inadequate for effective ART.

This company demonstrated vibration simulation for aircraft. One PAL 133-30P supports the aircraft nose. This isolator can support and lift 13,300 lb (6000 kg), 11,500 kg, and other loads.

The vertical and horizontal natural frequencies are 1.0 and 1.5 Hz. The jacking unit raises the fuselage approximately 21 in. (530 mm).

This type of vibration simulation is inadequate for ART because it utilizes separate types of mechanical field influences without taking into account wear and other interacting mechanical influences and environmental factors.

The exhibits included several types of wind tunnels. These wind tunnels simulate fluctuation and air pressure and are appropriate for experiments, development, and maintenance in areas such as aerodynamics, flight mechanics, energetic, structures, materials, and test instrumentation. These wind tunnels are also suitable for research disciplines such as optics, electro-optics, electromagnetism, automation, and information technology required to support the aerospace industry worldwide. This type of wind tunnel cannot provide reliability testing for obtaining initial information for quality and reliability prediction. It is only one of the components for ART.

In this case, one can use the title of the DNB Engineering, Inc. (an exhibitor at the Aerospace Expo) brochure: *Is Your Testing Solution Reliable?* The answer to this question lies in the contents of this company's work demonstrated at its booths and by other companies. The simulation is for their testing capabilities and includes the simulation for only the types of testing shown below

- Electromagnetic
- Electromagnetic compatibility (EMC)
- Lighting
- High-intensity radiated field
- Product Safety
- Environmental
- Acoustics

These types of testing are independent of each other; therefore, they are insufficient for reliability/durability testing.

L-3 Communications Cincinnati Electronics showed Simulated Environments that one uses to provide environmental and climatic stress screening. This company's Environmental Simulation Test Laboratory includes the following simulation capabilities:

- For climatic testing
  - •• Temperature and humidity
    - One 32-ft^3 chamber (38" × 38" × 38" working area)
    - Temperature
  - •• Altitude (low pressure)
    1. One 64-ft^3 chamber (4′ × 4′ × 4′ working area)
    2. Temperature: –73 to +177°C (–100 to +350°F)
    3. Controlled altitude to 10 mbars (100,000 ft)
    4. Temperature cycling capabilities
    5. Microprocessor controlled

- •• Temperature cycling
  - Three chambers available
    6. Two chambers up to 6′ × 4′ × 4′
    7. One chamber up to 54" × 54" × 42"
    8. Temperature: –73 to +177°C (–100 to +350°F)
    9. Temperature cycling capabilities
    10. Temperature change of 10°C/min
    11. Microprocessor controlled
- For environmental stress screening simulation and testing
  - •• Thermal shock
    12. Dual basket system (28" × 28" × 28" work space)
    13. Temperature: –73 to +177°C (–100 to +350°F)
    14. Microprocessor controlled
  - •• For vibration and shock simulation and testing
  - •• Electrodynamic vibration shakers provide the following maximum performance spectrum:
    15. T1000 Unholtz-Dickie shaker with slip plate
    16. D300 Ling shaker with slip plate
    17. A395 Ling shaker with slip plate
    18. Two EST 2648 fixed horizontal shakers with temperature dwell and cycling
    19. Shock testing (electrodynamic vibration shakers that provide the capability of classical shock and shock response spectrum shock)

The previous items represent the simulation of separate parameters and therefore are unsatisfactory for successful accelerated reliability (durability) testing.

Let us analyze how the demonstrated test equipment relates to reliability testing. Lockheed Martin demonstrated the High-Speed Wind Tunnel for

- Calibrations of balances and transducers
- Strain gage installation
- Six-component internal strain gage force balances, with or without flow-through capability
- Special-purpose balances
- Calibration equipment
- Complete models, components, and modifications
- Strings and support equipment
- Thrust and test strands

Similar solutions to problems were included in other presentations. These solutions are not relevant for accelerated reliability (durability) testing.

L-3 Communications Cincinnati Electronics provides high-intensity radiated field testing that includes frequencies from 10 KHz to 40 GHz with continuous scan capability. The testing services offered include the following:

- Climatic testing (altitude [low pressure], humidity, temperature, temperature cycling, and salt fog atmosphere)
- Electromagnetic interference/capability
- Electromagnetic pulse
- Electrostatic discharge (ESD)
- Environmental stress screening
- High-intensity radiated field
- High-power RF testing (power handling capability)
- Lighting indirect effects
- Nacelle attenuation testing
- Shock and vibration

One can conclude that L-3 provides a wide range of testing. These are separate types of testing and use only part of the components that one needs for reliability testing. Therefore, it cannot accurately predict quality, reliability, and durability.

CEL Aerospace Test Equipment Ltd. (Canada) demonstrated that this company has

- Test cell equipment
- A turbofan test cell
- A turbofan test cell—low/mid speed
- A turbofan test cell—high speed
- An APU test cell
- Fuel accessories test stands
- Component test stands
- Engine throttle controls
- An industrial engine test cell
- A turbo shaft test cell—tilt rotor

This test equipment operates separately and independently; thus, it is not sufficient for ART.

The description of the East-West Technology Corporation types of simulation and equipment shows that this equipment is also not intended for reliability testing.

RENK LABECO Test Systems Corporation's experience is in the manufacturing of transmission (dynamometer systems) and power transmission for testing, research, and development in aviation, automotive, aerospace, railway, and other areas of industry. This equipment independently simulates separate components of the field situation. Therefore, it also is not adequate for the reliability testing of the product.

Other industrial companies that demonstrated their test equipment in the Aerospace Test Expo 2006 also did not show equipment for ART. Therefore, the above as well as other companies' booths at the Aerospace Test Expo 2006 did not mention any equipment that could be used for ART as a source of the information for the accurate prediction of quality and reliability during the period of use (service life and guarantee period). As a result, problems continue with practical prediction. For example, when the research station was sent to Mars in 2001, the prediction was that the station would work for a minimum of 3 months. However, it only worked for 3 days.

In the chapter about testing services, only one company, Garwood Laboratories, Inc., mentioned that, in addition to shock and vibration, climatic, MIL/Aero and commercial hydraulics, EMS/EMI/RF lighting simulation, and product safety testing, it performs NEBS ESS/reliability testing. Nevertheless, it did not describe what it meant by the term "reliability testing." We can conclude from the above-mentioned analysis of Aerospace Testing Expo 2006 that existing techniques and equipment cannot help accurately to predict quality and reliability over time. The reason is the absence of an accurate simulation of the field situation. They independently simulate only the separate components of the field input influences or part of them in the real environment. They do not simulate most of the interconnected field input influences (changing during a given time of temperature, humidity, fog, air fluctuation and pressures, solar radiation, mechanical and chemical pollution, etc.) that act simultaneously and in combination.

This type of simulation will not provide a basis for ART that offers initial information for the accurate prediction of product quality, reliability, and durability.

The exhibit profile of Aero Test America 2008 contained more than 40 types of testing, including static and durability testing and reliability/life cycle testing. However, the last two types of testing used only the simulation of parts from the whole complex of real-life input influences. This is one more example of the misunderstanding of what "durability testing" or reliability testing actually means.

Similar results emerged for the automotive industry during Automotive Testing Expo North America 2008 (October 22–24, Novi, Detroit), which encompassed 43 categories of products including the following:

- Acoustic testing
- Aerodynamic and wind tunnel testing
- Automated test equipment

- Automatic inspection
- Calibration
- Component testing
- Crash test analysis
- Data acquisition and signal analysis
- Dynamometers
- Electrical system testing
- Electronics and microelectronics testing
- EMC/electrical interference testing
- Engine/emission testing
- Hydraulics testing
- Impact and crash testing
- Materials testing

One of the 43 categories was reliability/life cycle testing. In this group, 89 companies demonstrated their products. Only two companies, Weiss Environmental Technology Inc. and Seoul Industry Engineering Co, Ltd. R&D Center, demonstrated combined testing equipment for ART/accelerated durability testing (ADT). Several companies demonstrated equipment that combined simulation for only two to three types of input influences out of many types of field input influences.

ADT, especially ART, did not have an exhibit in the automotive industry during Automotive Testing Expo North America 2008. Thus, is it not possible to discuss the development of the accurate prediction and the accelerated improvement of reliability, durability, maintainability and, availability within this show.

Technical Program of SAE 2009 Aero Tech Congress

Aerospace Operations

- Aerospace Modeling and Simulation
- Airspace Systems Design

Automated Fastening/Assembly and Tooling (AeroFast)

- Change to Assembly Methodologies and Advanced Assembly Fixtures and Tooling
- Large Component Assembly and Subassembly
- Major Section and Final Assembly
- Automated Drilling and Fastening Systems
- Automated Robotic Drilling and Fastening Systems
- Advanced Portable Tools
- New and Enhanced Fasteners
- Composites Assembly

Aviation Cyberphysical Security

- Aeronautical Network and Application Security
- Product Life Cycle Management
- Security of Distributed, Integrated Software
- Intensive Systems

Avionics

- Display, Cockpit, and IFE Systems
- Flight Control Systems
- Avionics and Aircraft Control Systems IVHM
- Distributed Electric Systems, Controls, Platforms, Smart Sensors, and Actuators
- Application of Avionics and Software in Harsh Environments (Space/Defense)
- Integrated Model-Based System, Application, and Architectures
- Infrastructure Technologies—Network Technology
- Infrastructure Technologies—Multicore Processing and COTS Platforms
- Systems Integration Platforms and Architectures
- Integrated Modular Avionics and Distribution
- Computing Platforms

Business/Economics

- Aircraft for 2030 and Beyond
- Light Business Jets
- Next Generation Air Traffic Management
- Market Forecasts
- New Global Markets
- Business Models

Environment

- Aircraft ECS and Cabin Environment
- Airplane Design for Environment
- Emissions
- Energy
- Environmental Materials
- Environmental Materials and Processes
- Noise

Flight Sciences

- General Aerodynamics
- Aircraft Design
- Aircraft Design Methods

- Flight Control Technology
- Flying Cars
- Computational Fluid Dynamics (CFD)
- Aircraft Icing

Manufacturing, Materials, and Structures

- Advanced Robotics Applications
- Composites Fabrication and Joining
- Direct Digital Manufacturing
- Environmental Compliance, Green or Sustainable Applications
- High Output Composites
- Laser Applications
- Lean Manufacturing, Six Sigma, and Supply Chain
- Metals, Fabrication and Processing
- Metrology Automated Systems
- Product Design and Manufacturing Integration
- RFID Applications in Aerospace
- Structural Health Monitoring

Power Systems

- Advanced Power Systems
- Power Management and Distribution
- Energy Storage—Aircraft Batteries
- High-Temperature Electronics
- Thermal Management
- Modeling and Simulation
- Prognostics and Health Management

The above-mentioned program did not consider reliability and testing. This emphasizes the importance of developing the ART/ADT technology that forms the contents of this book.

5.2 COMBINED EQUIPMENT FOR ART/ADT AS A COMBINATION (INTEGRATION) OF EQUIPMENT FOR DIFFERENT TYPES OF TESTING

There are many types and designs of testing equipment. However, most of them cannot perform ART/ADT.

The current trend in the development of testing technology is the movement toward the design of equipment employing ART/ADT functionality. The first step in the process is the merging of various features into a single unit. As a result, the combined equipment is becoming more and more complicated. But professionals who are responsible for the purchase and use of these newly developed units understand that more accurate real-word simulation during the testing means less expense during the following steps of design, manufacturing, and usage.

Therefore, the types of testing equipment currently available in the market that are prime candidates for ART/ADT use are described next. The description and specification of advanced testing equipment will help the reader select useful testing equipment for the development of ART/ADT.

Chapter 1, as well as other chapters, describes that the most accurate simulation of the field situation is the simulation of random stress. One can see in section 1.4 of Reference 18 a description of the system of control for the physical simulation of the random input influences. This description includes

- Principles for the simulation of random input influences
- The mechanisms for the simulation control of the field random input influences
- An example of physical simulation of random input influences on the trailer

Universal equipment for random simulation control is also included: the generator of normal stationary random processes, as well as the more developed generator for nonstationary random processes, and the multichannel generator of random processes.

Successfully developing appropriate ART/ADT technology requires a multidisciplinary team to engineer and manage the application of this technology to a particular product. The team should include, as a minimum, the following:

- A team leader who is a high-level manager who can understand the strategy of providing this technology and understands the principles of accurate simulation of the field situation including what professional disciplines need to be included in the team
- A program manager to guide the team through the process and to remove any barriers that prevent the team from succeeding, as well as from understanding the design and technology needed for ART/ADT
- Engineering resources
  - to perform unit filtering (selection and elimination)
  - to conduct failure analysis
  - to solve chemical problems in simulation
  - to solve physical problems in simulation
  - to predict methodology
  - to guide the system of control development, design, and diagnostic and corrective action for mechanical, electrical, structural, and hydraulic problems

- to determine the need for hardware and software development and implementation

The team must work in close contact with departments responsible for design, manufacturing, marketing, and sales.

5.3 CONSIDERATION OF COMPONENTS FOR ART/ADT AND COMBINED (INTEGRATED) EQUIPMENT TESTING

5.3.1 General Consideration

ART/ADT equipment consists of components—groups of equipment. The basic components include

- Equipment for mechanical tests
  - Vibration testing
  - Dynamometer brake testing
- Equipment for multi-environmental testing and its components
  - Combined equipment for multi-environmental tests
  - Equipment for corrosion testing
  - Wind tunnels
  - Solar test chambers
  - Dust and sand test chambers
  - Ozone test chambers
  - Accelerated weathering test chambers
- Equipment for electrical tests

The number and type of components needed to conduct accelerated reliability (durability) testing depend upon the test subject (product).

In the mid-1960s, industrial companies, as well as research centers such as university research centers, began to provide the direction for future ART using a simultaneous combination of the types of appropriate testing equipment.

The development of ART/ADT was a systematic process. It began as a combination (integration) of two processes, then three, then four, and more as needed from the above-mentioned combinations.

Russia was the first to do this, and then it spread to Western countries. In Russia, it began with farm machinery in the Kalinin Governmental Test Center. To provide ART, the equipment was integrated for the simultaneous simulation of the road (field) conditions and other equipment, including the use of artificial materials to accurately simulate farmers' products (grain, corn, fertilizer, manure, tillage, and sand). This was the first such experience reported in

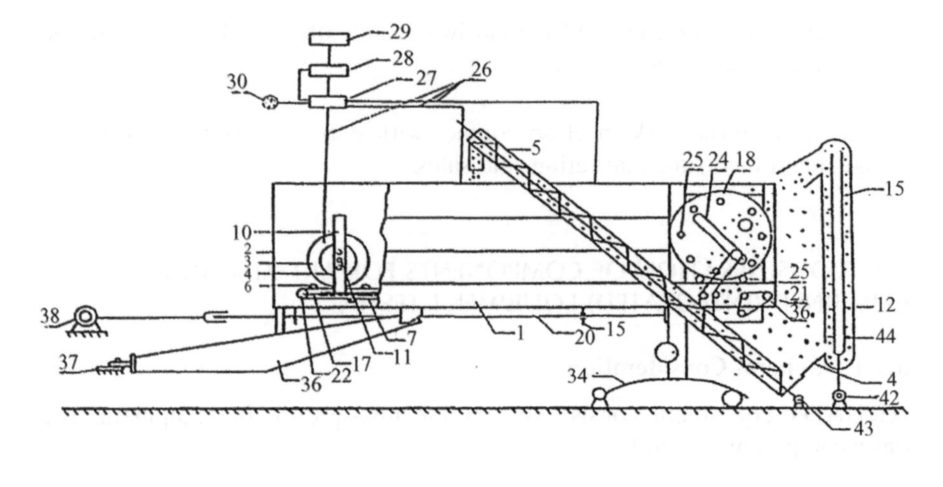

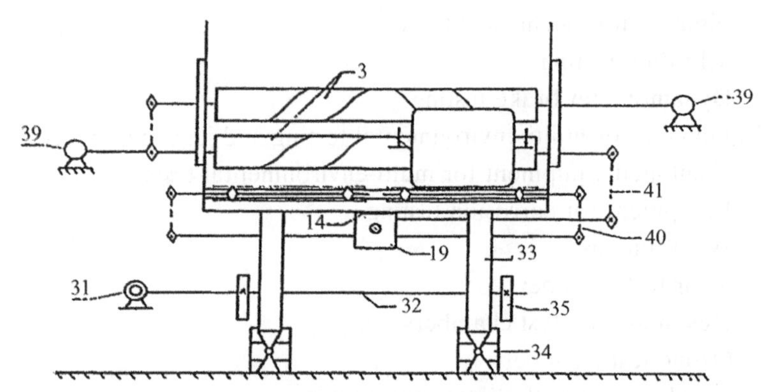

Figure 5.1. Scheme of the equipment for accelerated testing of the manure applicator's distributors [42]: 1, frame, 2, body, 3, spreading units, 4, bunker, 5, downhill elevator, 6, plank, 7, elevator of test subject, 8, wheel, 9, pneumatic tire, 10, stilt, 11, axle, 12, scraping elevator, 13, scraper, 14 to 18 are units of fastening, 19, reducing gear, 20, main driver shaft, 21, 22 shaft of test subject elevators , 23, roll, 24, groove, 25, stops, 26, pipeline, 27, control valve, 28 pressure valve, 29, compressor, 30, manometer of pressure, 31, 38, 42, and 43 are electric motors, 32, shafts, 33, prop, 34, buffer, 35, eccentric device. 36, beam, 37, bearing, 39, loading device, 40, 41, chain transmission, 44, substitution material.

the world. As an example, two types of the combined test equipment (which are still in use) are shown in Figures 5.1 and 5.2.

The equipment was developed in the 1990s. Figures 5.3 and 5.4 show two components of the above-mentioned equipment used for testing.

As an example of the quality of testing equipment simulation for field loading (Fig. 5.2), one can see that the normalized correlation functions of the upper shaft screw unit loading in the laboratory are similar to those for the field (Fig. 5.5).

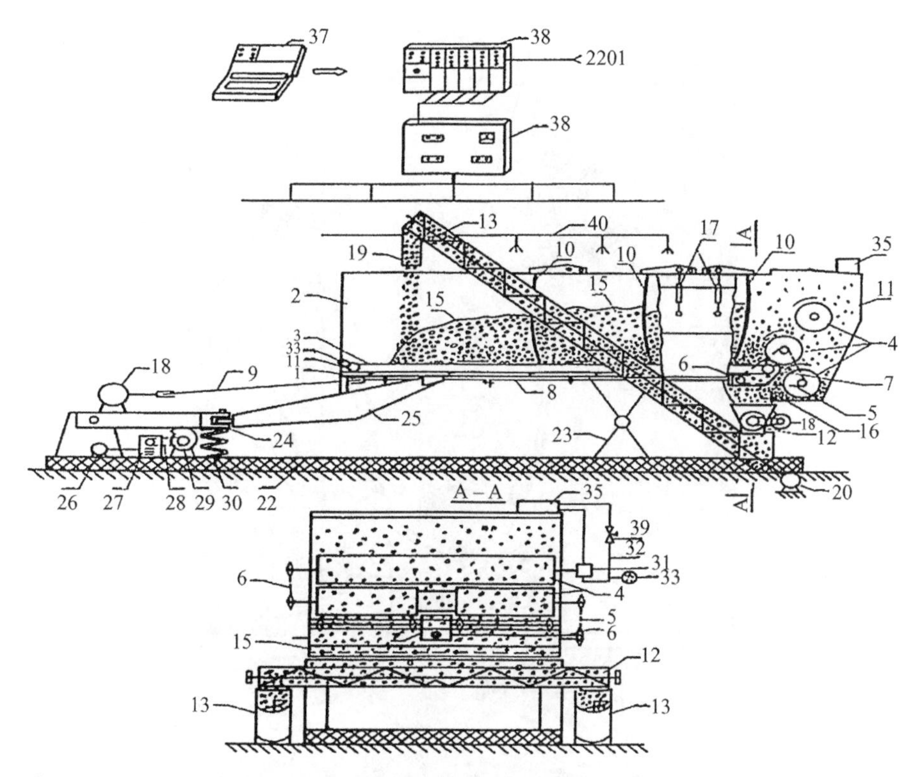

Figure 5.2. The scheme of equipment for testing the working heads and drives of manure applicators [42]: 1, frame; 2, body; 3, elevators; 4, spreader units; 5, chain drivers; 6, driver shaft of spreading units; 7, reducing gear; 8, common driver shaft; 9, cardan shaft; 10, barriers; 11, bunker; 12, horizontal screw; 13, inclined screw; 14, direct trough; 15, substitution material; 16, oven door; 17, regulated screw; 18, electric motor; 19, electric motor; 20, electric motor; 21, drive shafts of the elevator; 22, support plate; 23, hinge supports; 24, support; 25, beam; 26, electric motor; 27, reducing gear; 28, drive chain; 29, eccentric; 30, buffer; 31, hydraulic pinion pump; 32, hydraulic drives; 33, manometer; 34, throttle; 35, oil box; 36, programmer microprocessor controller; 37, apparatus for programming and adjustment; 38, control cupboard; 39, alarm sensors; 40, system of moisturizing.

The simulation of manure, grain, and other materials provided the correct substitute media for this special methodology.

5.3.2 How to Change a Natural Medium to an Artificial Medium

The equipment often contacts different media during operation. Most farm machinery contacts technological (natural) media: grain, glass, corn, and so on. This includes off-highway machinery, trucks, and many other types of equipment.

Figure 5.3. System of additional loading (11, body; 12, horizontal screw; 31, hydraulic pinion pump; 32, hydraulic drives; 33, manometer; 34, throttle).

Figure 5.4. The sensors of emergency signalization.

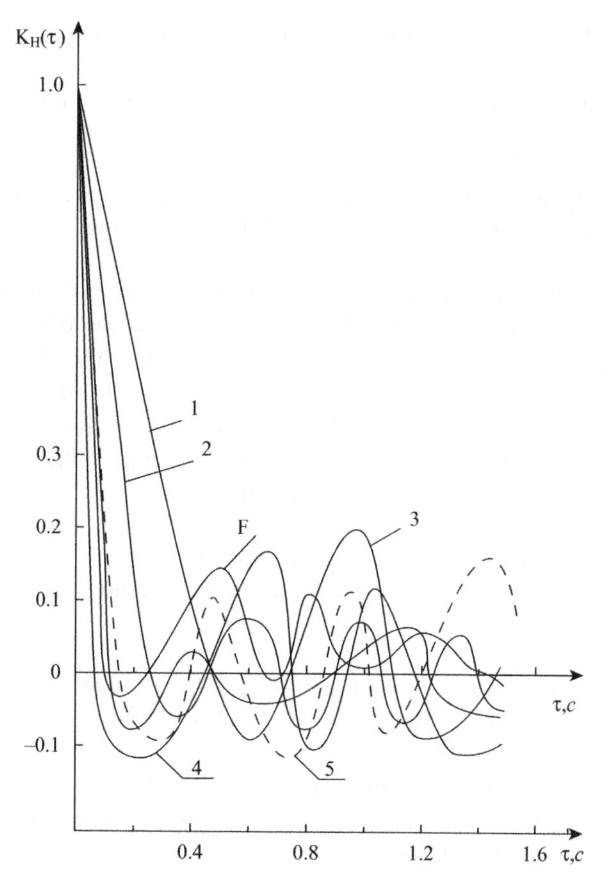

Figure 5.5. Normalized correlation functions of upper shaft screw unit loading processes: F—field; 2, 3, 4, and 5 are numbers of the various regimen in the laboratory.

One needs to simulate these contacts with natural media during ART/ADT because they influence the reliability and durability of the equipment. However, one cannot use identical measures for natural media (fertilizer, manure, and grain) because all natural media change their physical/chemical and mechanical qualities rapidly.

One problem that arises is how to substitute artificial media for natural media for ART/ADT. For example, how does the corrosive-abrasive influence of artificial media compare with natural media? For several years, the author and his colleagues studied this problem and developed principles that led to a satisfactory solution.

The author based his solution on the acceptance of the structure and composition of the artificial media that simulate the wearability mechanism with enough accuracy. This, in turn, sufficiently and accurately reflects the wearability mechanism of the machinery's work when in contact with natural technological media.

The criterion of the similarity of an artificial medium to a natural medium is based on the wearability that developed. For capacity, one can use the similarity of the rows of the materials' wearability.

After testing in the above-mentioned media, the basic steps of establishing equality between these two media are [147] the following:

- Testing samples of materials that consist of the product components (steel or plastic) and that come into contact with the natural medium
- Obtaining a series of measurements of the amount of wear of these samples resulting from the natural medium and various artificial media
- Establishing a correlation of this wear (or fatigue) between the natural and artificial media to determine if a particular artificial medium can effectively replace the natural medium

Consider the experimental research conducted in the search for natural media substitutes for a manure applicator. The methodological process includes five basic steps [147]:

1. Manufacture the test specimens from different steels (determinate the sample size and configuration) to study them in both a technological medium and in artificial media.
2. Study the effects of various natural media and the selection of an artificial medium and then identify identical results for physical-mechanical and chemical properties.
3. Determine the conditions and regimens for testing metallic specimens from different types of steel in natural and artificial media. The regimens of testing have to be similar for both. For this purpose, analyze the regimens of work for different components of machines (speed, surface of friction, distribution of the time between work processes, and breaks).
4. Provide testing of metallic samples from different types of steel in the natural and artificial media, obtaining wear rows for artificial and natural media
5. Analyze the wear rows of materials resulting from specimen testing in the above-mentioned media

An example using manure illustrates the above-mentioned process. To achieve the substitution process (not only for manure substitution), one can use the following device in Figure 5.6. The physical-mechanical quality of studied manure is included in Table 5.1.

This research determined the friction coefficients provided in Table 5.2. Table 5.3 includes the value of wear for each medium after 375 hours of testing. The measure for the value of wear is the loss of mass. The weight of each

Figure 5.6. Device for testing the wearability of materials [147]: 1, frame; 2, motor; 3, transmission.

TABLE 5.1 Physical-Mechanical Quality and Chemical Composition of Studied Manure

Dampness (%)	Dry Component (%)	Organic Component (%)	Strawness (%)	Abrasive Content (%)	Content in % to Gray Matter		
					N	P_2O_5	K_2O
65.00	35.00	31.90	6.10	1.82	0.45	0.05	0.78

sample of steel was checked after every 75 hours of testing. Table 5.2 shows that artificial medium 4 (quartz sand, 60.44%; turf, 21.36%; and water, 18.2%) is similar to the natural manure across all steel samples. The corrosive-mechanical wear of medium 7 is very different from the natural medium (manure), therefore it was decided that this medium would be removed from the next experiments. The following lines of corrosive-mechanical wear (Table 5.3) were determined using the results obtained from the above-mentioned media. In these lines, the values of corrosive-mechanical wear changed for the

TABLE 5.2 Friction Coefficient of Different Media by Steel [43]

Number	Media Composition	Friction Coefficient
1	Manure (moisture $W = 65\%$)	0.98
2	Manure (moisture $W = 70\%$)	0.90
3	Manure (moisture $W = 75\%$)	0.38
4	Quartz sand (dry) with sizes of samples from 0.1 to 0.4 mm	0.60
5	Quartz sand (moisture $W = 5\%$)	0.80
6	Quartz sand (moisture $W = 10\%$)	0.62
7	Quartz sand (moisture $W = 15\%$)	0.86
8	Quartz sand—60.44%, turf—21.36%, water—18.2%	0.92
9	Polymer granules (dry)	0.50
10	Polymer granules (wet)	0.44
11	Polymer granules (wet) and quartz sand (6%)	0.47

TABLE 5.3 Total Corrosive-Mechanical Wear of Steels by Manure and Artificial Media (g/m^2) [43]

Media	S1020	5140	T5140*	1045	E1045*	1070	E1070*	W108	LW108*
Manure	125.6	136.5	105.8	143.9	115.9	138.0	165.4	122.0	130.1
Artificial medium 1	133.3	153.7	142.3	144.0	176.1	137.3	167.7	127.1	170.1
Artificial medium 2	279.7	280.8	253.6	275.0	313.6	307.3	323.7	280.9	361.0
Artificial medium 3	139.4	100.9	112.3	101.2	150.2	110.3	138.1	86.9	153.1
Artificial medium 4	135.3	145.0	110.1	145.5	122.4	142.8	162.8	129.0	139.8
Artificial medium 5	253.4	260.6	188.8	285.4	183.4	389.5	358.6	281.0	236.7
Artificial medium 6	101.8	98.8	91.5	102.7	87.5	89.5	88.5	90.9	75.9
Artificial medium 7*	2.0	15.4	0.6	22.3	1.4	0.9	4.4	1.2	7.2

*Shows that the steel is heat treated.

previously mentioned types of steel. Here ξ_w, ξ_1, ξ_2, ξ_3, ξ_4, ξ_5, and ξ_6 are rows of corrosive-mechanical wear of steels that were obtained from their testing in a natural medium and six compositions of artificial media. The analysis of the lines of corrosive-mechanical wear showed that the mechanism of the above-mentioned wear of the studied steels differs for different media.

Evaluate the level of divergence of the lines of corrosive-mechanical wear with the formula

$$R_i = \frac{n_{mn}}{m_{um}}, \tag{5.1}$$

where n_{mn} is the number of materials in line ξ_i that must be displaced to obtain an analog of another line (ξ_i); n_{um} is the total number of tested materials.

If the lines are fully equal, $R_i = 0$, but they are fully divergent when $R_i = 1$. Therefore, R_i is a measure of the divergence of the corrosive-mechanical wear of steels after testing in artificial media compared to the level of steel wear after testing in a natural medium. In other words, how different is the mechanism of corrosive-mechanical wear in artificial media is from that in natural media. Table 5.4 shows the coefficients of lines of $\xi_1, \xi_2 \ldots \xi_6$ to line ξ_w. One can see from Table 5.5 that the mechanism of corrosive-mechanical wear in manure is closest to that of the fourth artificial medium (0.22).

In Table 5.6, one can see that the results of the abrasive indices show the relative capability of medium wear. The abrasive index R_a is

TABLE 5.4 Lines of Wearability of the Studied Steels [43]

ξ_w	5140*	1045*	W108	1020	W108*	5140	1070	1045	1070*
ξ_1	W108	1020	1070	5140*	1045	5140	1045*	1070*	W108*
ξ_2	5140*	1045	1020	5140	W108	1070	1045*	1070*	W108*
ξ_3	W108	5140	1045	1070	5140*	1070*	1020	1045*	W108*
ξ_4	5140*	1045*	W108	1020	W108*	1070	5140	1045	1070*
ξ_5	1045*	5140*	W108*	1020	5140	W108	1045	1070*	1070
ξ_6	W108*	1045*	1070*	1070	W108	5140*	5140	1020	1045

TABLE 5.5 Values of Inverse Coefficients (R_I) in the Ratio of Materials' Wearability $\xi_1, \xi_2 \ldots \xi_6$, and ξ_w

Compared Lines	ξ_1 and ξ_w	ξ_2 and ξ_w	ξ_3 and ξ_w	ξ_4 and ξ_w	ξ_5 and ξ_w	ξ_6 and ξ_w
Coefficient of inversion	0.89	0.89	1.00	0.22	0.89	0.89

TABLE 5.6 Indices of Abrasives R_a of Line ξ_i in Ratio to Line ξ_w

The Rows of Wearability	S1020	5140	T5140*	1045	E1045*	1070	E1070*	W108	LW108*
ξ_1–ξ_w	1.06	1.13	1.34	1.00	1.44	0.99	1.01	1.04	1.31
ξ_2–ξ_w	2.23	2.06	2.40	1.91	2.71	2.23	1.96	2.30	2.77
ξ_3–ξ_w	1.11	0.74	1.06	0.70	1.30	0.80	0.83	0.71	1.18
ξ_4–ξ_w	1.08	1.06	1.04	1.01	1.06	1.03	0.98	1.06	1.07
ξ_5–ξ_w	2.02	1.91	1.78	1.98	1.58	2.82	2.17	2.30	1.82
ξ_6–ξ_w	0.81	0.72	0.86	0.71	0.75	0.65	0.54	0.75	0.58

TABLE 5.7 The Spread in Values of Wearability

Line	W	1	2	3	4	5	6
Characteristic of line	1.56	1.34	1.42	1.76	1.48	1.96	1.35

$$R_a = \frac{W_{ji}}{W_{j1}},$$

where W_{ji} is the absolute value of the above-mentioned wear for the jth material and the first line.

An analysis of the characteristic line values shows the spread in values of corrosive-mechanical wear for the materials for all examined rows of steel wear. It also shows that the closest characteristics are between manure and the fourth artificial medium (Table 5.7).

The conclusion for the similarity of the wear mechanisms of steels in different artificial media and manure came after an analysis of the corrosive-mechanical results for various steel wear in these media. The following functions were determined to relate the artificial media and manure (Table 5.8).

Analysis of the research results show that the value of corrosive-mechanical wear of steels in manure has linear time dependence. The data from the fourth medium also have linear dependence between time and the value of corrosive-mechanical wear in these steels (Fig. 5.7). For other media, this dependence is not as linear. Therefore, we can conclude that the fourth artificial medium is suitable to simulate manure for ART.

5.3.3 Combined Equipment Development

Development of the combined equipment took place after the 1970s, continually increasing the number of components. The author, along with his colleagues, obtained dozens of patents for this combined testing equipment. Each new combination of testing equipment for simulation field input influences, as well as for each test subject, required a new technical solution. This equipment simulated the real-world situation—an interconnected combination of interacting field input influences.

This work continued in Moscow at the All-Soviet Union Institute of Agricultural Mechanization (VIM). The results of this work are provided in Reference 42.

The same process was used for the automotive industry and for other industries.

To facilitate the development, the Russian government organized a special engineering center, TESTMASH, at AUTOELECTRONICA in Moscow. The next step was an organization called State Enterprise TESTMASH, which became the base for this engineering center. The engineering center and state enterprise developed combined testing equipment for the ART and ADT of

TABLE 5.8 Calculated Functions of Corrosive-Mechanical Wear of Steels by Manure and Artificial Media

Steel	Manure	Artificial Media				
		1	2	3	4	5
1020	1.4 + 0.34t	15.9 + 0.34t	0.96 + 0.9t	–0.9 + 0.38t	–7.2 + 0.37t	–6.6 + 0.7t
5140	10.7 + 0.34t	20.6 + 0.38t	10.0 + 0.75t	0.2 + 0.28t	–3.3 + 0.41t	–15.0 + 0.74t
5140*	4.8 + 0.28t	16.0 + 0.37t	–2.0 + 0.7t	2.4 + 0.3t	–1.4 + 0.31t	–9.0 + 0.5t
1045	6.6 + 0.37t	20.7 + 0.36t	11.4 + 0.74t	0.7 + 0.28t	–11.4 + 0.42t	0.97 + 0.96t
1045*	4.6 + 0.29t	27.6 + 0.43t	6.7 + 0.86t	0.91 + 0.69t	–4.2 + 0.34t	0.96 + 0.87t
1070	9.0 + 0.35t	17.6 + 0.35t	6.7 + 0.88t	–1.3 + 0.3t	–5.2 + 0.4t	–21.1 + 1.08t
1070*	17.7 + 0.4t	21.6 + 0.42t	1.01 + 0.89t	0.82 + 0.38t	–2.3 + 0.46t	–7.0 + 0.98t
W108	–0.5 + 0.32t	0.81 + 1.2t	0.98 + 0.88t	–0.6 + 0.24t	–2.5 + 0.35t	–11.0 + 0.78t
W108*	8.4 + 0.33t	29.0 + 0.43t	1.02 + 0.99t	0.89 + 0.8t	–3.3 + 0.39t	0.93 + 0.97t

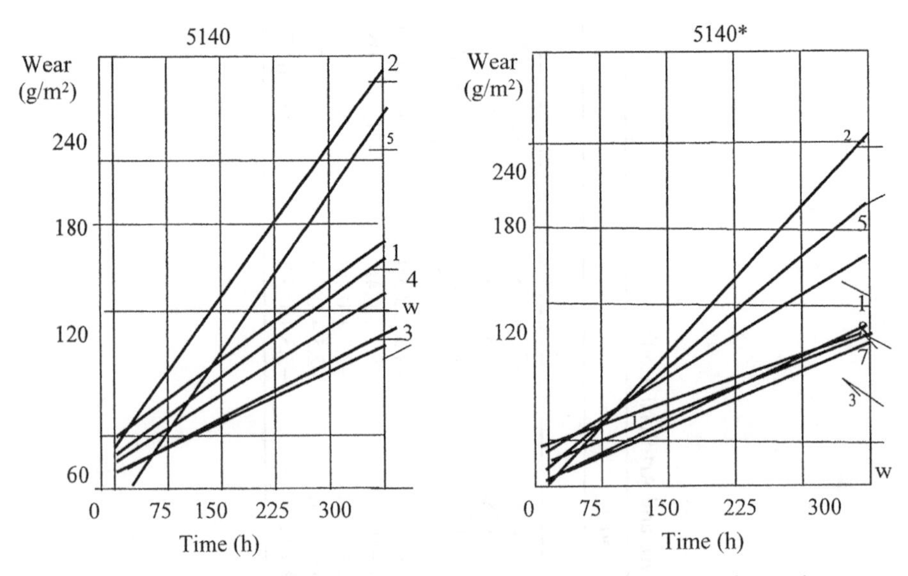

Figure 5.7. Dynamic wear of steels 5140 and 5140* by corrosive-mechanical wear in manure (W) and artificial media [43].

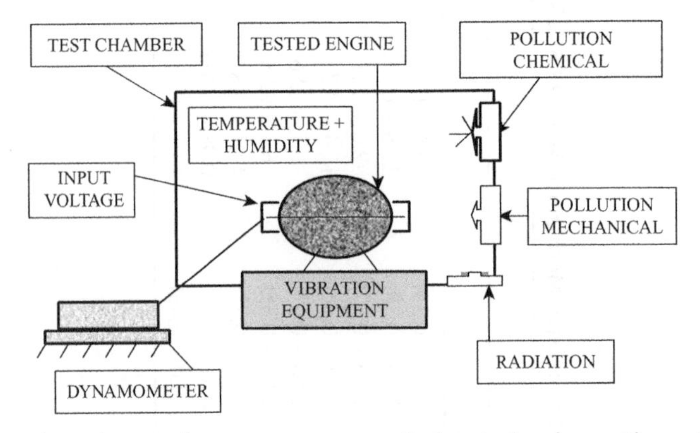

Figure 5.8. The scheme of a computer-controlled test chamber with equipment for engine (and complete equipment) accelerated reliability/durability testing.

cars, trucks, and heavy trucks. These universal test chambers provided the simulation of the environmental input influences (temperature, humidity, salt spray, and radiation) that worked simultaneously, as well as with mechanical, electrical, and other groups of influences. In the example in Figure 5.8, one can see the computer-controlled equipment for engine combination testing. There are sets of equipment for ART/ADT composed hardware and other items.

In Western countries, especially in the United States, combined test chambers began development in the 1980s–1990s. The electronic industry was the first to follow.

Figure 5.9. Combined test chamber with vibration and input voltage [102].

These included combined equipment [102] for physical simulation by the following:

- Test chamber (temperature and humidity)
- Vibration (electrodynamic) equipment inside the test chamber
- A device for the simulation of input voltage

Figures 5.9 and 5.10 show combined testing equipment for an electronic product.

When identifying a thermal chamber for environmental testing, key points to specify are [102] the following:

- Mechanical refrigeration versus liquid nitrogen (LN_2) cooling
- Operating temperature range
- Temperature rate of change

Most suppliers of combined environment thermal/vibration chambers use LN_2 as a standard method of cooling.

Weiss Umwelttechnik GmbH (Germany) is now the world leader in the design and manufacture of serial testing equipment for the combined (integrated) simulation of a part of the environmental field influences, as well as the mechanical influences, including vibration.

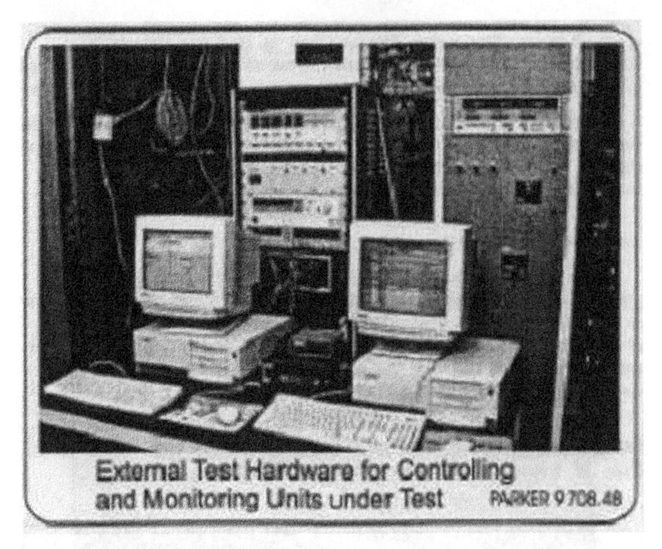

Figure 5.10. Test hardware for controlling and monitoring units under the test chamber, which is shown in Figure 5.9 [102].

Several types of the combined test equipment that Weiss Umwelttechnik GmbH produces are listed next.

This company is proposing advanced testing equipment that is close to the integrated equipment required for ART and ADT. For example,

1. *Climate Test Chamber with Road Simulator.* This system combines humidity, heat, cold, solar radiation, and vibration.
2. *Combined Corrosion Test System.* This system is composed of two operating reliability test beds and enables the combination of the following test parameters: temperature, humidity, and corrosion with NaCl, CaCl, and MgCl solutions while simultaneously vibrating along three axes.
3. *Global Ultraviolet (UV) Testing System.* The simultaneous effects of UV radiation, temperature, humidity, (acid) rain, and their diurnal variations are decisive factors for the premature aging of many types of materials.

Weiss Umwelttechnik GmbH's equipment (Figs. 5.11–5.17) [148] is used for combined testing in Europe, America, and Asia.

Specification (Fig. 5.11)

Test space volume: 500 m^3

Temperature range: −20 . . . +50°C

Climate range: +10 . . . +35°C

Humidity range: 5.5 . . . 15.5 gH_2O/kg dry air, max 75% relative

Figure 5.11. Climate test chamber on rolling roads (MTS Systems Corporation) for the whole product (Windshear Inc.).

Figure 5.12. Cold-heat climate test chamber with road simulator and sunlight simulator [148].

Figure 5.13. Cold-heat climate test chamber with multiaxis vibration table [148].

Figure 5.14. Combined testing system: vibration, climate, and corrosion [148].

Specification (Fig. 5.12)

Test space volume: 480 m^3
Temperature range: –40 . . . +90°C
Climate range: +10 . . . +80°C
Dew point temperature range: +5 . . . +60°C

Figure 5.15. Climate test chamber combined with dynamometer [148].

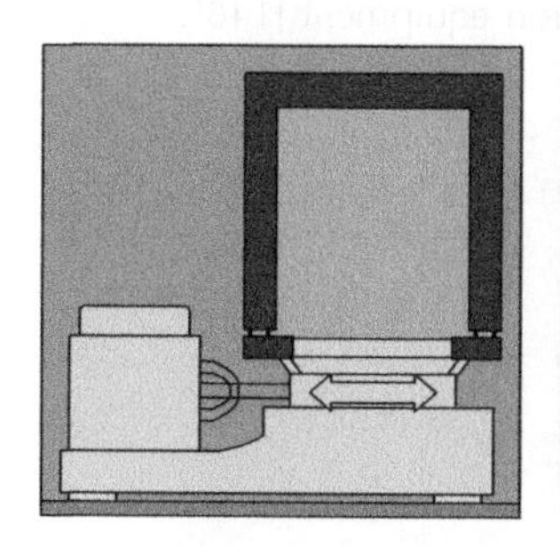
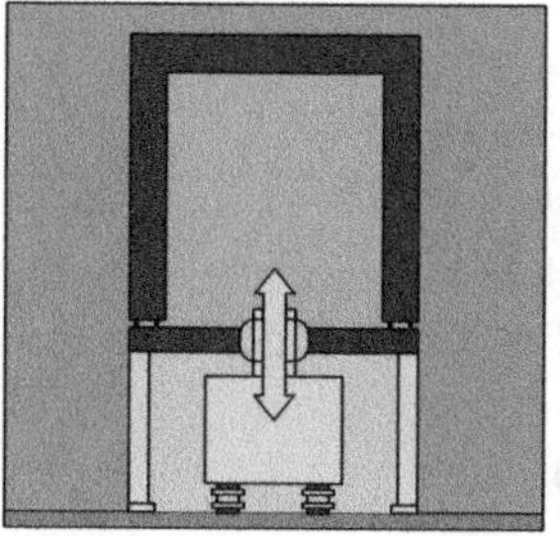
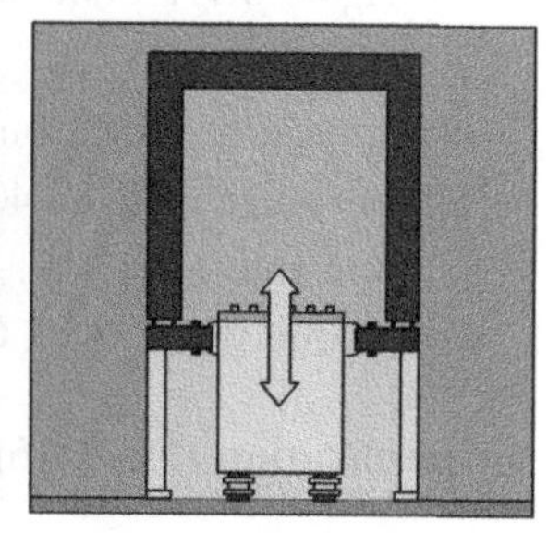

Figure 5.16. Vibration test chamber with different configuration [148].

Specification (Fig. 5.13)

Test space volume: 14 m^3
Vibration: in three linear axes
Temperature range: –30 . . . +90°C
Climate range: +20 . . . +60°C
Dew point temperature range: +10 . . . +55°C

Specification of One More Unit

Test space volume: 13.6 m^3
Vibration: in three linear axes

Figure 5.17. Combined test chamber with vibration equipment [148].

Amplitude of vibration: max ± mm
Temperature range: –40 . . . +130°C
Climate range: +13 . . . +90°C
Corrosion: with NaCl, CaCl, and MgCl solutions

Specification (Fig. 5.15)

Test space volume: 579 m^3
Temperature range: –20 . . . +35°C
Climate range: +10 . . . +35°C
Humidity range: 5.5 . . . 12 gH_2O/kg dry air

Weiss Umwelttechnik GmbH designs and manufactures combined vibration test chambers with different configurations (Fig. 5.16).

Other world-class companies that produce test equipment are now using the same process described in Chapter 4 of this book, particularly in the section "ART/ADT Methodology as a Combination of Different Types of Testing." Presently, they are not producing equipment for accelerated reliability (durability) testing, but trends in their development show that it would be possible for them to do it in the future.

A good example of this situation is ESPEC Corporation. Approximately 10 years ago, this company produced mostly temperature/humidity chambers.

Now they state in their brochure "Testing more than temperature and humidity." To provide real-world simulation, ESPEC designed an insulated floor system that allows a road vibration simulator to integrate with a drive-in chamber. This floor moves to allow the simulator to adjust for vehicles with different wheelbases, from subcompact cars to extended-cab pickups.

In another instance, ESPEC created a drive-in chamber out of special fiberglass for corrosion testing with road salt. The system includes an undercarriage with salt spray capability and an easy-maintenance brine tank made by ESPEC.

Chamber Integration with

- Single and multiaxis vibration systems
- Road vibration simulator
- Salt or rain spray
- Dynamometers
- Emission test stands
- Test-buck fixtures
- Light simulation (infrared [IR], UV, and visible)
- Video recording (for airbag testing)
- Measurement and data acquisition systems

Other advanced companies are using the above-mentioned process to develop complex equipment for accelerated reliability (durability) testing. For example, the Seoul Industry Engineering Co., Ltd. R&D Center [149] demonstrated a bus climatic wind tunnel (Fig. 5.18).

Brief Features

- Bus, truck, and commercial vehicle test
- 4WD vehicle test
- Full-spectrum solar simulation
- Sunlight area and adjustable angle
- Frontal nozzle area (4.8–10.5 cm^2)
- Idle state simulation bypass airflow
- Throttle actuator
- Tire burst checking system
- Exhaust gas pressure control system
- Roller heating steam injection system
- Driver's aid system
- Vehicle climatic wind tunnel for bus, truck, and commercial vehicle
- Air conditioner cool-down test
- Heater warm-up test
- Cold and hot start test

Figure 5.18. Bus climatic wind tunnel [149].

Specifications

Temperature control range: –40 → 60°C
Temperature drop time: 20 → –30°C (in 3 hours)
Humidity control range: 10 → 90% relative humidity (RH) (at 10 m–60°C)
Wind speed control range: 0 → 160 km/h (0 → 100 mph)
Wind flow uniformity: 3.0 at 80 nozzle area
Solar light control range: 0 → 1400 W/m^2 (0 → 1200 kcal/m^2·h)
Chassis dynamometer speed control range: 0 → 200 km/h (0 → 124 mph)
Chassis dynamometer power control range: 0 → 373 kW
Chassis dynamometer roller size: 1600 mm (diameter)
Makeup air: provided at –40°C dew point
Test vehicle max weight: 20,000 kg
Vibration available

Figure 5.19. Unholtz-Dickie Corporation vibration test equipment T2000 series [150].

The Unholtz-Dickie Corporation uses combined testing equipment [150]. One can see an example of the combined (integrated) equipment for this company in Figure 5.19. However, this company needs more time to develop equipment suitable for ART/ADT. Specifications for this equipment are in Table 5.9.

Other companies are also working in this direction.

The information in this section demonstrates that there is a real and practical way to use this equipment to provide actual accelerated reliability (durability) testing technology in the near future.

The Thermotron Company also produces combined specific equipment for different applications [151]. Figure 5.20 displays a temperature/humidity test chamber combined with electrodynamic vibration equipment in one axis. This company is also producing equipment for accelerated stress testing that uses extreme temperature and repetitive shock vibration, mechanical or liquid nitrogen cooled options, and high-volume airflow. Thermotron will need

TABLE 5.9 T2000 3-in. System Performance and Configuration [150]

System Model	SA 1180-T2000-44-3	2XSA1120-SA T2000-64-3	2XSA 1180-T2000-44-3	2XSA 1240-T2000-44-3
Sine Force (pk)	20,000 lbf (89 kN)	25,000 lbf (111 kN)	25,000 lbf (111 kN)	25,000 lbf (111 kN)
Random Force (rms)	20,000 lbf (89 kN)	23,000 lbf (102 kN)	23,000 lbf (102 kN)	23,000 lbf (102 kN)
Half Sine Shock[6] 4 ms	50 g: 980 lb 100 g: 420 lb 250 g: 60 lb	50 g: 1200 lb 100 g: 580 lb 300 g: 70 lb	200 g: 240 lb 300 g: 150 lb 450 g:10 lb	200 g: 400 lb 300 g: 200 lb 500 g: 50 lb
versus Payload lb (kg) 11 ms	60 g: 620 lb 100 g:130 lb 100 g: 500 lb	60 g: 800 lb 100 g: 450 lb 120 g: 160 lb	100 g: 700 lb 120 g: 300 lb	100 g: 800 lb 120 g: 400 lb
Usable frequency range		5 Hz to 3000 Hz		
Maximum Acceleration Sine (pk)	180 g (1766 m/s^2)	220 g (2158 m/s^2)	220 g (2158 m/s^2)	220 g (2158 m/s^2)
Random (rms) ISO 5.344	100 g (981 m/s^2)	170 g (1668 m/s^2)	180 g (1766 m/s^2)	180 g (1766 m/s^2)
Maximum Velocity (pk) Sine Sweep[3] Shock	70–80 in./s 143 in./s	80 in./s 180 in./s	80 in./s 235 in./s	80 in./s 280 in./s
Shaker Displacement (pk-pk)[8.15]	3.0 in. (76 mm)	3.0 in. (76 mm)	3.0 in. (76 mm)	3.0 in. (76 mm)

Armature weight 110 lb (50 kg).
Armature diameter 17.5-in. (445 mm) overall diameter, with 16-in. (406 mm) diameter, outer bolt circle.
Armature resonance (f_R) (typical) 2300 Hz.
Shaker body isolation standard <4 Hz with air mounts and damping tanks.

additional time to design and manufacture equipment capable of performing ART/ADT.

In the aerospace area, the Arnold Engineering Development Center (AEDC), Moog Inc., PFW Aerospace AG, and other advanced companies are working on the development of combined testing equipment. For example, AEDC supports a robust and versatile Test Technology Branch focused on three primary disciplines: modeling and simulation, instrumentation and diagnostics, and facility and testing technology. A team of engineers, scientists, and craft and support personnel provides expertise to develop, adapt, and apply complex computational models, test techniques, and others.

The turbine engine ground test complex at AEDC is responsible for propulsion testing in the Engine Test Facility test cells, which are used for develop-

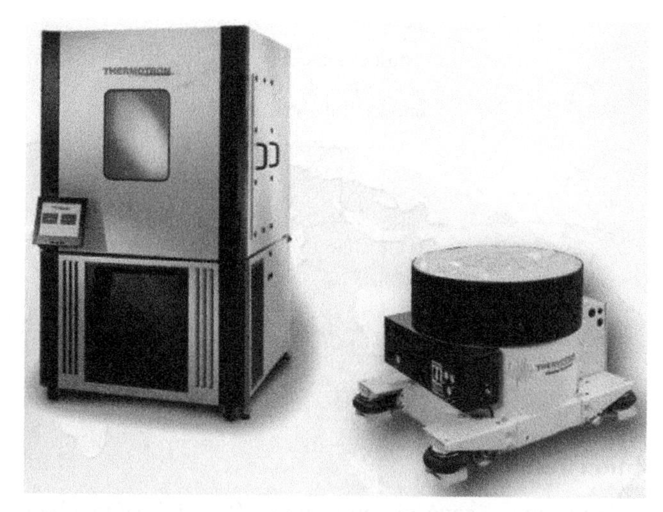

Figure 5.20. A high-performance Thermotron SE Series environmental test chamber and a Thermotron DSX Series Electrodynamic Vibration Test System [151].

ment and evaluation testing of turbine-based propulsion systems for advanced aircraft. AEDC operates eight active test cells for atmospheric inlet and altitude testing. Altitude Test Cells C-1 and C-2 comprise the Aero Propulsion Systems Test Facility, part of the Engine Test Facility. C-1 and C-2 are each 28 ft in diameter and approximately 45 ft in length. Either cell can provide engine inlet temperatures of up to 350°F and accommodate engines producing up to 100,000 lb of trust.

The Aerodynamic and Propulsion Test Unit is a blow-down test facility designed for testing the true temperature performance of propulsion, materials, structures, and aerodynamics of supersonic and hypersonic systems and hardware.

The National Full-Scale Aerodynamics Complex (NFAC) wind tunnel facility, located at NASA's Ames Research Center at Moffett Field in California, is composed of two large test sections and a common, six-fan drive system. A wide range of available support systems combine with this facility to allow the successful completion of aerodynamic experiments.

5.4 EQUIPMENT FOR MECHANICAL TESTING

There are many types of mechanical testing (Fig. 4.7). The most complicated types of mechanical testing are vibration testing and dynamometer brake testing (Fig. 5.21) due to their use for the testing of units and entire vehicles. There are other types of mechanical testing (Fig. 4.7), but their use is mostly for materials and parts. Let us describe the current testing capability and advanced testing equipment for both types of mechanical testing (vibration and dynamometer).

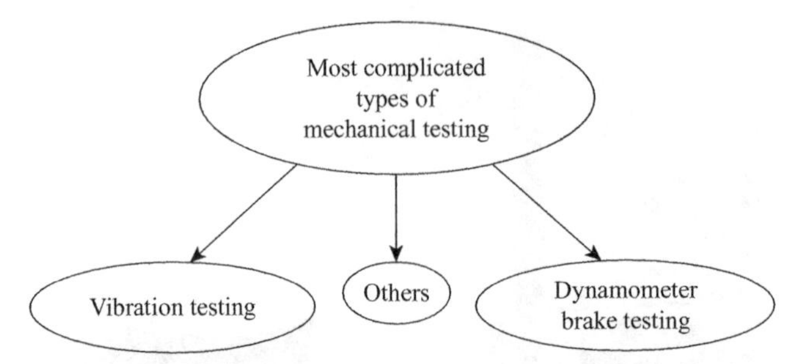

Figure 5.21. Most complicated types of mechanical testing of units and complete equipment.

5.4.1 Vibration Testing

Vibration testing is one of several components (Fig. 4.8) of mechanical testing. It is as one component in a combination of components for ART. Vibration testing may also occur as an independent type of accelerated testing. It depends on the goal of the testing.

One of the basic criteria for the use and development of the equipment as a component of ART is the accurate simulation of real-life input influences on the actual product.

5.4.1.1 The Current Situation. The vibration of a product is an output parameter. The input influences for mobile products that result in the vibration of this product consist of the road features (type of road, profile, density, hardness, and moisture), design and quality of the wheels, coupling of the wheels with the surface of the road, stability of the road surface due to weather changes, speed of the test subject's movements, speed and direction of the wind, and other influences. The types of road surface may be concrete, asphalt, cobblestone, gravel, dirt, or other materials. These influences have a random character.

They influence the stiffness, inertia, elastic, and damper qualities of the product. The result of this interconnection is the vibration of a mobile product on a particular surface under specified conditions and the product's design (Fig. 4.23). Therefore, the mobile product is a dynamic system that vibrates due to the interactions of these interconnections.

There are two general approaches for physical simulation of field vibration: proving grounds and laboratory testing equipment. Both evolved in the 20th century and continue development in the 21st century.

As stated in Reference 43, the basic negative aspects of proving grounds include the following:

- It is often too expensive for middle-sized and small companies to use proving grounds.

- The vibration process cannot be controlled in the same way that it is controlled in the laboratory.
- There is a dependence on the weather.
- It is not convenient for the separate divisions and departments of large companies.

The basic trends for the use and development of this testing include the following factors:

- More accurate simulation of the real-life vibration of the product
- Reduction of the simulation cost
- The possibility of combining vibration testing with electrical, environmental, and other types of testing to conduct ART/ADT

Initially, vibration testing developed with uncontrolled devices like simple shakers. Then, electrohydraulic, electrodynamic, and pneumatic vibration devices were developed. Multiaxial shakers have eliminated most of the negative aspects of previous laboratory test equipment and proving grounds.

Simple mechanical shakers, especially single-axis shakers, cost less than other varieties. However, they cannot be used successfully for ART/ADT because they cannot simulate the random character of field input influences, as well as real multiaxis vibration experienced in the field.

Therefore, in the 1950s, the companies designing vibration equipment began to produce servohydraulic, electrodynamic, and pneumatic equipment with control systems to solve this problem.

Some laboratory vibration testing equipment (VTE) has a negative aspect compared to proving ground testing. It provides a much lower quality of road simulation (Fig. 5.22). The use of servohydraulic, electrodynamic, and pneumatic equipment for vibration testing in the laboratory with block-programmed control began in the1950s. The accuracy of simulation was much lower (Fig. 5.23) than on the proving ground, but the laboratory simulation was much less expensive.

The companies that design servohydraulic, electrodynamic, and pneumatic equipment continue to improve their equipment. Each additional step of improvement (which includes mostly multiaxis vibration and a system of control) increases the cost of this equipment. However, the quality of road simulation improved slowly, because the principles of simulation were unchanged. Therefore, this approach has not been proven to be useful for users of this type of equipment. For many companies, the basic negative aspect (simulation of profile instead of road simulation) has not improved until recently. Presently, most companies producing vibration equipment continue to improve the profile simulation or other equipment designs instead of real road simulation. Nevertheless, advanced companies are making progress. Figure 5.24 shows the dynamics of improvement in vibration equipment by advanced companies. Before the 1980s, the total effectiveness (including technical and economic

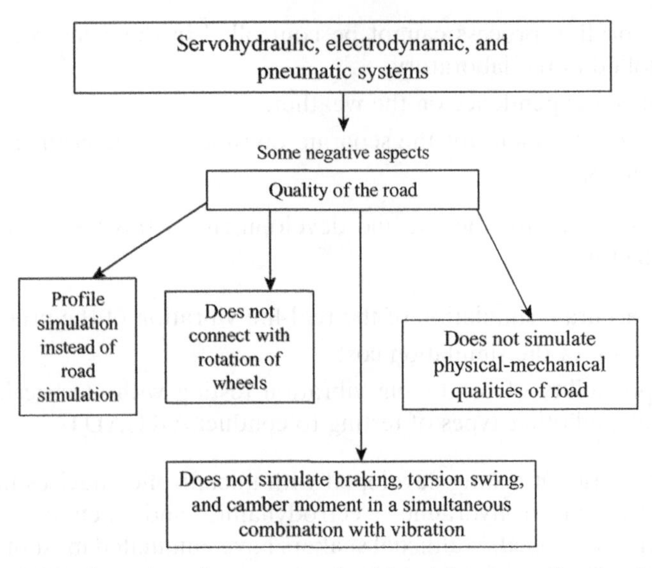

Figure 5.22. Some negative aspects of vehicle laboratory vibration testing.

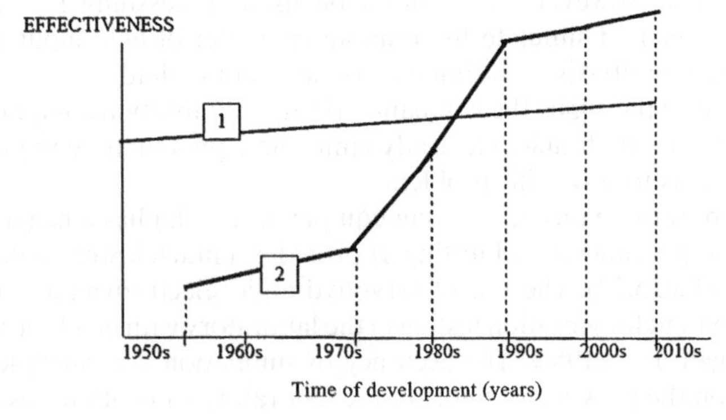

Figure 5.23. Development of vibration testing quality: 1– proving ground, 2—vibration.

parameters) of testing with this equipment was lower than proving ground testing. From the mid-1980s, this effectiveness increased. The speed of improvement in the effectiveness of the vibration equipment in the laboratory (Fig. 5.24) was highest in 1970–1990, when advanced companies created new designs for vibration equipment, especially with the working functioning unit. As an example, the components of the new design are shown in Figures 5.34–5.40. From the 1990s forward, the speed of improvement of this product is moving slowly, with the exception of improvements in the control systems.

Figure 5.24 shows the trends for advanced companies. The product effectiveness for other companies that are producing mechanical or pneumatic shakers in the vertical axis is currently worse than proving grounds. Therefore,

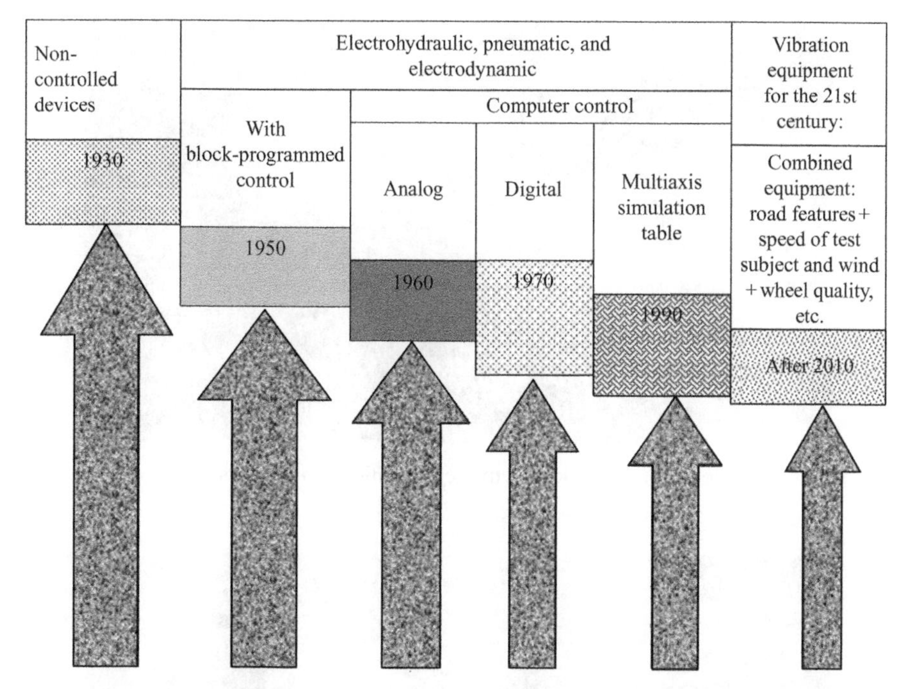

Figure 5.24. Stages of basic vibration testing equipment development.

achieving accurate simulation of real road vibration is still one of the basic problems that need development for vibration testing techniques.

There are two basic aspects of VTE in the technical area. The first is the type of vibration, that is electrodynamic, electrohydraulic, pneumatic, or mechanical shakers. The second involves the system of control: block-programmed or different types of computer controlled (analog, digital).

Vibration equipment has applications for different purposes. For example, in aviation, the certification process requires structural and dynamic testing as part of the process. This includes vibration testing. Figures 5.25 and 5.26 demonstrate pneumatic vibration testing where the vibration for mobile machinery is multiaxial.

In another example, the 190 type aircraft is supported by a Fabreeka soft support system (SSS) [152] when undergoing ground vibration testing (GVT). The 190 type aircraft weighs approximately 111,000 lb (55.5 tons). The Fabreeka SSS includes jacking units to raise the aircraft off the landing gear.

The vibration equipment with block-programmed control was designed and manufactured 60 years ago, but we still see this equipment, especially mechanical shakers, in the market today offered by both industrial companies and research centers.

One knows that mobile products vibrate in the field with 6 degrees of freedom and have a random character. Therefore, for accurate simulation of

Figure 5.25. Vibration in the test certification process [152].

Figure 5.26. The 190 aircraft and vibration equipment [152].

vibration, the testing equipment for laboratory testing has to be similar in character. In the current market, the electrodynamic shakers are the less advanced vibration equipment that mostly act along only one (vertical) axis and cannot accurately simulate the field situation. The positive side of electrodynamic shakers is the ability to work with high frequency (up to thousands of hertz). This type of vibration equipment began usage in combination with test chambers. This combination is not useful right now because the vibration along one (vertical) axis does not accurately simulate the field situation. Some industrial companies are producing electrodynamic vibration equipment that acts in 3 degrees of freedom (see Fig. 5.27, second series), but this is a very complicated design.

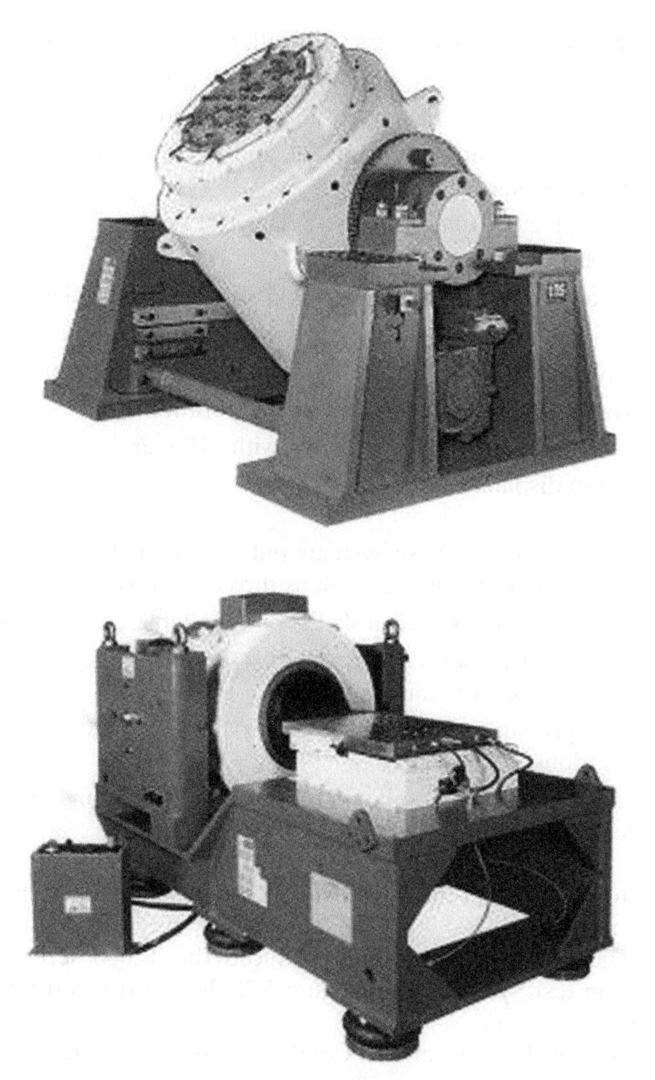

Figure 5.27. Example of LDS shakers for the vibration test produced by Brüel & Kjær—(first is High Force Range; second is Combos—up to e axes) [153].

One can see in Figure 5.27 electrodynamic vibration equipment LDS shakers for the vibration test produced by Brüel & Kjær [153]. Use these systems for the following:

A. Medium Force Range

V800–V8 Electrodynamic Shakers: This test system family offers

- 2000–14,800 lb
- High "g" with lightest high-performance armatures

- High cross-axial stiffness with roller suspension design
- Large displacement: ±1.5 in. on the V875LS
- Test loads up to 1540 lb with integral load support
- Ultraquiet running for "squeak and rattle" applications

B. High Force Range

V900–V9 Electrodynamic Shakers: This test system family offers

- 20,000–65,000 lb
- Patented resin bonded, carbon fiber armature design
- Leading performance for displacement and peak velocity
- Payloads of up to 11,000 lb of mass, for example, satellites
- Water cooled for quiet and clean operation
- High velocity for shock testing with 118 in./s
- Large displacement up to ±1.5 in.

Combos

Slip Table Systems: These systems put test items through their paces along two horizontal axes as well as through the vertical axis:

- Easy alignment between shaker and slip table, easy shaker rotation
- Hydrostatic bearing tables (HBT series) provide ultimate pitch, roll, and yaw restraint.
- Low-pressure tables (LPT series) provide more cost-effective solutions.
- Lin-E-Air™ body isolation for low distribution and low-frequency test

The first form shown in the next test system is High Force Range, and the second is Combos.

Unholtz-Dickie Corporation also produced the T2000 series electrodynamic vibration test Systems [154]. Figure 5.28 demonstrates the R-Series of this equipment.

Multiple System Configurations: Combining the R16 or R24 Shaker with SAI-Series Amplifiers from 30- to 90-kVA output

- 100% air cooled
- Choice of armature sizes—17.5"/25.75" (51/63/76 mm)
- Auto load support and centering (1500 lb [680 kg])
- SAI-Series Class D Amplifiers (with IGBT output circuitry)
- Three base configurations (low profile, pedestal/slip table)

Data Physics Corporation demonstrated at the Automotive Testing Expo 2010 that recently acquired the SignalForce Vibration Test System for higher force vibration. Water-cooled electrodynamic shakers producing 20,000 lbf (90 kN) to 50,000 lbf (222 kN) were mainly found in defense and military-

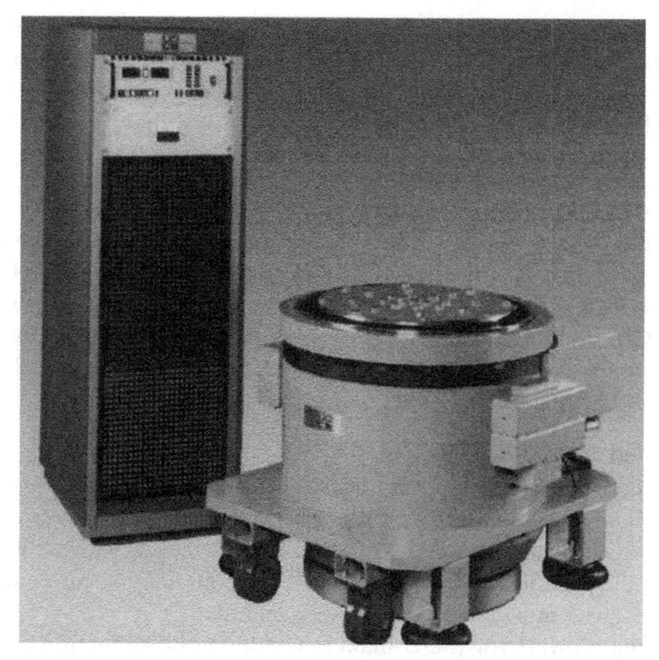

Figure 5.28. R-Series electrodynamic vibration equipment produced by Unholtz-Dickie Corporation [154].

related testing facilities but are used now in the automotive research and testing laboratories, due mainly to the large and heavy hybrid components and battery packs. SignalForce Vibration Test Systems (both air and water cooled) provide dynamic testing solutions including shaker tables, power amplifiers, slip tables, shaker head expanders and fixtures for environmental stress screening, modal excitation, and AGREE testers.

Design Groups of Electrodynamic Shakers. Data Physics Corporation provides a broad range of electrodynamic shaker systems for use with SignalStar vibration controllers. Four design groups are offered:

Inertial Shakers: Cover the force range –6.7 lbf (30 N) to 56.2 lbf (250 N)

Modal Shakers: Cover the force range –11 lbf (49 N) to 475 lbf (2.1 kN)

Air-Cooled Shakers: Cover the force range –2 lbf (9 N) to 12,000 lbf (53.4 kN)

Water-Cooled Shakers: Cover the force range –18,000 lbf (80 kN) to 50,000 lbf

Inertial Shakers

Testing of large structures can often present considerable difficulties, so when the payload cannot be mounted on the shaker, Data Physics can provide a shaker that can be mounted on the payload. Inertial shakers are fully enclosed, permanent magnet shakers that can be mounted onto structures at any angle—they are entirely self-supporting. Inertial shakers have found applications testing car chassis, building structures, ships' flight decks, helicopters, submarines, geophysical surveys, and vibration cancellation systems.

Modal Shakers

The modal shaker series exhibits almost zero axial stiffness while offering very high radial stiffness for stability. This performance combination is achieved through the use of a linear guidance bearing. There are six models of modal shakers with a total of 10 variants. Previous applications include the testing of airframes, space laboratory structures, automotive chassis, road surfacing materials, and artificial limb joints.

Air-Cooled Shakers

Air-cooled shakers are the workhorses for many component testing applications such as automotive assemblies and consumer electronics equipment. The Data Physics range includes many shaker models with axial guidance bearings and 2 in. (50 mm) displacement, and our largest air-cooled model offers 2.5 in. (63.5 mm) continuous displacement.

All cooling blowers for these shakers include noise reduction silencers; all models feature high levels of efficiency.

Water-Cooled Shakers

Water-cooled shakers cover the force range 18,000 lbf (80 kN) to 50,000 lbf (222 kN). These shakers all produce 2 in. (50 mm) displacement and all are water cooled.

All models may be supplied on high-quality pneumatic mounts, for vertical vibration isolation, in trunnions for horizontal and vertical operation, or in monobases for three-axis testing using a slip plate. PDF data sheets may be obtained by clicking on the required model.

All water-cooled shakers have axial guidance using dual hydrostatic bearings and pneumatic load support; they may also be supplied with automatic load support systems to ensure the armature remains in its central position when payload is added or removed. By using hydrostatic bearings, the SignalForce water-cooled shakers are able to handle large overturning moments without requiring external guidance in the head expander.

Lear Corporation demonstrated at the Automotive Testing Expo North America 2010 that recently acquired two SignalStar Matrix multishaker vibration controllers for use with Team Cube™ shaker systems to reproduce 6-degrees of freedom vehicle vibration environments. The Matrix multishaker controllers are used for closed-loop control of random and sinusoidal vibration in addition to replication of measured vibration time histories recorded in vehicles.

One other type of VTE is servohydraulic (electrohydraulic) vibration equipment. The advantage of this type of equipment is the possibility to simulate multi-axis vibration in three axes and, what is most important, in six

Figure 5.29. Scheme of vibration testing equipment, which simulates vibration in three linear axes (MTS Systems Corporation).

axes—three linear and three angular. This provides the most accurate simulation of the field vibration. The negative aspect of this type of vibration equipment is that it cannot provide vibration in high frequencies.

The modern systems of this type of equipment are multiaxis simulation table (MAST). MTS Systems Corporation, Carl Schenck, Instron Structural Testing Systems (IST), and other companies develop and produce this type of equipment.

Vibration input can be generated in up to 6 degrees of freedom: three linear—vertical, lateral, and longitudinal, and three angular—pitch, roll, and yaw.

Figure 5.29 demonstrates an arrangement of MAST that simulates vibration along three axes. MTS Systems Corporation is producing several types of MAST. The performance specification for some of these types is included in Table 5.10.

Notes

1. Performance figures are at the maximum payload with the center of gravity (CG) located at the centroid of the vertical actuator attachment points to the table.
2. Performance is based on a single degree of motion and will decrease if more degrees of motion are active.
3. Lateral motion has two actuators.
4. IST has a policy of continuous product development and all specifications are subject to change without notice.

IST has developed several standard design MAST tables that can provide vibration with 6 degrees of freedom: maximum angle—pitch, and angle—yaw

TABLE 5.10 Performance Specification of MAST Systems [192]

	Unit	MAST-9710	MAST 9720	MAST 9725	MAST 9730	MAST 9735
Payload used for calculation	kg	300	450	450	800	800
	lb	660	990	990	2200	2200
Table mass	kg	300	300	300	545	545
	lb	660	660	660	1200	1200
Table work—area dimensions	mm	1200 × 1200	1700 × 1500	1700 × 1500	2100 × 1800	2100 × 1800
	in	47 × 47	67 × 60	67 × 60	83 × 71	83 × 71
Peak acceleration—vertical	m/s^2	127	60	102	64	106
	g	13	6.1	10.3	6.5	10.8
Peak acceleration—lateral	m/s^2	100	48	77	50	80
	g	10.1	4.8	7.8	5.1	8.1
Peak acceleration—longitudinal	m/s^2	49	40	40	26	40
	g	5	4	4	2.6	4
Peak velocity—vertical	m/s	1.5	1.7	2.1	2	2
	in./s	59	66	82	78	78
Peak velocity—lateral	m/s	1.25	1.5	1.9	1.8	2
	in./s	48	59	74	70	78
Peak velocity—longitudinal	m/s	1	1.1	1.5	1.3	1.3
	in./s	39	43	59	51	51
Stroke—vertical*	mm	±75	±75	±75	±75	±75
	in.	±3	±3	±3	±3	±3
Stroke—vertical*	mm	N/A	±125	±125	±125	±125
	in.	N/A	±5	±5	±5	±5
Stroke—lateral*	mm	±75	±75	±75	±75	±75
	in.	±3	±3	±3	±3	±3
Stroke—longitudinal	mm	±75	±75	±75	±75	±75
	in.	±3	±3	±3	±3	±3
Operating frequency range	Hz	0–60	0–50	0–50	0–60	0–60
Table resonant frequency	Hz	250	250	250	285	285

*For most of our standard MASTs, an optional stroke extension to ±125 mm (±in.) is available.

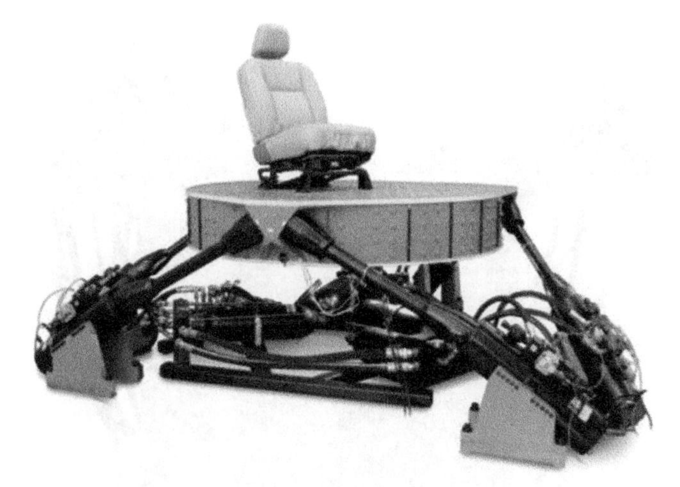

Figure 5.30. MTS Model 353.20 Multi-Axial Simulation Table (MAST) for vibration testing [155].

±8.5°, maximum angle—row ±6.8°, operating frequency range 0.1—50//60 Hz; stroke is ±16 mm.

MTS Systems Corporation [155–157] also works on the development of electrohydraulic vibration equipment. Currently, this company provides this type of equipment that works in 6 degrees of freedom (six axes). Let us show the MTS Model 353.20 Multi-Axial Simulation Table (MAST™), which offers precise, repeatable 6-degrees of freedom simulation of automotive vibration environments, at frequencies of up to 100 Hz (Fig. 5.30). It is the system for component vibration, squeak, and rattle testing in temperature-controlled or acoustics-controlled environments.

Figure 5.31 demonstrates the MTS system of vibration equipment that works in 6 degrees of freedom.

The MTS Model 836 multiaxial test system can utilize the MTS FlexTest GT Controller that provides direct digital control of all dynamic system functions. Featuring multichannel and multistation capabilities, this flexible controller can be expanded to accommodate up to four test stations, and it can be set up to allow test management and control with multiple PCs (Fig. 5.32).

Technical Specifications

Range of Motion (Nonsimultaneous)

$X = Y = Z = \pm 25$ mm

$R_x = \pm 15°$

$R_y = R_z = \pm 5°$

Force Capacity (Nonsimultaneous)

$F_x = 25$ kN

$F_y = 50$ kN

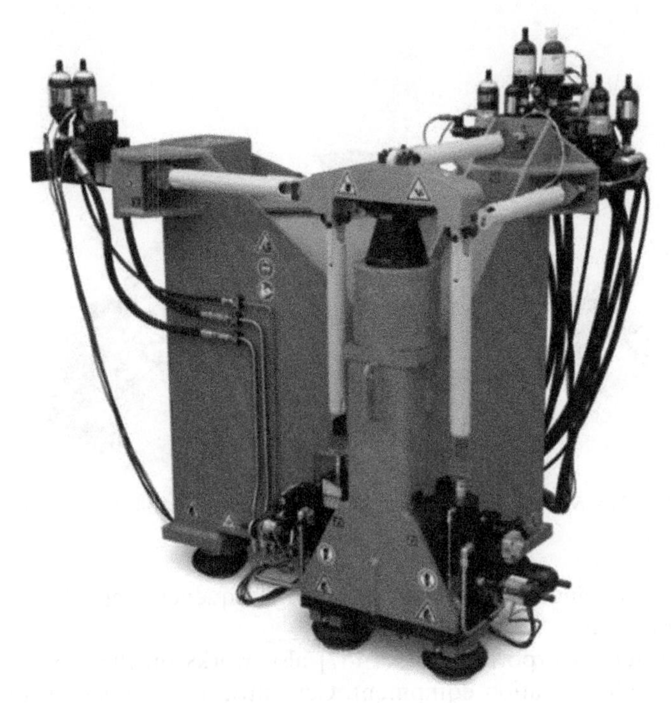

Figure 5.31. MTS Multiaxial Elastomer Test System—Model 836 [157].

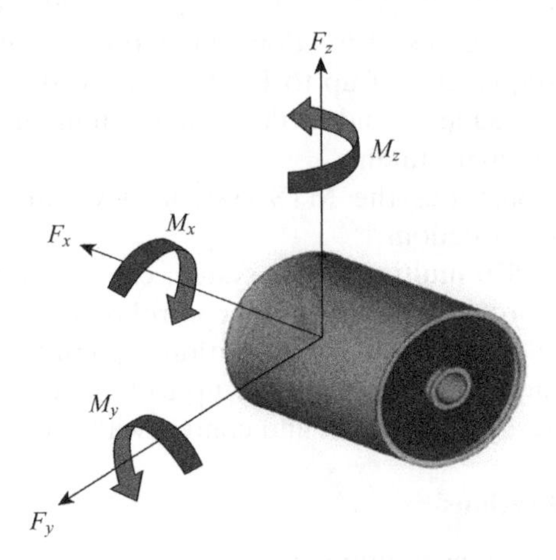

Figure 5.32. MTS System Corporation Typical Bushing Orientation [157].

$F_z = 50$ kN
$M_x = 2$ kN · m
$M_y =$ Mz $= 12$ kN · m

The current pneumatic vibration system does not show a big difference from the system used a dozen years ago [18].

One of the basic negative aspects of the present electrohydraulic and electrodynamic vibration testing (and pneumatic types of testing) is insufficient methodology for the real-life simulation of input influences in the laboratory.

Usually, this testing consists of three basic steps: a profile of the road and load (acceleration, tension, etc.) data of the test subject; signal processing and analysis; and laboratory test setup.

The basic negative aspects of current electrodynamic, electrohydraulic, and pneumatic vibration equipment are inferior simulation of

- Physical-mechanical quality of features of the roads (see Fig. 4.24)
- Smooth movement of the test subject wheels across any road obstacles
- Impact influences on the wheels and structures

5.4.1.2 Vibration Testing in the Space Industry. The space industry has specific requirements for vibration testing because one cannot easily repair damage to a deployed satellite. Testing the equipment is a major part of the development program, with multiple different parameters simultaneously tested.

Vibration testing validates the integrity of the manufactured system. As computers have become more powerful, it is now possible to simulate the effect of different frequencies that vibration or chocks will impose on a structure. However, this analysis runs on a structurally perfect model and cannot simulate the response caused by material weaknesses or manufacturing faults.

Vibration systems vary in force from air-cooled permanent magnet shakers delivering 9 N of sine force to water-cooled shakers running into hundreds of newtons. The space industry favors water-cooled shakers to maintain a clean environment where the satellites are tested. Water-cooled shakers typically range in force from 80 kN to around 300 kN (sine) [158]. Despite this high force, multiple-shaker test systems are becoming increasingly common as the size and weight of the payloads increases. Up to four shakers have been coupled for a common platform to mount the payloads for testing. They can then deliver forces greater than 1000 kN. The LDS water-cooled V900 Series shakers are the de facto standard shakers in the space industry.

It is important to consider testing the defined test profile, but it is more important not to damage the product by overtesting it. Such an error would lead to untold cost and time delays. The control element of the vibration test system is as important as the shakers themselves. Some controllers such as LDS Laser USB offer integrated control of the vibration test, power amplifier, and environmental chamber to maximize the test effectiveness.

Accurate data acquisition is an important element of vibration testing, particularly for satellites. Rather than needing miles of signal cables, it is advisable to place distributed data acquisition systems close to the sensors. Then how do you accurately synchronize these units if they are sampling hundreds of data points per second? In its genesis system, LDS Nicolet uses a special technique that automatically compensates for the time skew caused by different lead lengths in order to guarantee precise in-phase results in such an environment.

However, for field applications in hot, cold, wet, or dusty conditions and for vibration up to 100 g, the battery-operated Liberty would be a good solution. Both systems run the user-friendly perception software designed for maximum test productivity. It allows the user to display any selection of channels even from different mainframes within a single window on the PC screen. A huge time and cost saver is the off-line setup feature that allows the user to perform a test setup using virtual hardware while the real hardware performs actual tests on a different application.

5.4.1.3 Development of Advanced Equipment for Accelerated Vibration Testing. History shows that vibration equipment has been in use for 40–50 and more years.

As was shown in Figure 5.26, the greatest progress in the development of equipment for vibration testing was from the 1970s to the 1990s. During this time, the author and his colleagues also created a new piece of vibration technology equipment. This technology is based on theoretical and experimental research results on testing equipment, electronics, car trailers, trucks, farm machinery, and other items. The scientific-technical level of this equipment is higher than most of the current vibration equipment. The improved VTE has unique features such as working heads that provide assigned controlled random values for the test subject's vibration amplitude and frequency. The system of control consists of a computer with a peripheral network system and vibroacceleration and tension sensors. Vibration is applicable for both mobile and stationary products. It is applicable as universal vibration equipment as well as sensitive equipment for special products. This technology covers wheeled and stationary applications acting along up to one to three linear and one to three angular axes (from 1 to 6 degrees of freedom). The design of this VTE is simpler than many current electrohydraulic, pneumatic, and electrodynamic equipment systems. It also has no deterioration of performance, and it expands the functional applications to a wider range of markets. Table 5.11 presents the typical specifications for the new vibration equipment.

The basic advantages of this equipment in comparison with other equipment include the following:

- More accurate simulation of field vibration for the test subject
- Lower cost, because it is simpler
- It is applicable as a component of ART/ADT in combination with equipment for other mechanical types of testing (dynamometer brake), multi-environmental, electrical, and other factors
- This equipment is easy to mount for operation and easy to add to or combine with other machinery elements, such as a transmission, engine, and other equipment, as well as with a completely different type of industrial product, for example, cars and trucks.

TABLE 5.11 Example of Typical Specification for New Vibration Equipment

Names of Indexes	Test Subject	
	With Wheels	Without Wheels
Weight of test subject (ton)	Up to 30	From 0.0001 to 2.00
Weight of vibration equipment (ton)	38	From 0.001 to 4.00
Axles excited	Three linear	Three linear + three angular
Table sizes		From 20" × 20" to 100 × 100
Frequency of vibration of the tested product (Hz)	0.1–120.0	0.1–300.0
Amplitude of vibration of the tested product (mm)	From 0.1 to ±90 (wheel swing axis)	From 0.1 to ±90.0
Necessary power (kW)	120	From 1.6 to 105.0

The above-mentioned advanced type of equipment was developed for KAMAZ Inc. (Section 4.6), one of the largest Russian truck companies. Figure 5.33 (KAMAZ Inc., Engineering Center, Block No. 3) shows the plan of a room for vibration equipment in 3 degrees of freedom for a completed truck, as a first step for ART/ADT. This project took into account that this room is available for development in the next steps into a complex test chamber to provide ART/ADT of complete trucks.

This testing equipment is the first in the world to provide this capability.

Two types of vibration equipment for different test subjects are shown in Figures 5.34–5.40. Figures 5.34–5.38 show VTE with a car's trailer as a test subject. This is the first type of equipment for testing wheeled transport along three linear axes (degrees of freedom). The second type of VTE is for testing a different wheel less product along six axes, three linear and three angular (Figs. 5.39 and 5.40). Figures 5.36–5.38 demonstrate the working units of this equipment.

This equipment is used on car trailers, trucks, cars, farm machinery, off-highway machinery, defense equipment, and other equipment.

Figure 4.25 shows that the characteristics of random processes for advanced vibration equipment are also similar to these characteristics under field conditions.

The results of testing show that the developed advanced technology of VTE allows eliminating the basic negative aspects of the current vibration equipment.

The above-mentioned vibration equipment has the following advantages when compared with current equipment:

- Simulation
  - Different physical-mechanical qualities of roads
  - Torsion swing in the simultaneous combination of linear and angular vibration

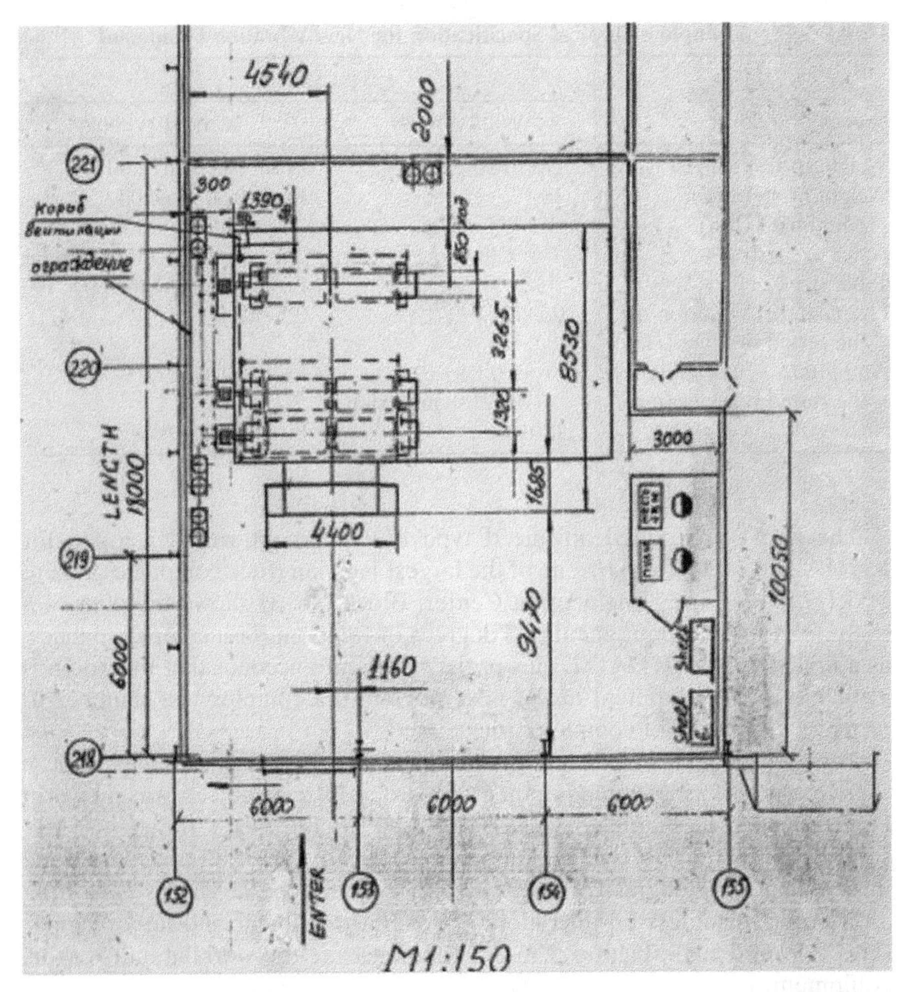

Figure 5.33. Plan of room for vibration equipment in 3 degrees of freedom for completed truck. KAMAZ Inc. (Russia), Engineering Center, Block No. 3.

- It can bridge most of the components of combined equipment for ART/ADT, the connection between vibration testing, dynamometer testing, corrosion testing, sun radiation testing, electrical and electronic testing, and other types of testing.

The second type of the advanced vibration equipment (Figs. 5.39 and 5.40) is simpler and less expensive. One may use it for many types of stationary and mobile products such as those that are mechanical, electrical, and electronic, and for products that are more complicated. One may use it for components like seats in the automotive industry, aircraft, aerospace, and others, as well as different whole products.

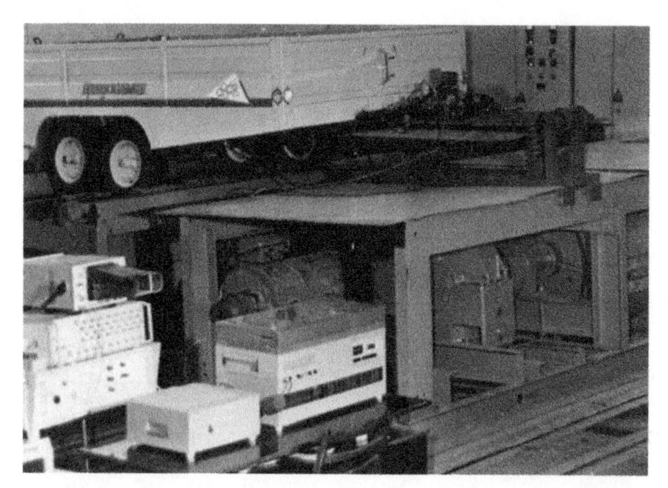

Figure 5.34. Equipment for vibration testing with test subject—car's trailer (from front-back side).

Figure 5.35. Equipment for vibration testing with test subject—a car's trailer (from backside).

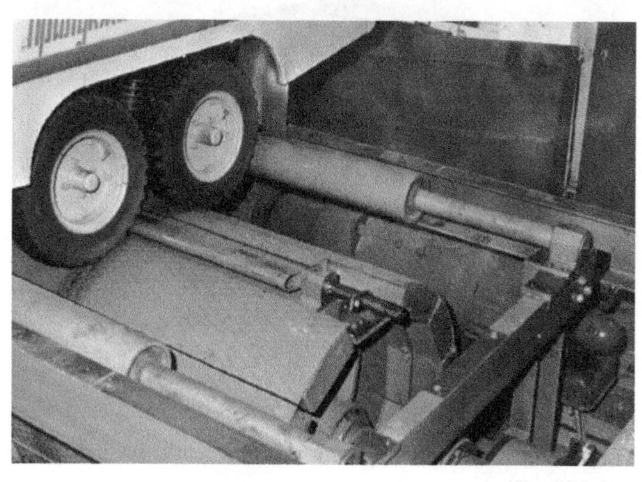

Figure 5.36. Working heads of vibration equipment with test subject (car's trailer).

Figure 5.37. Working heads of vibration equipment with wheel of the car's trailer.

Figure 5.38. Another photo of working heads of vibration equipment with a car's trailer's wheel.

Figure 5.39. Advanced vibration equipment for testing in 6 degrees of freedom (three linear and three angular).

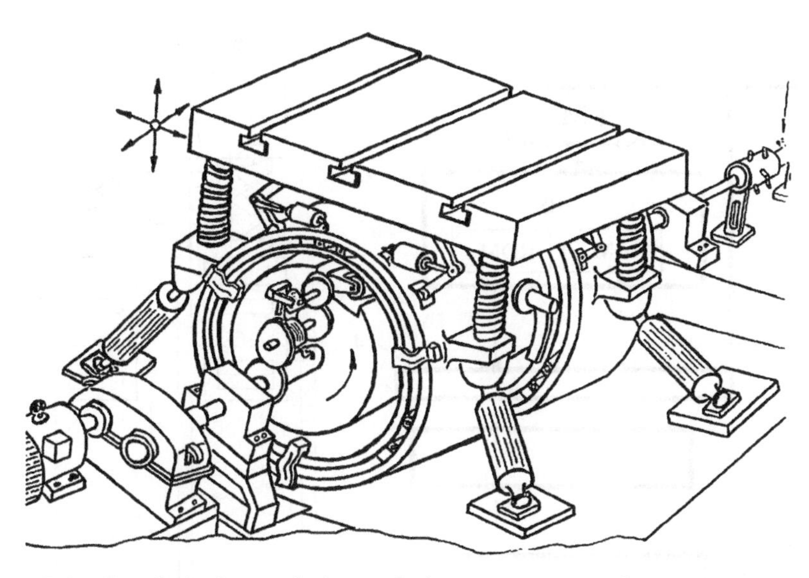

Figure 5.40. Detailed scheme of advanced vibration equipment for testing in 6 degrees of freedom (three linear and three angular).

The results of testing this equipment show that it is much simpler and less costly than vibration equipment in the market.

The equipment specifics offer the possibility to use them instead of electrodynamic vibration equipment, with an important advantage such as the simulation vibration along 6 degrees of freedom (one to three linear and one to three angular axes).

5.4.1.4 System of Control for Vibration Testing. To implement the new technology previously mentioned practically for vibration testing, one needs a corresponding control system that simulates the random character of input influences. One possible control system is discussed in Reference 101. As a component of this system, the structural scheme of the control system for advanced vibration equipment is shown in Figure 5.41.

The computerized control block of the system consists of a processor (1); an external memory (EM) (2); a block of real storage (RS) (3), which includes a random access memory (RAM) and read-only memory (ROM); and a bus for control commands and data transmission (4). Additionally, the control system functions use an adapter (24), an expansion card (23) (with modules and interfaces for input [20] and output [22] signals blocks), and an analog-to-digital converter (14) (ADC) with a normalized amplifier (15).

The adapter makes it possible to join the expansion card to the computer and ensures the conformity of information exchange between different level channels.

The expansion card (23) is able to use several blocks of input and output signals (22) from the controlled subject (5) and ADC for the transformation of analog signals from sensors of the controlled subject to digital signals.

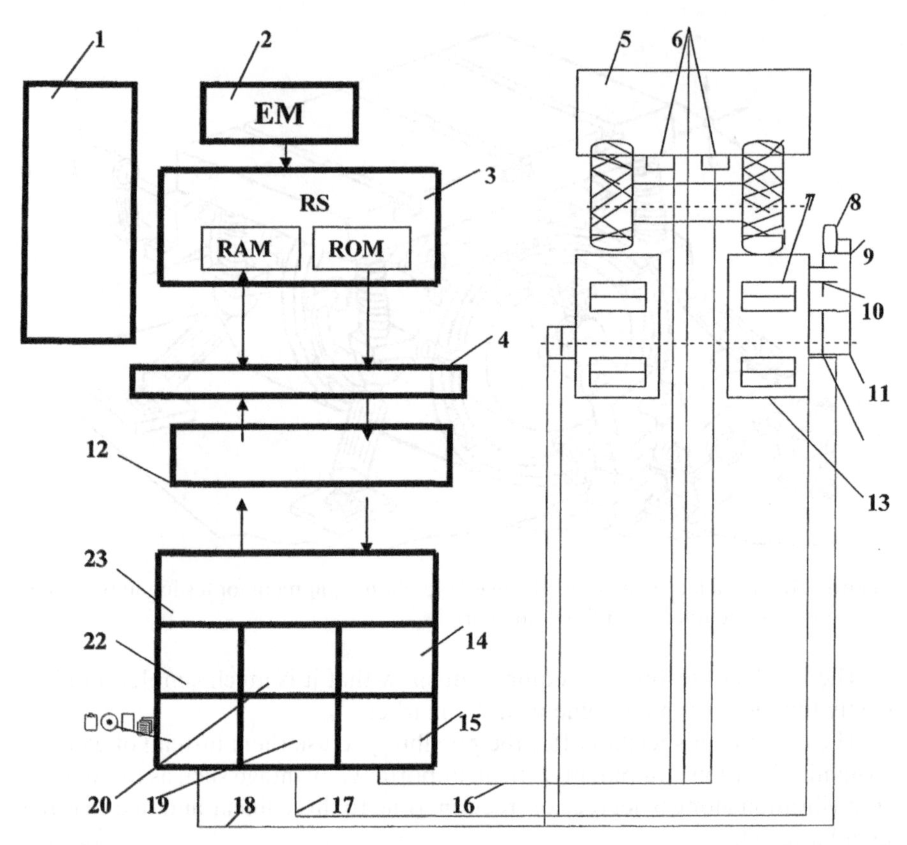

Figure 5.41. Structural scheme of drum vibration equipment control system [43]: 1, computer processor; 2, external memory (EM); 3, block of real storage (RS); 4, control command and data transmission buss; 5, test subject; 6, loading measurement sensor; 7, controlled simulators; 8, feedback sensor; 9, feedback sensor signal transformer; 10, feedback sensor roller; 11 and 12, feedback sensor current collectors and controlled simulator filters; 13, vibration drum; 14, analog-to-digital converter (ADC); 15, normalized amplifier; 16, group sensors collection cable with ADC; 17, output control signal interface with control mechanisms of simulator position; 18, feedback sensor interface with input signals; 19, output signal interfaces; 20 and 22, input and output signal blocks; 21, input signal interfaces; 23, expansion card; 24, adapter.

The mechanical part of the equipment consists of vibrodrums (13) with height-controlled simulators (7) of the road profile. Each simulator with a control mechanism mounts inside the drum, separate from the other simulators. The signals for changing the simulators are transferred from the computer by a group of cable connections through the current collector. A feedback sensor (8) that connects with the axis of the simulator determines the position of each simulator.

The deformation and vibration of the test subject are measured with sensors that are mounted on the test subject. Analog signals are received and transferred by a cable connection to the amplifier and are then converted in the ADC to digital form.

Operation of the test equipment is in real time. For the simulation of the random rate of loading (or product degradation), it is necessary to have individual control of each simulator position.

The plan for the road simulation influences the position control that is in Figure 5.42. It is necessary to define the numerical characteristics for the program used in preparing the control for the test equipment. The program may be composed by random signal generators, which possess the following characteristics:

1. Probability distribution of the random input influences on the test subject
2. Mathematical expectation of the distribution m

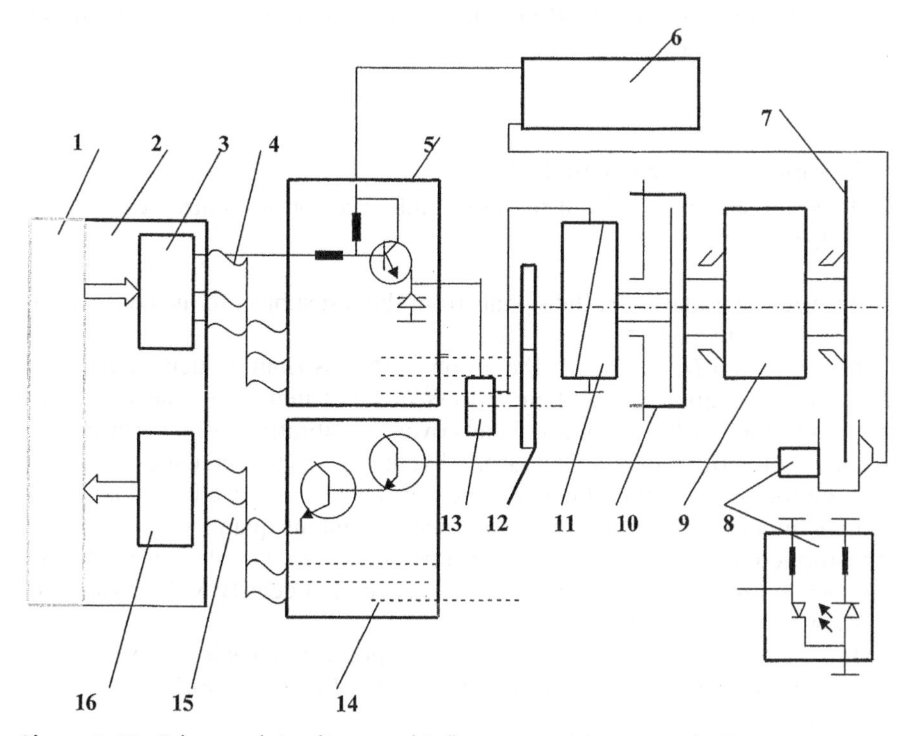

Figure 5.42. Scheme of simulator road influences position **control** [43]: 1, computer; 2, block of expansion; 3, output signal block modules; 4, block output signal group collection cable; 5, control signal interface block; 6, power supply unit; 7, simulator position sensor disk; 8, simulator position sensor recorder; 9, controlled simulator; 10, electromagnetic clutch coupling disks; 11, electromagnetic coupling coil; 12, simulator controlled drive mechanism; 13, current collectors; 14, feedback signal interface block; 15, block input signal group collection cable; 16, input signal module block.

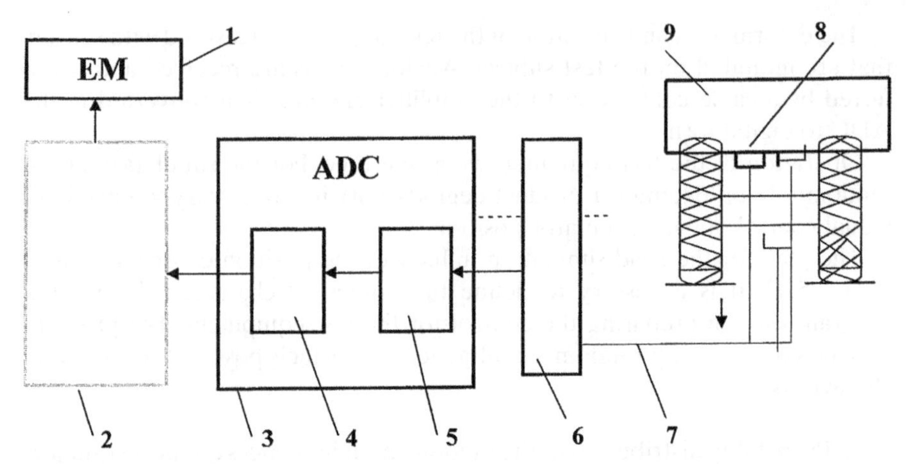

Figure 5.43. Scheme of transfer degradation of the signal from test subject to the computer: 1, external memory; 2, computer; 3, ADC; 4, ADC module digital unit; 5, ADC module analog unit; 6, normalized amplifier; 7, group connection cable; 8, degradation sensors; 9, test subject.

3. Variance D or standard deviation α
4. Power spectrum $S(\omega)$ of the input influences or normalized correlation $K(\tau)$

Transfer degradation of the signals from the test subject to the computer is shown in Figure 5.43.

Given m and D, one builds the graph of the distribution density $f(b)$ and $F(a)$ for the continuous random process of input influences that are transformed to the height of working heads over the vibration drum surface or to an angle of turn, a, of the simulator's blade. These characteristics show the distribution of the input influences. The turn angles are since in real and continuous numbers. Therefore, the continuous random quantity $f(a)$ must be transformed to discrete points, n, in groups over the input influence range. For this step, represent the characteristics of distribution density with a histogram as shown in Figure 5.44.

It is necessary to calculate the number of points n in each group.

The profile and other input influences often follow a normal probability distribution.

Now, in order to find the sequence of generated random signals η_{2i-1} and η_{2i}, we use a set of random numbers generated from the normal distribution $\xi_I(I = 1, 2, 3, \dots)$ [105]:

$$\begin{aligned} \eta_{2i-1} &= (-2\ln\xi_i)^{0.5}\cos(e^9\xi_i) \\ \eta_{2i} &= (-2\ln\xi_i)^{0.5}\sin(e^9\xi_i). \end{aligned} \tag{5.1}$$

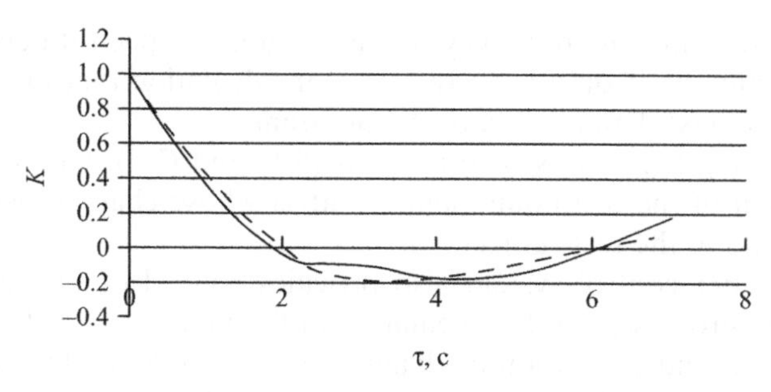

Figure 5.44. Load correlation of test subject [43]; calculated directly,———; approximated,------------

To prepare a program, it is necessary to provide a predetermined power spectrum. For example, the power spectrum of input influences on the mobile product's wheels may be

$$S(\omega) = D/2\pi\left[\frac{\alpha}{\alpha^2+(\omega+\beta)^2}+\frac{\alpha}{\alpha^2+(\omega+\beta)^2}\right], \tag{5.2}$$

where α and β are parameters that can be determined from random signals of real-life input influences, and ω is the circular frequency of the input influence.

This characteristic of the power spectrum corresponds to the Fourier series of the correlation function $K(\tau)_p$:

$$K(\tau)_p = De^{-\alpha(\tau)}\cos\beta\tau, \tag{5.3}$$

where τ is the parameter time.

It is necessary to provide the frequency of decomposition for the power spectrum's characteristic in order to compile a digital program for vibration equipment control. For this purpose, the power spectrum $S(\omega)$ is the n harmonics—regions of equal width $\Delta\omega$. The variance D_k of each harmonic and α_k wave amplitude corresponding to each selected region is calculated. The frequency of these waves is equal to the mean of the frequency of the wave for the region under consideration.

The variance D_k is equal to the sum of the variances of all the harmonics of the spectral decomposition. Calculate the wave amplitude of each harmonic with the following formula:

$$\alpha_k = (2D_k)^{0.5}. \tag{5.4}$$

As a result, we have an equation for the evaluation of the control program numerical series $y(t)$, to obtain the given power spectrum

$$y(t) = \alpha_1\cos(\omega_1 t) + \alpha_2\cos(\omega_2 t) + \ldots + \alpha_n\cos(\omega_n t), \tag{5.5}$$

where ω_n is the circular frequency of each region of spectral decomposition and t is an interval of time from the start of calculation that one can use to determine the next digital value of the program.

By combining the results of Equations (5.1) and (5.5), one may obtain a digital version of the test equipment's control whose characteristics depend on the real life of the test subject.

The computer control system also provides control of the test subject's technological process, periodic calculation of the indices of work through the requested time, and evaluation of changes in these indices, which provide the means to evaluate any deterioration in time, the technical conditions, or vibration distribution on the test subject.

To obtain the correlation function, one can use the equation

$$K(\tau) = D^{-1}(T-\tau)^{-1}\sum_{1}^{T-\tau} n_0(i)n_0(I+\tau), \tag{5.6}$$

where

$n_0(i)$ is the centralized value of the random magnitude of the degradation at moment I and

$n_0(i+\tau)$ is the same with the time displacement of τ.

Figure 5.44 shows how to obtain the correlation function from Equation (5.6) based on the experimental data of variable input influences on the test subject.

This method may find the deviation from theoretical characteristics if τ is large.

Therefore, one must choose the parameters α and β such they will approximate this curve, where α measures the degree of decrease in the elementary branch and β is the frequency of branch waves. Different methods are available for the approximation of this correlation function. One possibility is using the method of least squares to evaluate α and β, so that the deviation of the sum of the squares, which may be calculated by using Equations (5.3) and (5.6), has a minimal difference.

Next, consider the examples of a typical system of control specification for the vibration testing of a complete product (truck) and of the components of a complete product (see Table 5.12).

The system of vibration control (SVC) consists of

- Block of control computer (CC)
- Block of analog signal transformation (ADC)
- System of vibration influence sensors
- Three blocks of control with electromagnetic clutches
- Three blocks of turn angle transformation
- 27 sensors of turn angles

TABLE 5.12 Typical Specifications of Drum Vibration Equipment System of Control for Whole Product (Truck) Testing [43]

Speed of Central Processor, Short Operations in Second	1 Million
1. Capacity of ROM	24 K
2. Capacity of RAM	4 K
3. Number of the clutch control command	48
4. Number of angle sensors	27
5. Number of acceleration sensors	64
6. Number of degradation sensors (deformation, wear, etc.)	88
7. Frequency range of measured acceleration (Hz)	From 0.5 to 200
8. Number of digital–analog transformers	3
9. Channel of connection with outside computer	1
10. Mass (kg)	30

The control program includes

- A program of vibration equipment control in real time
  - A program for the selection and control of vibration testing regimens corresponding to the regimen occurring in the field
  - A program of correction and selection of the regimens for obtaining necessary statistical characteristics of vibration as a random process from each vibrosensor
- A prepared program of the control file of SVC that provides special road simulation
- A control program with a database of the product's component degradation during accelerated testing
- Database degradation information for components from the field
- Test programs for SVC with built-in means
- Service programs checkout for the vibration equipment

The SVC must be able to function in a temperature range of –40 to +50°C, and in a vibration range of 0.1–500.0 Hz, with a single road acceleration of no more than 15 m/s^2 with multiple road induced loads resulting in a combined acceleration ranging from 80 to 100 m/s^2, and a 2- to 20-ms duration.

5.4.2 Dynamometer Brake Testing

A dynamometer is a machine used to measure the torque and rotational speed to calculate the power produced by an engine, motor, transmission, or other rotating prime.

A dynamometer can also determine the torque and power required to operate a driven machine such as a pump.

In addition to determining the torque or power characteristics of a machine under test (MUT), dynamometers operate in a number of other roles.

5.4.2.1 Principles of Operation. An absorbing dynamometer acts as a load driven by the prime mover that is undergoing testing. The dynamometer must be able to operate at any speed and must load the prime mover to any level of torque that the test requires. A dynamometer is usually equipped with some means of measuring the operating torque and speed.

The dynamometer must absorb the power developed by the prime mover. Usually, the power absorbed by the dynamometer is dissipated to the ambient air or it is transferred to cooling water. Regenerative dynamometers transfer the power to electrical power lines.

Dynamometers can be equipped with a variety of control systems. If the dynamometer has a torque regulator, it operates at a set torque while the prime mover operates at whatever speed it can attain to develop the torque that has been set. If the dynamometer has a speed regulator, it develops whatever torque is necessary to force the prime mover to operate at the set speed.

A motoring dynamometer acts as a motor that drives the equipment undergoing testing. It must be able to drive the equipment at any speed and develop any level of torque that the test requires.

Only torque and speed are measured. Calculate power from the torque and speed figures according to the formula

$$\text{Power} = \text{torque} \times \text{speed} / K,$$

where K is determined by the units of measure used.

To calculate power in horsepower (hp), use

$$\text{HP} = \text{torque} \times \text{rpm}.$$

Torque is in foot-pound and the rotational speed is in revolution per minute.

To calculate power in kilowatts, use

$$\text{KW} = \text{torque} \times \text{rpm} / 9549.$$

Torque is in newton-meter and the rotational speed is in revolution per minute.

5.4.2.2 How Dynamometers Are Used for Engine Testing. Dynamometers are useful in the development and refinement of modern-day engine technology. The concept is to use a dynamometer to measure and compare power transfer at different points on a vehicle, thus allowing modification of the engine or drivetrain for more efficient power transfer. For example, if an engine dynamometer shows that a particular engine achieves 400 N-m (300 ft-lb) of torque, and a chassis dynamo shows only 350 N-m (260 ft-lb), one would know to look to the drivetrain for the major improvements. Dynamometers are typically very expensive pieces of equipment and are reserved for certain fields that rely on them for a particular purpose.

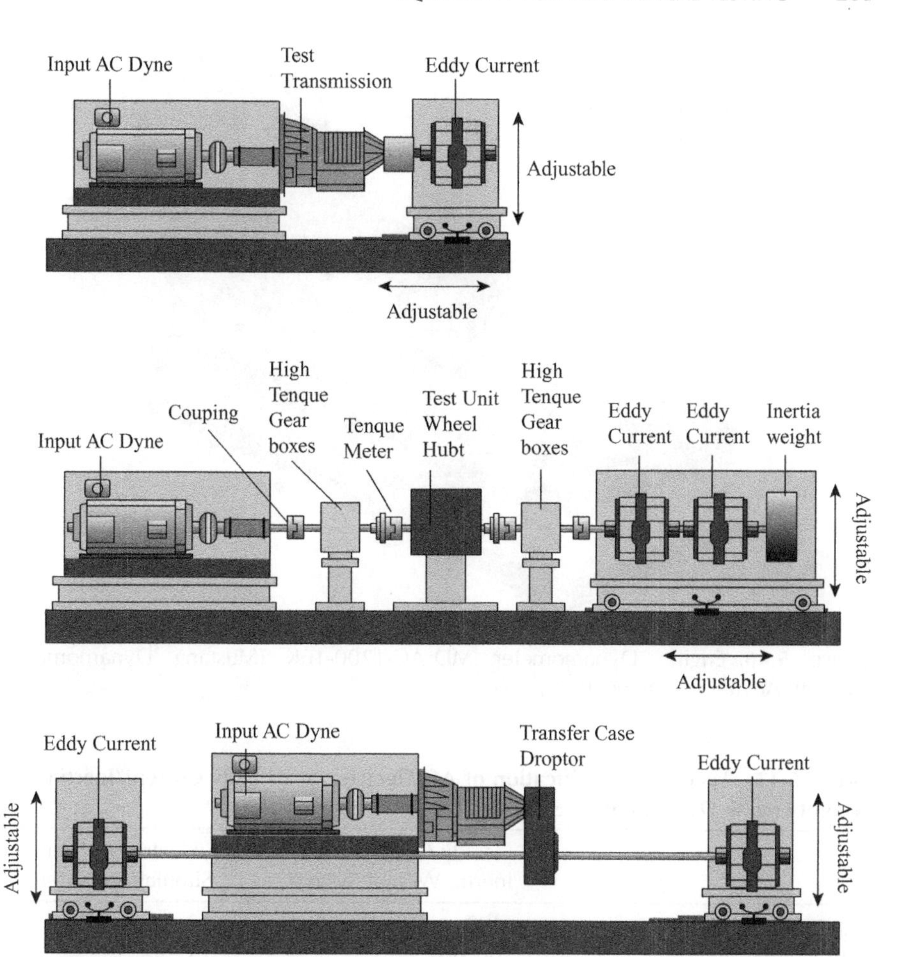

Figure 5.45. HD Heavy Duty Series Mustang Dynamometers/Mustang Advanced Engineering [187].

Mustang Dynamometer/Mustang Advanced Engineering is the manufacturer of different types of dynamometers for various areas of industry. Figure 5.45 demonstrates a Heavy Duty HD Series.

By utilizing high-speed AC motors and an advanced control system, the MD-AC-1200-10K (Fig. 5.46) makes it possible to perform tests such as vehicle and track simulations that are simply impossible for the competing systems employing older technology. Table 5.13 compares the Mustang's AC/electric system to its competitor's eddy current and inertia weight systems (AC Engine Dynamometers, Mustang Advanced Engineering).

The brake dynamometer applies a variable load on the engine and measures its ability to move or hold the revolution per minute as related to the

Figure 5.46. Engine Dynamometer MD-AC-1200-10K (Mustang Dynamometer/ Mustang Advanced Engineering [187].

TABLE 5.13 Technical Specification of AC/Electric versus Eddy Current/Inertia Weight Engine Dynamometer Systems [187]

	Eddy Current and Inertia Weight System	AC/Electric with Inertia Simulation System
System response	–100 ms	<10 ms
Shift simulation	Too slow	Yes
Inertia simulation resolution	±200 lb	±1 lb
System setup for different axle and transmission ratios	Manual inertia weight changes request	Automatic
Water requirements	80 gal/min at 800 hp	None, air cooled
Dyne power recovery	None	Near full recovery
System configuration	Complicated	Simple and compact
Overload capacity	None	150%

"braking force" applied. The engine is usually connected to a computer that records the applied braking torque and calculates the power output of the engine based on information from a "load cell" or "strain gauge" and tachometer (speed sensor).

An inertia dynamometer provides a fixed inertial mass load and calculates the power required to accelerate that fixed, known mass. It uses a computer to record revolutions per minute and the accelerated rate of calculated torque.

Engine testing generally runs from somewhat above idle to the engine's maximum revolution per minute with the output measured and graphed.

There are essentially only two types of dynamometer test procedures:

- *Steady-State (Only on Brake Dynamometers).* The engine operates at a specified revolution per minute (or series of usually sequential revolutions per minute) for 3–5 seconds by the variable brake loading as provided by the PAU.
- *Sweep Test (Inertial or Brake Dynamometers).* The engine is tested under a load (inertial or brake loading) but is allowed to "sweep" up in revolution per minute in a continuous fashion, from a specified lower "starting" revolution per minute to a specified "end" revolution per minute.

Types of sweep tests include the inertia sweep test that is an inertial dynamometer system, providing a fixed inertial mass flywheel that computes the power required to accelerate the flywheel (load) from the starting to the ending revolution per minute.

Issues with inertial sweep power testing include the fact that the actual rotational mass of the engine or engine and vehicle in the case of a chassis dynamometer is unknown and the variability of the entire mass will skew the power results.

The inertia value of the flywheel is "fixed," so low power engines are under a load for a longer time and the internal engine temperatures are usually too high by the end of the test, skewing optimal "dyno" tuning settings away from the outside world's optimal tuning settings. Conversely, high-powered engines usually complete a common "fourth-gear sweep" test in less than 10 seconds, which is not a reliable load condition compared to the operation in the outside world. By not providing enough time under a load, internal combustion chamber temperatures are unrealistically low and power readings, especially past the power peak, skew low.

Loaded sweep tests on brake dynamometer systems consist of two types:

- *Simple Fixed Load Sweep Test.* Apply a fixed load during the test that is somewhat less than the output of the engine. The engine accelerates from the starting revolution per minute to its ending revolution per minute, varying in its acceleration rate, depending on power output at any particular revolution per minute point. Power is calculated using torque × rpm/5252 + the power required to accelerate the dynamometer and engine's/vehicle's rotating mass.
- *Controlled Acceleration Sweep Test.* Similar in basic usage to the simple fixed load sweep test, but with the addition of active load control, that targets a specific rate of acceleration. Commonly, acceleration of 20 fps/ps is used.

The advantage of the controlled acceleration rate is that the acceleration rate used is relatively common from low power to high-power engines avoiding unnatural overextension and contraction of "test duration." This provides more accurate and repeatable test and tuning results.

There is still the remaining issue of a potential power reading error due to the variable engine/dynamometer/vehicle's total rotating mass. Most modern computer-controlled brake dynamometer systems are capable of deriving that "inertial mass" value to eliminate the error. Interestingly, a "sweep test" will always be suspect because many of the sweep users ignore the inertial mass factor and prefer to use a blanket "factor" on every test, on every engine or vehicle.

Inertial dynamometer systems are not capable of deriving inertial mass and use the same inertial mass on every test.

Using steady-state testing eliminates the inertial mass error because there is no acceleration during a test.

The dynamometer tests are performed on the brake pads and brake linings for light passenger vehicles. However, scale tests are also conducted on brake friction material for larger applications.

The following is an example of technical specification ECESO [159]:

Maximum power: 35 kW
Maximum inertia: 50 kgm^2
Maximum speed: 1400 rpm
Maximum torque: 2000 Nm
Air over hydraulic actuation
Fully computer controlled with constant torque capacity

5.4.2.3 Passenger Car Inertia-Type Brake Dynamometer. Some companies provide laboratory facilities with the development and performance of testing of brake assemblies including special emphasis on their requirements and testing in accordance with International Organization for Standardization (ISO), SAE, and other standards.

This machine is a unit designed to be both ergonomic and economical, and its rugged cast iron machine base houses the drive motor that creates a compact tabletop design. A vertical brake cooling air system accommodates the brake setup by creating a large working area.

The typical dynamometer has the following features:

- Available in single-ended, dual-ended, and "twin" dynamometer configurations
- A rotational inertia system provides the stored energy for deceleration conditions. The dynamometer drive motor provides the required torques for drag conditions and for simulation of grade effects.

- Modern digital computer technology simplifies the test setup and dynamometer conditions with or without grade simulation.
- The system incorporates special features to maximize test data accuracy and reliability. Automatic signal calibration, simultaneous acquisition of time and distance frame data, and digital/adaptive control "outer loops" are some of the major features.
- Full inertia simulation, inertia "trimming," and uphill/downhill grade simulation add to the test sequence flexibility.
- The computer-based control system provides for data acquisition recording, display, and selective printout of analog "in-stop" and computed test data and test status information. All data and status information are available for further analysis using an off-line computer via an interface with a local area network (LAN).
- The machine can be set up by one person and can operate automatically on a 24-hour continuous-duty cycle basis.
- The brake applies pressure or torque controls without overshooting a programmed level even at high apply rate conditions.
- Dual-end dynamometers simulate front and rear brake combinations or a complete axle. They may be tested simultaneously providing data on the "load sharing" and brake balance behavior of the two brake assemblies.

5.4.2.4 Engine Dynamometer. An engine dynamometer measures power and torque directly from the engine's crankshaft or flywheel, when the engine is not in the vehicle. These dynamometers do not account for power losses in the drivetrain, such as the gearbox, transmission, or differential.

5.4.2.5 Chassis Dynamometer. A chassis dynamometer measures power delivered to the surface of the "drive roller" by the drive wheel. Position the vehicle on the roller or rollers so that the car then turns and the output is measured.

Modern roller-type systems use the Salvisberg roller, which improved traction and repeatability over smooth or knurled drive rollers.

On a motorcycle, typical power loss at higher power levels mostly through tire flex is about 10%, and gearbox chain and other power transferring parts are another 2–5%.

Other types of chassis dynamometers that eliminate the potential wheel slippage on old-style drive rollers and attach directly to the vehicle's hubs for direct torque measurement from the axle are available. Hub-mounted dynamometers include units made by Dynapack and Rototest. These dynamometers should read about 10–15% higher than a "rear wheel" chassis dynamometer.

Chassis dynamometers can be fixed or portable.

Modern chassis dynamometers can do much more than display, horsepower, and torque. With modern electronics and quick reacting, low-inertia dyne systems, it is now possible to tune to the best power and the smoothest runs in real time.

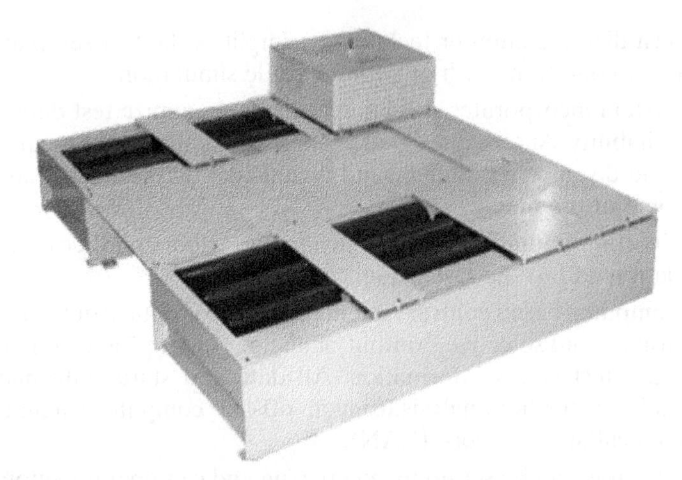

Figure 5.47. Hybrid Series Chassis Dynamometer AWD-AC/EC.

It is also common on a retail level to "tune to an air fuel ratio" with a wideband O_2 sensor in step with revolution per minute.

Because of frictional and mechanical losses in the various drivetrain components, the measured rear wheel brake horsepower is generally 15–20% less than the brake horsepower measured at the crankshaft or flywheel on an engine dynamometer [160].

One can see a Hybrid Series Chassis Dynamometer AWD-AC/EC in Figure 5.47 (Mustang Dynamometer).

Standard Features

- Hybrid and electric vehicle development
- Precise road load and inertia simulation
- Test regenerative braking systems
- 100- to 600-hp AC motor
- 900-hp Eddy Current PAU
- Mechanically linked front/rear roll sets
- Wheelbases 66.5–146.0 in.
- Road load and federal drive cycles

5.5 EQUIPMENT FOR MULTI-ENVIRONMENTAL TESTING AND ITS COMPONENTS

5.5.1 Current Situation in Environmental Test Chamber Design

Accelerated multiple environmental testing (AMET) requires the simulation of complex environmental influences acting simultaneously and in combination with each other. AMET is one of the basic components of accelerated reliability (durability) testing.

Each product may be tested at an extreme range of environmental parameters (temperature, fluctuation, pollution, radiation, etc.) to confirm the continuation of the design and production process compliance. Testing can simulate the environment the products will encounter in transportation and operation. Nevertheless, professionals who plan, manage, and provide AMET must realize that environmental influences do not act independently on the product in real life. They always act in combination with other types of influences such as road features, air conditions, input voltage, and other factors depending on the product's use conditions. Product reliability (durability) changes due to the interaction of these types of influences.

Unfortunately, most publications, including those of the more highly developed technologies such as aircraft, electronics, and aerospace, use only the simulation of separate environmental input influences or two to three of them, such as temperature, humidity, chemical or mechanical pollution, or radiation (often only UV). Literature and practice in environmental testing only address the following types of separate testing:

- Thermal shock
- Thermal cycling
- High-temperature burn-in
- Low temperature
- Salt fog
- Humidity
- Solution spray
- Rain spray
- High-pressure spray
- Cold rooms/warm rooms
- Low humidity
- Spray bar assembly
- Temperature/humidity
- Dust
- Light

These types of separate environmental testing are used for reliability or durability evaluation and prediction.

The title "Environmental Testing Chambers" frequently appears in literature, but concerned production companies are referring to the contents of their temperature/humidity test chambers, or corrosion test chambers, or solar radiation test chambers, or very occasionally to separate solar radiation test chambers or dust chambers as environmental test chambers.

Many industrial companies then use these types of test results as the basis for the reliability and durability evaluation and prediction of their product. This is a misconception. In real life, environmental factors act only in combination and act simultaneously with mechanical, electrical, and other factors. Therefore, after such incomplete environmental testing, one cannot accurately predict reliability and durability.

The basic reasons for these deficiencies are the following:

- Industrial companies, as well as service companies, want to reduce the expense of testing, including environmental testing. Therefore, they use lower-cost test technology and buy lower cost, less capable testing equipment. Temperature or temperature/humidity test chambers are less expensive than complex (multi-environmental) test chambers, but this does not reflect the product life cycle cost during the design, manufacture, and usage stages. They also do not take into account during manufacturing the cost of complaints, recalls, and additional product development. As shown in Chapters 2–4, high quality, reliability, durability, and maintainability testing requires larger expenditures of funds but leads to reduced recalls and complaints. Minimal expenses result only if one takes into account the above-mentioned factors during the design phase.
- Many national and international organizations for standardization, where international experts work as volunteers, include the representatives of production companies. The companies using the above-mentioned products very infrequently send their representatives to these organizations; therefore, the standards often reflect mostly the producers' interests.
- The situation described earlier is the basic reason for slow technical progress in the development of AMET equipment and standards.

Correctly predicting product reliability and durability requires accurate simulation of real-life environmental influences to achieve accurate reliability/durability testing. The rapid and correct solution of reliability and other quality problems also requires this approach. If the simulation is not accurate, the results are not accurate. AMET is applicable to a separate type of stress testing for solving specific problems.

The second approach is using it as a component of reliability (durability) testing. Additionally, it applies to other components of ART (mechanical and electrical).

In practice, when carrying out the fatigue testing of a product (steel specimens and pieces), one has to take into account that the results of fatigue testing of metals or nonmetallic materials are different from the results of testing pieces of these materials because local tensions concentrate on these pieces. Therefore, corrosion of the product that destroys its surface results in a local concentration of strength-reducing tensions that decrease the total product's fatigue resistance decreases.

The names of test chambers frequently do not correspond to their content and true purpose. For example, "chamber for durability accelerated testing" is usually referred to as the chamber for temperature and humidity without condensation. The "chamber middle volume for environmental testing" is the only chamber that tests for temperature changes, or for changes in temperature and humidity, or for changes of separate air pressure, salt spray, temperature, and humidity, but it does not test for the full set of influencing factors.

As was mentioned earlier, the use of test chambers for the independent simulation of one to three environmental input influences characterizes the current situation, not chambers for multi-environmental testing.

Many companies are designing and manufacturing equipment to test one to three environmental input influences. For example, consider the high-performance Thermotron chambers that are used in the automotive industry to test vehicles under extreme temperature and humidity conditions (Fig. 5.48).

The situation is similar for smaller environmental test chambers. Many of these chambers correspond to American Society of Mechanical Engineers (ASME) standards, but it is necessary to take into account that ASME standards apply mostly to materials, not to machines and their units.

Many companies continue to design and manufacture temperature and temperature/humidity test chambers that are useful for the solution of separate problems but are not useful for ART/ADT. For example, the Weiss/Envirotronics Company [161] is producing many types of these test chambers. See their ET and EH Series test chambers in Figure 5.49.

The ET/EH chambers can be customized to meet specific needs in terms of size, horsepower, or configuration:

- Temperature range: –73 to +177°C
- Humidity range: 10–98%

Figure 5.48. High-performance Thermotron chamber for automotive industry to test vehicles under extreme temperature and humidity conditions.

Figure 5.49. Weiss/Envirotronics typical temperature and temperature/humidity chamber [161].

- High-/low-temperature testing (ET)
- High-/low-temperature and humidity testing (EH)
- Custom design

Of course, this equipment is not appropriate for ART/ADT technology.

A good example that is not appropriate for ART/ADT technology is the series of test chambers produced by other companies. These companies design, manufacture, and sell a series of the test chambers for separate and different components of the real environment. They demonstrate test chambers and state that the customer can evaluate the product's performance in a simulated real-world environment, whether the customer needs to mimic rain, a shower, spraylike conditions, or a combination of one to three parameters of the environment. In addition to the standard integrity test, this series is also available in configurations that will allow the testing of parts while they are under power. Each chamber can be customized to include automatic controls with a sophisticated data collection or with simple on/off controls.

5.5.2 Combined Equipment for Multiple Environmental Testing

There is currently no equipment on the market for full multiple environmental testing. Multiple environmental testing needs a combination of test equipment as shown in Figure 5.50, including equipment for accelerated weathering

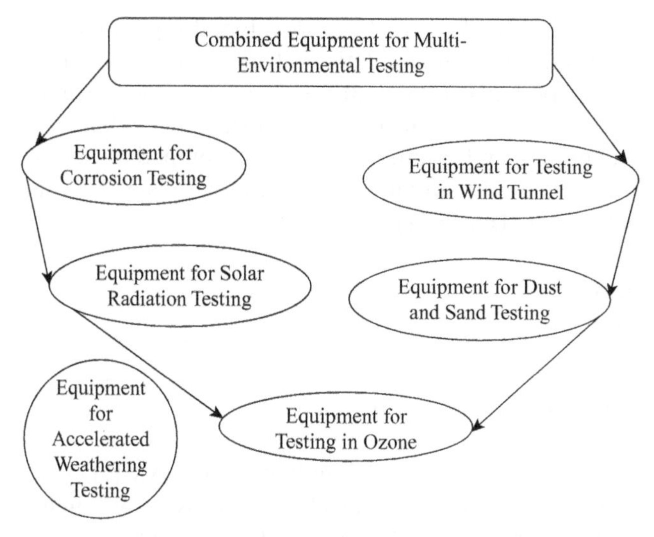

Figure 5.50. Scheme of combined equipment for multiple environmental testing.

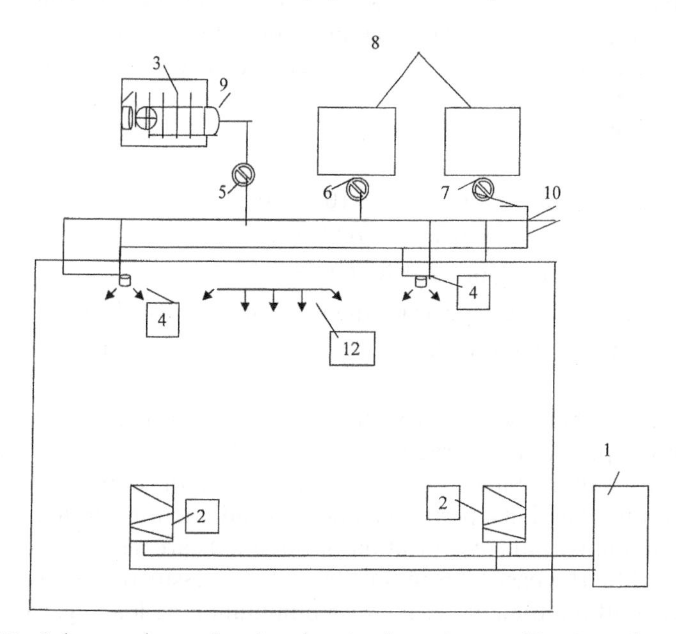

Figure 5.51. Scheme of test chamber for simultaneous combination of temperature, humidity, pollution, and radiation (1, system of control; 2, heat exchanger; 3, compressor; 4, sprayer; 5–7, valves; 8, capacities; 9, receiver; 10 and 11, pipelines; 12, lamps).

testing. A step toward the solution of integrated design and manufacturing problems is to consider the combined equipment for multi-environmental testing that includes the complex of components shown in Figure 5.51.

State Enterprise TESTMASH was the first to create and design combined equipment for multi-environmental testing.

The author and his colleagues have developed equipment as a first step toward the accurate physical simulation of multi-environmental input influences. They developed this equipment for a variety of organizations and companies from aerospace to the automotive industry to academia.

Several techniques and equipment were developed to integrate climate chambers that simulate the simultaneous combinations of the four basic environmental parameters (Fig. 5.51):

- Temperature
- Humidity
- Pollution
- Solar radiation

This equipment is a series of test chambers for the AMET of separate parts and units and complete machinery. The chambers have a system for simulation, regulation, and the control of environmental parameters.

The chambers also have ventilation and cooling systems, wastewater, an aggressive evaporation drainage system, a sprinkler system, and a power and alarm system. The chamber design is such that it can integrate with other testing equipment.

Specification

1. Temperature range (°C): from –10 to +60 ± 1
2. Humidity range (%): from 40 to 97 ± 3
3. Temperature changing velocity (°C/min): 0.7
4. Sun radiation range spectral range: 280–400 nm (42 W/m^2)
5. Pollution: sprinkling with salt water, ammonia solutions, and dust

This test chamber does not provide a high acceleration coefficient or a high change of the physics-of-degradation process. Therefore, the time for failures is faster (two to three times) than time for failures of the optimum life regimen of temperature, humidity, pollution, and sun radiation in the field. The temperature, humidity, and concentration of pollutants are also a little bit higher than in real life. The basic reason for this is the necessity of compressing work time (taking off the time of breaks with minimum time loading, which has no influence on the product's reliability).

When one uses the accelerated coefficient and wants to use the testing results for reliability/durability evaluation/prediction, he needs to remember that a more accelerated coefficient means less correlation between the accelerated reliability/durability indices in the field and in the laboratory.

One must also remember that the accuracy of ART/ADT results determines the value of the initial input for resolving quality, reliability, and durability problems during design and manufacturing.

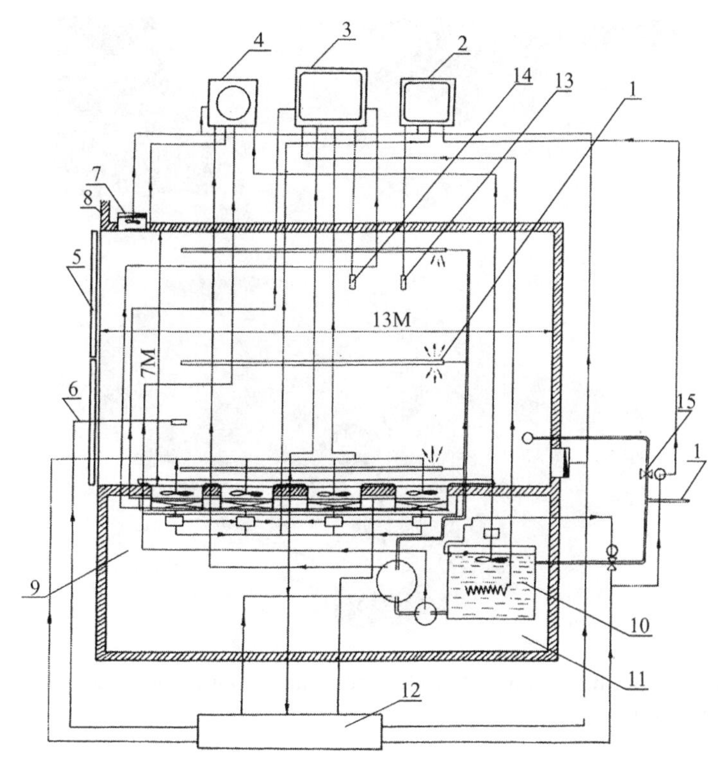

Figure 5.52. Scheme of test chamber with simultaneous combination of temperature, humidity, pollution, and radiation.

Figure 5.52 shows a second type of test chamber developed by the author with colleagues for the simultaneous combination of temperature, humidity, salt spray, and solar radiation.

This test chamber consists of a

- System of humidity regulation
- System of temperature regulation
- System of sprinkling aggressive medium
- System of solar radiation
- System of blocking, signalization, and control
- System of power equipment

The author's previous publications [18, 42, 43, 104, 105] show the combined integrated equipment. Other companies at that time were involved only in the development of equipment for separate components for multi-environmental testing. Other advanced companies then followed this path.

Figure 5.53. Cold heat climate test chamber with a four-poster system and sunlight simulation [148].

Weiss Umwelttechnik GmbH is one of the most advanced companies in the design and manufacturing of combined testing equipment for multi-environmental testing. Listed next are several types of this equipment (Figs. 5.53 and 5.54):

Specification (Fig. 5.53)

Test space volume: 304 m^3

Temperature range: –40 . . . +80°C

Climate range: +5 . . . +80°C (limited by sunlight simulation)

Humidity range: 20 . . . 80% RH

Solar radiation

Specification (Fig. 5.54)

Test space volume: 138 m^3

Temperature range: –40 . . . +50°C

Pressure range: 715 . . . 1100 mbar

Altitude range up to 3000 m above sea level

Figure 5.54. Altitude simulation chamber [148].

5.5.3 Equipment for Corrosion Testing

Currently no equipment can accurately simulate the corrosion process because the current corrosion chambers that are available for materials corrosion only simulate part of the field environment. This equipment cannot simulate a simultaneous combination of multi-environmental influences that influence corrosion (chemical pollution, mechanical pollution, moisture, temperature, etc.) with mechanical influences (vibration, deformation, friction, etc.), as shown in Figure 4.11. In addition, these chambers are unable to simulate the interconnection of different units and parts of the product.

Only combined equipment for accelerated reliability (durability) testing can provide the necessary information for the accurate prediction of the corrosion of cars, trucks, aircraft, and other machines.

Researchers are working toward creating corrosion test chambers for the simulation of a simultaneous combination of factors that exert an influence on the corrosion process. There will be further information on this subject later in the book.

Let us demonstrate a typical series of the current corrosion test chambers.

5.5.3.1 Cyclic Salt Corrosion Test Equipment. Spray and sprinkle corrosion testing was demonstrated earlier. The **cyclic corrosion test chamber** can provide a range of controlled cycles in the automotive industry among others. This testing equipment is capable of multiple automatic cycles including

- Salt fog
- Humidity fogging
- Dry cycle
- Dwell cycle
- Controlled humidity
- Solution spray
- Wet bottom RH
- Immersion cycle
- Temperature ranging from –20 to 90°C

Multipurpose Fog (MPF) Test Equipment. The equipment is designed to meet American Society for Testing and Materials (ASTM) International, SAE, and most other automotive test requirements for accelerated corrosion testing, such as ASTM B117, D1735, B368, MIL-STD-883, Method 1009.5, B368 (CASS),and Mil Std 810D, Method 509.2. With the optional cycling control package, MPF test equipment will also meet standard specification *ASTM G85 Modified Salt Spray (fog) Testing Annex A5 Dilute Electrolyte Cyclic Salt Spray (fog) Testing (Prohesion*™).

Base Cabinet All-Plastic Construction. This chamber is made of fiberglass using the most advanced techniques available to ensure longevity and strength. The plastic construction of this unit protects the chamber from the corrosive solutions used, in addition to providing a naturally insulated exposure zone for consistent and repetitive testing. The chamber lid is made of a transparent polymer-based material to allow the viewing of the ongoing testing while keeping the corrosive fog in the chamber. Electric heaters heat the chambers located under a diffuser plate below the exposure zone.

Microprocessor-Based Controller (MBC). The controller for the MPF Series Chambers is a set of microprocessor-based temperature controllers. They provide the chamber with simple and accurate operation and control. The MPF Chamber is designed to meet most automotive specifications for accelerated corrosion testing including salt fog, humidity, acetic acid, CASS, and simple cyclic tests such as ASTM G85.

Base Cabinet. The interior lining is inert white PVC sheet lining. The exterior is painted with a blue finish coat. Smoke-gray transparent PVC lid with gas cylinder lifting (Models 24 and 410 are equipped with air cylinder lifting). A water jacket on all four sides and the bottom heats the cabinet. The electrical enclosure, pilot lights, selector switches, and wiring components all comply with NEMA 12 standards.

5.5.3.2 Noxious Gas Test Chambers. Noxious gas test chambers are used to test the corrosion resistivity of metals, to determine the aging speed of plastics

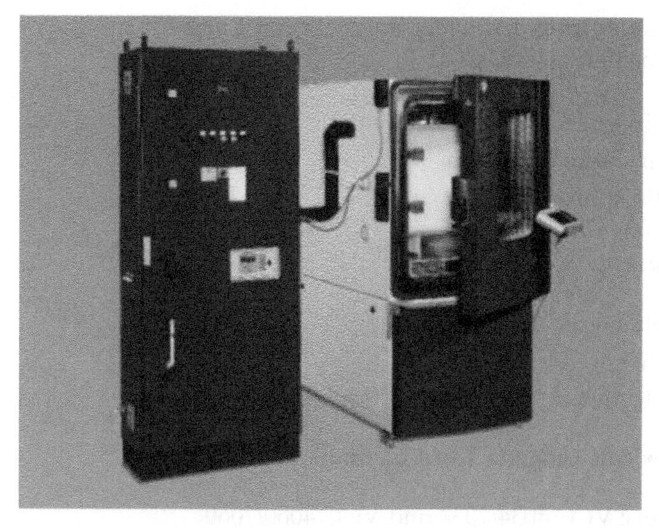

Figure 5.55. Noxious test chamber produced by Weiss Umwelttechnik GmbH [188].

and glass, and to test coatings and coverings for leaks and tightness. This type of gas test chamber is useful for accelerated corrosion testing. Weiss Umwelttechnik GmbH produces noxious gas test chambers.

Basic units of the test chambers produced by Weiss Umwelttechnik GmbH (Fig. 5.55) consist of the well-known climatic test cabinets of the WK series. The noxious gas test chamber can be included in these test chambers.

These test chambers also have the following specifications:

- Design details of the climate test chambers are in the WK Series leaflet
- Noxious gas test chamber made of acrylic glass or polyvinylidene fluoride (PVDF)
- Dosing pump (PTFE) or mass flow controller
- Monitor of low pressure in interior test chamber
- Thinning of the discharge air containing noxious gas

Temperature range:

- With acrylic glass: +15 . . . +40°C
- With PVDF: +15 . . . +80°C

Humidity range: 40 . . . 75% relative
Dew point temperature range:

- With acrylic glass: +10 . . . +38°C
- With PVDF: +10 . . . +60°C

Flow rate noxious gas: adjustable between 1 and 250 mL/h
Volume: approximately 641 L

- Usable height: approx. 400 mm
- Width: approx. 400 mm
- Depth: approx. 400 mm

Volume: approximately 1251 L

- Usable height: approx. 500 mm
- Width: approx. 500 mm
- Depth: approx. 500 mm

Noxious Gas Climatic Test Chambers

- Models: VCC 4034/7034 and VCC 4060/7060
- Test space volumes: 340 and 600 L
- Temperature range: +15 to +60°C
- Dosing of up to four gases
- Noxious gas test chamber: stainless steel or Plexiglas
- Thermal mass flow controller

TechnoLab (Germany) [162] produces noxious gas test/corrosion test chambers with different gas concentrations and gas compositions.

5.5.4 Wind Tunnels

A wind tunnel is a research tool for studying the effects of air moving over or around a test subject.

Wind tunnel testing is the technical support of any major development process involving aerodynamics. In this book, wind tunnel testing is a component of ART. It is applicable for aircraft, helicopters, cars, trains, and laboratory research.

Wind tunnels may be classified by their basic architecture (open circuit or closed circuit), according to their speed (subsonic, transonic, supersonic, or hypersonic), the air pressure (atmospheric or variable density), or their size (full scale).

There are a number of wind tunnels (meteorological tunnel, shock tunnel, plasma jet, hotshot tunnel, and tunnels with moving ground) that fall in a special category of their own.

The main components of a tunnel are the entrance cone, test section, regains passage, propeller, motor, and return passage. Flow straightness, corner vanes, honeycomb layers for reduced turbulence, air exchangers, and diffusers are other common features.

Measurement equipment and testing procedures include instrumentation for the measurement of pressure, temperature, forces, moments, and turbulence intensity.

A short history follows for wind tunnels. Leonardo da Vinci, ca. 1500 [163], was the first to propose the idea behind the wind tunnel. He wrote that the effects of air flowing on an object are the same as the object moving in still air. Although Francis Herbert Wenham designed and used the first practical wind tunnel in 1870, the Wright brothers, in 1901, performed the first systematic experiments on airfoil characteristics in a well-designed wind tunnel [164].

In a classic set of experiments, Osborne Reynolds (1842–1912) of the University of Manchester demonstrated that the airflow pattern over a scale model would be the same for the full-scale vehicle if a certain flow parameter were the same in both cases. This factor, now known as the Reynolds number [165, 166], is a basic parameter in the description of all fluid flow situations, including the shapes of flow patterns, the ease of heat transfer, and the onset of turbulence. This comprises the central scientific justification for the use of models in wind tunnels to simulate real-life phenomena.

The Wright brothers' use of a simple wind tunnel in 1901 to study the effects of airflow over various shapes while developing their Wright flyer was in some ways revolutionary. They were simply using the accepted technology of the day, although this was not yet a common technology in America.

The subsequent use of wind tunnels proliferated as the science of aerodynamics and the discipline of aeronautical engineering were established and air travel and air power were developed. Wind tunnels were often limited by the volume and speed of airflow that could be delivered.

Wind tunnel tests in a boundary layer wind tunnel simulate the natural drag of the earth's surface. For accuracy, it is important to simulate the mean wind speed profile and turbulence effects within the atmospheric boundary layer. Most codes and standards recognize that wind tunnel testing can produce reliable information for designers, especially when their projects are in complex terrain or on exposed sites.

Air is blown or sucked through a duct equipped with a viewing port and instrumentation where mounted models or geometric shapes are studied. Typically, the air is moved through the tunnel using a series of fans. For very large wind tunnels several meters in diameter, a single large fan is not practical, and so an array of fans operates in parallel to provide sufficient airflow instead. Due to the sheer volume and speed of air movement required, stationary turbofan engines often power the fans rather than electric motors.

The airflow entering the tunnel is itself highly turbulent due to the fan blade motion, and it is not directly useful for accurate measurements. The air moving through the tunnel needs to be relatively turbulence free and laminar. To correct this problem, a series of closely spaced vertical and horizontal air vanes smooth the turbulent airflow before reaching the subject of the testing.

Due to the effects of viscosity, the cross section of a wind tunnel is typically circular rather than square. In addition, there will be greater flow constriction in the corners of a square tunnel that can make the flow turbulent. A circular tunnel provides a much smoother flow. In general, the wind tunnel provides a controlled safe test environment on the ground for engineers to evaluate new concepts and ideas for flight. In new aerospace concepts, it is now common for elaborate and sophisticated computer codes to first optimize the vehicle (aircraft, helicopter, or launcher) design. Nevertheless, wind tunnel experiments will always contain all the physics of flight. Consequently, engineers minimize the risk (both financial and physical) and cost associated with their project's development by performing ground tests before flight. In effect, the wind tunnel tests are critical to ensure that the vehicle's performance is optimal before proceeding to the expense of flight-testing the prototype.

The wind tunnel tests validate and improve computational predictions for upcoming projects. Moreover, wind tunnel simulations are an indispensable element in preparing a new vehicle for flight. Take the German–Dutch Wind Tunnels (DNW-LLF) in Figures 5.56–5.58. We can see here

- Engine and ground simulation tests on a model of the Airbus A400M above a moving belt in the 8 × 6 m test section of the DNW-LLF (Fig. 5.56)
- Force and pressure measurements on a model of the Embraer Lagacy 500 in the transonic wind tunnel DNW-HST
- Acoustic measurements on a full-scale wing of the Airbus A320 in the open jet test section of the DNW-LLF.

It operates both the aeronautical wind tunnels of the German Aerospace Center (DLR) and the Dutch National Aerospace Laboratory (NLR). These wind tunnels are located in Germany and in The Netherlands.

Figure 5.56. Engine and ground simulation tests on a model of the Airbus A400M above a moving belt in the 8 × 6 m test section of the DNW-LLF.

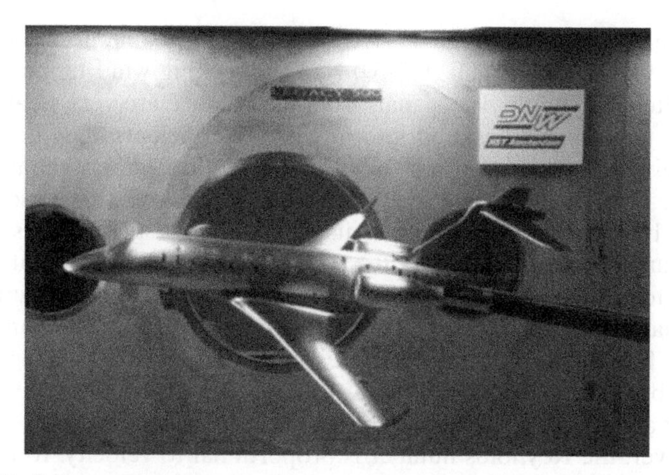

Figure 5.57. Force and pressure measurements on a model of the Embraer Legacy 500 in the transonic wind tunnel DNW-HST.

Figure 5.58. Acoustic measurements on a full-scale wing of the Airbus A320 in the open jet test section of the DNW-LLF.

The main objective of these tunnels is to provide a wide spectrum of wind tunnel tests and simulation techniques.

5.5.4.1 Wind Tunnel Testing of Microair Vehicles (MAVs). MAVs and small unmanned air vehicles (SUAVs) represent a new challenge for aerodynamics and controls. Typical MAVs have a wingspan of 15 cm or less, and they fly relatively slowly at around 13–15 m/s. The aerodynamic forces generated by the air vehicles are small (0.5–5.0 N), thus making wind tunnel measurements quite challenging. The signals from the balance are very weak, about a microvolt, and when translated into forces or coefficients, they need proper analysis, giving special attention to many factors that can potentially jeopardize the results. Some of these factors include the low signal-to-noise ratio, the use of the balance at its sensitivity limits in particular test conditions (typically low lift), and the high sensitivity to some parameters like ambient pressure. Other factors are the critical operational area of the Reynolds number and the peculiar design of the wing with a very low aspect ratio (AR) and flexible structure. The Reynolds number relates wind tunnel results to flight values. This number relates the model scale to the viscous properties of the wind tunnel air and varies over the entire model.

Moreover, the Reynolds number is proportional to velocity. To best simulate flight conditions, the Reynolds numbers of a model must be as close as possible to the flight values. The results are appropriate for general aerodynamic results, performance assessment, and the generation of a system characterization report with the main aerodynamics characteristics of the tested vehicle. The results are useful for the configuration selected, to create a numerical model of the vehicle, and for the design of its control system.

A wind tunnel model is smaller than full scale, so without changing the wind tunnel air parameters, the test Reynolds numbers would always be lower than in flight. Since this number is also proportional to air density, we can offset the deficit to some degree by increasing the stagnation air pressure in the wind tunnel and by lowering the test gas temperature (although this technique is not used at Modene).

5.5.4.2 Experimental Apparatus and Procedures. The facility used in the work described next [166] is a horizontal, open-circuit low-speed wind tunnel. The wind tunnel has an overall length of 10 m and is designed to provide low turbulence levels in the test section, less than 0.5% at 10 m/s. The test section is 914 × 914 mm with a length of 2 m. It uses a centrifugal-type fan and a variable frequency controller operated by a PC with the dedicated data acquisition system that remotely regulates its speed. The maximum velocity for test purposes is approximately 15 m/s, with a subsequent maximum Reynolds number of approximately 200,000.

One of the improvements in the capabilities of the existing wind tunnel setup has been the acquisition and implementation of a six-component sting balance. The balance is an internal five-force/one-moment sting balance. Two forces are normal to the sting axis in one plane (lift); two forces are normal to the sting axis on a plane perpendicular to the previous (side); one force is axial; and the moment is around the roll axis. Usually, the design of a wind tunnel balance is of necessity a compromise between the required maximum load capability of all components and the accuracy required for minimum loads. One can see the rate loads in Table 5.14.

TABLE 5.14 Sting Balance Rated Loads [187]

Component	Rate Loads
Normal force	±44.5 N
Side force	±44.5 N
Axial force	±17.8 N
Pitch *M*	±2.63 kN·m
Yaw *M*	±2.63 kN·m
Roll *M*	±8.77 kN·m

A typical test procedure starts with the input of the model position, the angle-of-attack (AOA) sequence, and the model characteristic dimensions. The acquisition of the tares with the wind off follows. A set of zeros are acquired at the start and at the end of the run for further analysis and eventual correlations resulting from drifting of the signals.

The signals from the balance, during both tares and tests, are sampled and are averaged in a preselected way. Data collection is typically in blocks by sampling 500 points at 1000-Hz frequency. The software performs all data conversions in forces and displays the six coefficients (C_L, C_D, C_S, C_l, C_m, and C_n) versus AOA. At the end of the run, the data are stored in a file for further analysis.

Different publications [165–168] describe the results of testing airfoils at low R_e numbers. One can investigate the effect of systematic variation for some design parameters, such as AR, using this setup, but this was not the intention of the work. One can find some results and analysis in publications [165,169,170]. The results from testing three complete aircraft are in Reference 165.

The main objective of the experiments was to perform an aerodynamic characterization of a specific aircraft (aerodynamic coefficients and static stability derivatives). The power was off so the propeller was wind milling freely. This wind tunnel provides the capability to test aircraft and other systems while operating.

One testing program completed in References 171 and 172 shows the following:

- The vehicle was swept through angles of attack ranging from –4° to 18° at beta (yaw) angles. Data were collected on a 6-degrees of freedom force balance, measuring lift, drag, and side force, as well as moments around all three axes.
- The tests confirmed the predictions of the vortex–lattice code used in the design process. With proper CG placement, the transition is aerodynamically stable and the stall characteristics are benign. As the AOA increases, the canard stalls first, followed by the wing root, and then finally the wing tip—providing a natural recovery or "safe stall."

- With the elevator on the tail of the transition and not on the canard, the pilot maintains the pitch authority allowing him to "mush" the stall a bit more than is possible with a typical canard design. Additionally, the sequence of flow separation from the lifting surfaces will allow maintenance of the roll authority throughout the stall.

In Reference 173, another wind tunnel test occurred. A small origami plane, just 8 cm in length, was mounted inside the hypersonic wind tunnel facility at the Kashiwa Campus of the University of Tokyo. The plane experienced a hypersonic flow at Mach 7 (5300 mph) for 10 seconds. It withstood massive aerodynamic temperatures of 200°C and aerodynamic pressure.

This experiment may stimulate children all over the world—who can directly relate to an original plane and fold one for themselves—to take a greater interest in science and technology.

5.5.4.3 Large Wind Tunnels. There are many types of large wind tunnels. For example, Lockheed Martin's home base for low-speed wind tunnel testing is in Marietta, Georgia. For aircraft customers, the testing provides a verification of mathematical models and hypotheses on the aerodynamic performance of the aircraft during its flight mission. For automobiles, important aspects such as the level of road noise, the amount of drag, and/or how much down force is present can be examined. Solutions can then be determined in an effort to improve upon an existing design.

Lockheed Martin High Speed Wind Tunnel Test Systems Design Products and Services [174] include the following:

- Calibrating of balances and transducers
- Strain gage installation
- Force measurement consulting
- Six-component internal strain gage force balances, with or without flow-through capability
- Special-purpose balances
- Calibration equipment
- Complete models, components, and modifications
- Stings and support equipment
- Thrust and test stands

Support facilities of this wind tunnel include the following:

- High-volume static flow facility
- High-pressure nitrogen system (6000 psi)
- Complete on-site machine shop

The unique capabilities of this tunnel include

- Captive trajectory
- Automated roll
- Inlet throttle
- High-angle remote roll sting
  α –10° to +110°
  φ ±180°

The Lockheed Martin Aerodynamic Group can provide any combination of the following services for customers' test programs:

- Design definition
  - Preliminary through final configuration
  - Surface sizing analysis
  - Drag predictions
  - Trade studies
  - Airload predictions
- Model design support
  - Model requirements
  - Design oversight
- Aeroprediction
  - Empirical
  - Analytical
  - CFD
- Test support
  - Planning
  - Predictions
  - Conduct
- Posttest data analysis
  - Result presentations
  - Aerodynamic database creation
  - Support 3-DOF simulation
  - Support 6-DOF simulation

As we can now see, the wind tunnels mentioned earlier and other wind tunnels are not appropriate for ART/ADT. Nevertheless, one may use them for this purpose, as a component of ARD/ADT. ONERA [175], the French aerospace research establishment, was founded May 1946 as a public organization under the supervision of the French Ministry of Defense. Now they are performing simulations and testing in order to design and build tomorrow's aerospace vehicles. Two of the four wind tunnels operate continuously using closed circuits, one at atmospheric pressure (0.9 bar), and the other pressurized to 2.5 bar. Powerful fans and compressors generate the necessary wind velocities.

The other two wind tunnels have a very different drive system and operate in a blow-down mode. The release of compressed air generates the runs upstream of the test section into the atmosphere or a vacuum chamber situated downstream of the test section.

The best quality airflow is maintained in the test section. The wind passes through the test section (where the model is located) for a duration that can vary from a few seconds to several hours. During this time, instruments acquire pressures, forces, and temperatures, while cameras can film the airflow on and around the model to help better understand the flow field characteristics. These measurements are transmitted to a powerful central computer that processes the information, renders the images of the airflow, and generates the test results in real time. Thus, one can modify the test parameters. The intervention can increase the value of the test and reduce the inherent cost.

The test section (Fig. 5.59) [175] is 26 ft in diameter, 46 ft long, and can accommodate very large test models. The large model size permits the instal-

Figure 5.59. The test section of the wind tunnel ONERA—The French Aerospace Lab [175].

lation of numerous automation devices. The wind tunnel is equipped with two contrarotating fans 49.2 ft in diameter that are driven by water-powered Pelton turbines developing up to 88 MV of power. A regulator controls the fan speed from 25 rpm (Mach 0.05) over 200 rpm (Mach 1).

The Mach number is the ratio of the local speed of the airflow to the local speed of sound.

Therefore, at Mach 1, the airflow is traveling at the speed of sound and is sonic. For a Mach number below 1, the airflow is subsonic, and for a Mach number above 1, the airflow is supersonic. The speed of sound means that sound travels at 750 mph when the air is dry, and its temperature is 0°C.

As the temperature increases, the speed at which the sound travels also increases. Depending on the speed reached in flight, the flow field has different characteristics from the point of view of fluid mechanics, density, and mathematical models. Airflows below Mach 0.8 are subsonic, while airflows from Mach 0.8 to Mach 1.2 are transonic. Airflows at Mach 1 are sonic; airflows from Mach 1.2 to Mach 5 are supersonic; and those above Mach 5 are hypersonic (Table 5.15).

For the latest Dassault Aviation business jet, over 2500 testing hours were required at the ONERA facilities, to confirm the performance and to validate computational predictions for the best design optimization (Fig. 5.60).

The different units of (DNW) wind tunnel facilities are in Tables 5.16–5.18.

The operations performed in the wind tunnel include the following:

- Drag/lift measurements on aircraft, helicopters, missiles, and racing cars
- Drag/lift moment characteristics of airfoils and wings
- Static stability of aircraft and missiles
- Dynamic stability derivatives of aircraft
- Surface pressure distributions on nearly all systems
- Flow visualizations (with smoke, oil, and talcum)
- Propeller performances (torque, trust, power, and efficiency)
- Performance of air-breathing engines
- Wind effects on buildings, towers, bridges, and automobiles
- Heat transfer properties of engines and aircraft

TABLE 5.15 Operating Ranges of the High-Speed Wind Tunnel

Indices	Number
Speed	Mach 0.2–5.0
Dynamic pressure	250–5000 psf
Temperature	100°F nominal
Altitude	−10,000 to 75,000 ft
Reynolds number	2.0 to 38.0×10^6

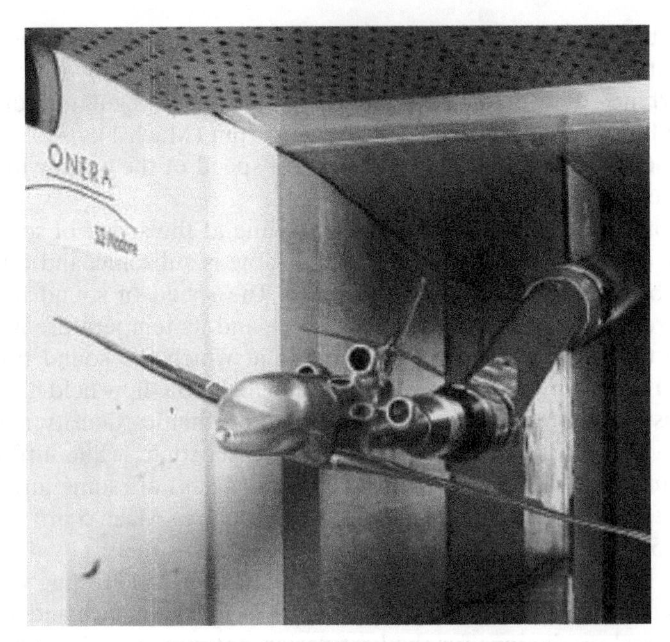

Figure 5.60. Testing in ONERA—The French Aerospace Lab [175].

A water tunnel can also be used to perform some of the above-mentioned operations.

A new wind tunnel was to open in early 2008. It is a commercially available full-scale, single-belt, rolling road wind tunnel. This tunnel is located near Charlotte, North Carolina, in the United States. According to company officials, Windshear Inc. [176] will operate the one-of-a- kind facility 24 hours a day, 7 days a week, with a staff of 25. The facility will be available for hire by all motor sports teams and the auto manufacturers.

The Windshear design accommodates 100% full-scale vehicles and provides constant airspeeds of up to 180 mph, with temperatures controlled to within ±0.55°C (1°F). The high-tech rolling road accelerates from 0 to 180 mph (290 km/h) in less than 1 minute. The "road" is actually a continuous stainless steel belt just 1 mm thick and is designed to last up to 5000 operational hours. During testing, "through-the-belt" sensors measure the aerodynamic down force under each tire while a sophisticated onboard data acquisition system collects other test-critical data (Fig. 5.61).

5.5.5 Solar Radiation Test Chambers

Solar simulators and light sources produce a simulated solar spectrum that is a part of multi-environmental testing. Like other types of mechanical and multi-environmental testing, testing in solar test chambers is a component of

TABLE 5.16 Facilities of the DNW Business Unit NOP [193]

	Type	Test Section $w \times h$	Maximum Velocity (m/s)	Mach Range	Reynolds No. (Based on $0.1\sqrt{S}$)	Maximum Power (MW)
Large low-speed facility (LLF)	Continuous, atmospheric wind tunnel	6×6 m closed or slotted walls	152	$0 \div 0.44$	6.0×10^6	12.5
		8×6 m closed or slotted walls	116	$0 \div 0.34$	5.3×10^6	12.5
		8×6 m open jet	85	$0 \div 0.25$	3.9×10^6	12.5
		9.5×9.5 m closed walls	62	$0 \div 18$	3.9×10^6	12.5
Low-speed Wind Tunnel Braunschweig (NWB)	Continuous, atmospheric wind tunnel	3.25×2.8 m closed or slotted walls	90	$0 \div 26$	1.8×10^6	1.4
		3.25×2.8 m open jet	75	$0 \div 22$	1.5×10^6	1.4
Low-speed tunnel	Continuous, atmospheric wind tunnel	3.0×2.25 m	80	$0 \div 23$	1.4×10^6	0.7

TABLE 5.17 Facilities of the DNW Business Unit ASD [193]

	Type	Test Section $w \times h$	Mach Range	Pressure Range (10^5 Pa)	Temperature Range	Reynolds No. (Based on $0.1\sqrt{s}$)
High-speed tunnel (HST)	Continuous pressured wind tunnel	2.0×1.8 m (2.0×1.6 m)	$0.1 \div 1.35$	$0.2 \div 4.0$	300 k ÷ 310 k	9×10^6
Supersonic tunnel (SST)	Block-down wind tunnel (t10 s ÷ 40 s)	1.2×1.2 m	$1.2 \div 4.0$	$1.5 \div 15$	290 K	15×10^6
Continuous supersonic tunnel	Blow-down wind tunnel (t = hours)	0.27×0.27 m	$0.3 \div 4.0$	$1.3 \div 40$	290 K	4×10^6

TABLE 5.18 Facilities of the DNW Business Unit GUK [193]

	Type	Test Section $w \times h$	Mach Range	Pressure Range (10^5 Pa)	Temperature Range	Reynolds No. (Based on $0.1\sqrt{s}$)
Transsonischer windkanal gottingen (TWG)	Continuous wind tunnel	1×1 m	$0.3 \div 0.9$ with adaptive walls $0.3 \div 1.2$ with perforated walls $1.3 \div 2.2$ with flexible Laval nozzle	$0.3 \div 1.5$	293 k ÷ 315 k	1.8×10^6
Kryo-Kanal Koln (KKK)	Continuous wind tunnel	2.4×2.4 m	$0 \div 0.38$	Ambient	100 k ÷ 300 k	9.5×10^6
Kryo-Roch Windkanal Gottingen (RWG)	Intermittent Ludwieg tube ($t = 0.6$ s ÷ 1.0 s)	0.4×0.35 m	$0.3 \div 0.95$ with adaptive walls	$2.0 \div 10.0$	100 k ÷ 300 k	15×10^6
Rohr Windkanal Gottingen (RWG)	Intermittent Ludwieg tube with two legs ($t = 0.4$ s)	Leg A: 0.5×0.5 m	3.0; 4.0	10.0	300 K max	3.5×10^6
		Leg B: $\varnothing = 0.5$ m	5.0;6.0;6.8	36	700 K max	2.2×10^6
Hochdruck Windkanal Gottingen (HDG)	Continuous wind tunnel	0.6×0.6 m	$0 \div 0.1$	$1.0 \div 100$	Ambient	12×10^6

Figure 5.61. Wind tunnel rolling road machine [176].

ART and ADT. Solar radiation test chambers also conduct plant growth, material-degradation simulations, photovoltaic (PV) cell testing, sunblock evaluations, and other testing.

The following text includes typical representative equipment for this type of testing. Solar simulation cabinets and systems simulate the energy of global radiation. They range from small, single-luminary exposure configurations to large-scale, multiple-luminary systems capable of exposing automobiles and other industrial products. The spectrum of sunlight (so-called global radiation) is reproduced during testing, facilitating the systems' performance decisions. The basic components of the solar constant unit are the radiation unit, the power supply, and the control system.

Different companies in the United States, Germany, India, China, and other countries design and produce this testing equipment.

One of the most famous companies in this area is Atlas Material Testing Technology (MTT) LLC [177, 178].

The company produces test chambers for high-speed lighting. The boost capability of this technology allows it to provide a light output 10 times greater than that of a tungsten system of a comparable installed lamp power.

The radiation unit uses a special metal halide lamp system (metal halide global [MHG]). The electronic power supply (EPS) drives the lamps with square wave currents that stabilizes and optimizes spectral characteristics. A dedicated computer called SolarSoft controls the solar constant unit, using a menu-driven software program that provides ease of use and fast test setup allowing a maximum utilization of the test facility.

Atlas produces the SC series of solar radiation test chambers.

As an example, look at the test chamber SC2000 MHG (Figs. 5.62 and 5.63):

- Test space volume: 3.4 m^3
- Test space dimensions, $W \times D \times H$: 2000 × 1150 × 1510 mm

Figure 5.62. Test chamber Atlas Material Testing Technology (MTT) SC2000 [178].

Figure 5.63. Inside components of test chamber Atlas Material Testing Technology (MTT) SC2000 [178].

- Irradiation unit type of irradiation: 2 × 4-kW MHG lamps
- Irradiation intensity: 800–1200 W/m^2 with reference to the test area, steplessly adjustable
- Uniformity: ±5% with reference to the test area
- Test area: 1700 × 800 mm, at a distance of at least 600 mm below the selling glazing

- Special radiation distribution global radiation: 280–3000 nm, recommended for aging tests

GAGA Instruments Pte Ltd [179]. produces a climatic chamber with a global sunlight spectrum. An economic alternative to large solar simulation equipment with a capacity of 6000 L, the test chamber is designed for testing medium-sized components. The Sun/Solar simulator S-13, 13" diameter, and 6000 K (Optical Energy Technologies Inc.), has the following features:

- 250-W metal halide arc lamp, 6000-K color temperature
  - 575-W lamp optional at $1500 additional cost
- Three times the luminous efficiency of an argon arc lamp
- Parabolic reflector produces 13" diameter beam
- Less than 3.5° angular subtense of the simulated sun
- 1.0 solar constant in space (adjustable)
- Useful in vacuum chamber (with feedthrough)
- Useful for testing sun sensors and solar panels
- Built-in autocollimation function option

Applications of this chamber, DIN 75220, include "Aging of automotive components in solar simulation units," SAE International testing methods, MIL-STD-810, Environmental Protection Agency (EPA), or others.

Pacific Laboratory Products [180] produces a series of the solar climatic test chambers with the following specifications:

- Models: SC XXX MH—four different types
- Test space volumes: 340, 600, 1000, and 3400 L
- Temperature range without radiation: –30 to +100°C
- Temperature range with radiation: –20 to +100°C
- Humidity range without radiation: 10–90% RH
- Humidity range with radiation: 10–80% RH
- Radiation intensity: approximately 1000 W/m^2

Nogix Co. Ltd. [181]. designed and produced a solar radiation test chamber that can create a continuous and instant light source for products and includes environment temperature and normal temperature conditions.

Illumination level: 60,000 ± 5000 lux
Temperature range: 0 to +100°C

5.5.5.1 Solar Simulation Systems. A car is a true representation of products that have to withstand various types of continuously changing environmental conditions. Atlas MTT (K.H. Steuermagel [KHS]) Solar constant solar simulation [40] systems provide solar energy for the testing needs of the automotive industry.

The radiation of the sun together with the ambient temperature and RH are essential factors. Solar radiation is global or total radiation that is the sum of direct and diffused (sky radiation) solar energy.

Regarding correlation, where it is important to achieve a comparison with the quantitative (irradiance) and qualitative (spectral distribution and uniformity) characteristics of natural global radiation, the simulation of effects, which are known by practical experience to be critical in achieving test results, need to be considered.

General requirements are similar for most solar simulation applications, whether it is for the examination of organic materials, the optimization of ventilation and air conditioning systems, monitoring the thermal stability, and interaction of components or emission testing.

The solar simulation system is also important for the consideration of other fields such as solar energy convergence systems, PV, agricultural, and other applications.

As for other solar radiation simulation systems, the basic components of the solar constant systems are the radiation unit, the power supply, and the control system.

A mechanical positioning system is often used to allow the solar constant system to simulate various natural solar conditions effectively. This enables motorized movement of the solar array within all axes for easy adaptation to various test configurations or to simulate natural solar day cycles. The control of the positioning can be a manual push-button or integrated into the SolarSoft program that will then offer automated control of the radiation along with the simulation of various sun positions in the sky. The positioning system is often unique to the application and test facility.

The companies that use the solar constant solar simulation systems utilize systems specifically designed to comply with established test methods, such as the DIN 75220 "Aging of automotive components in solar simulation units," SAE International methods, MIL-STD-810, EPA methods, or other methods.

Figures 5.64 and 5.65 demonstrate a solar simulation system application for automobiles and other products.

The large test chambers in Figures 5.64 and 5.65 are components of systems for ART/ADT.

ESPEC ENX112 Solar Panel Compact Walk-In Chambers. Compact Walk-In Chambers for Testing Solar Panels

The ENX112 chambers provide a test chamber large enough for testing full-size PV modules in the least amount of floor space. They can be used for panels 1.2 × 2.0 m or larger.

These models (Fig. 5.66) provide a variety of performance capabilities to meet maximum or minimum change rates in the International Electrotechnical Commission (IEC) or UL test standards. Incorporated are the features of

Figure 5.64. Large solar simulation system for testing automobiles and other products [40] (Atlas Material Testing Technology [MTT] LLC).

Figure 5.65. Sample drive-in chamber for full vehicle testing (Cincinnati Sub-Zero).

smaller Platinum and Global-N lines. The refrigeration systems of these models utilize modern, high-performance scroll compressors that allow a small footprint while allowing unparalleled service access.

The models come in different performance versions, depending on the customer's application. More powerful performance allows the customer to complete the temperature cycling test faster, or with a maximum number of modules. There is also a "damp heat"-only model that saves capital cost and energy.

Figure 5.66. Walk-in chamber for testing solar panels (ESPEC).

Features

- Performance to meet IEC 61215, 61646, UL-1703, and similar test requirements
  - 10.11 Temperature cycling
  - 10.12 Humidity freeze
  - 10.13 Damp heat
- Large interior volume
- Touch-screen controller
- Viewing window
- Compact footprint of 5′ × 11′
- Optional module rack with rollers
  - ENX112 rack can hold 1.2 × 2.0 m panels
  - ENX133 rack can hold 1.2 × 2.4 m panels
  - ENX164 rack can hold 1.8 × 2 m panels

EWSX282 Solar Panel Large Walk-In Chambers. Solid Walk-In Chamber Designed for Testing up to 20 PV Modules

ESPEC's solid construction walk-ins are desirable for testing large loads of PV modules or panels at 85°C/85% as well as humidity freeze and temperature cycling tests. The hermetically welded seams and stainless steel interior ensure integrity under extreme conditions required by IEC.

The models come in different performance versions, depending on application. More powerful performance allows faster completion of the temperature cycling test, or with a maximum number of modules. The "damp heat"-only model saves capital cost and energy.

5.5.6 Dust and Sand Test Chambers

Dust is everywhere. It seems to enter from fissures and cracks and settles on almost everything in sight. One can expect to find fragments of plants, molds, insects, pollen, spores, microorganisms, mineral dust, and many other substances. Dust influences the brake process, friction, wear, corrosion, and many other processes that influence a product's quality, reliability, and durability.

To simulate multi-environmental factors, one uses dust and sand chambers. Different companies design and manufacture different series of sand and dust chambers. Cincinnati Sub-Zero (CSZ) [182] designs and produces the USD Series units to expose automotive, electronic, and other product components to concentrated levels of dust in order to validate the products' seal integrity.

These units blow pulses of compressed air through a manifold located in the bottom of a collection trough, forcing dust up and over the product. Air pulses are time programmed for "on" and "off" cycles. Changing the air pressure and amount of the blowing time easily varies the dust concentration. The design of all dust chambers meets common test methods such as SAE J575, subparagraph "G," J.I.S., D-0207-1977, and ASTM C-150-77. Approved dust types are Arizona Dust, Portland Cement Dust, and 213 Silica Sand.

Standard features (Fig. 5.67) include

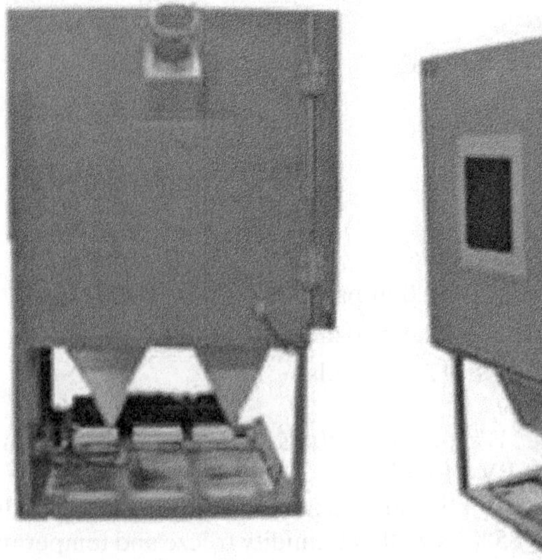

Figure 5.67. USD Series sand and dust chambers [182].

- Two digital timers
- Coalescing and particulate filters
- Exhaust with cleanable filter
- 16-gauge stainless steel interior
- Full-operating front-loading door
- Three-inch thick foam insulation
- Casters

Optional accessories include

- Access ports
- Viewing window
- Shelving
- Text fixture support rails
- Self-contained air compressor
- Interior light with switch

Workspace volume: 10 ft^3 (282 L) 27 ft^3 (762 L)

Workspace dimensions: 24" $H \times$ 30" $W \times$ 24" D 36" $H \times$ 36" $W \times$ 36" D (61 × 76 × 61 cm) (91 × 91 × 91 cm)

Exterior dimensions: 60" $H \times$ 55" $W \times$ 35" D 72" $H \times$ 61" $W \times$ 45" D (152 × 140 × 89 cm) (183 × 155 × 114 cm)

Weiss Umwelttechnik GmbH Dust Chambers allow for reproducible testing of the resistibility of electrotechnical products to dust. These chambers also determine the degree of ingress protection (IP) type for housings of electrical devices and equipment of road vehicles (Table 5.19).

There is a series of dust test chambers (ST 600, ST 1000, and ST 2000) designed and produced by Weiss Umwelttechnik GmbH.

Dust Test Chamber Type ST 600 (Fig. 5.68)

Features include

- Dust test with wind simulation
- Housing made of galvanized sheet steel, lacquered
- Ring dust with integrated test space

TABLE 5.19 Parameters of WEISS Dust Test Chambers (Type ST) [189]

Test Chamber Dimensions			Overall Dimensions			Door Opening	
Height (mm)	Width (mm)	Depth (mm)	Height (mm)	Width (mm)	Depth (mm)	Height (mm)	Width (mm)
1000	950	950	1900	1250	1050	850	850

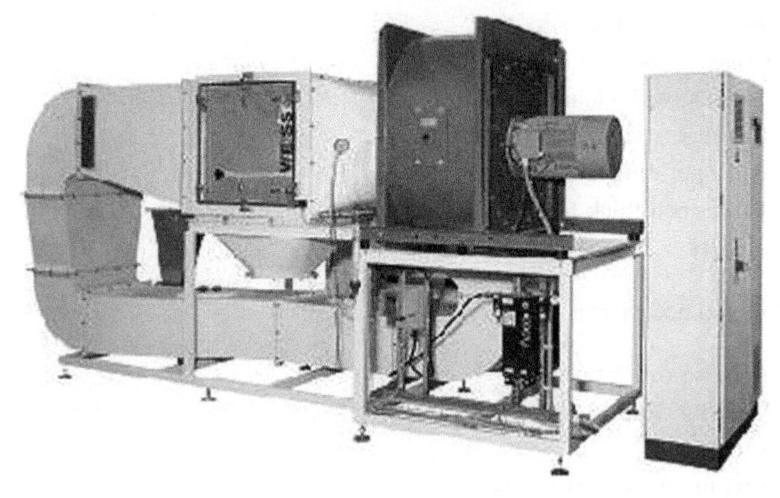

Figure 5.68. Dust test chamber type ST 600 [189].

- Horizontal airflow with temperature conditioning system
- Air velocity control (speed controllable recirculating air fan)
- Abrasion resistance of all components in contact with the dust
- Large doors for easy loading with surrounding special gaskets
- Dust measurement and control by means of a dust dosage device
- Dehumidification system (compressed air dryer)
- Operating hour counter

The specification includes

Temperature range: +5°. . . +40°C
Specimen weight: max 100 kg
Air velocity: 1.5, 3.0, and 5.0 m/s
Dust reservoir: 20 kg
Dust concentration: 2- to 10-g/m^2 dust

5.5.6.1 Dust Test Chamber Type ST 1000. According to DIN VDE 0470 T1 or EN 60529 for testing of the protected classes IP 5 X and IP 6 X, the ingression of dust into enclosures is tested. IP 5 is dust protected. IP 6 is dust tight. The system can be equipped with an underpressure device so that the dust not only settles on the entire specimen but also can ingress into the specimen.

The main features include

- Waste air exhaust via dust filter
- Abrasion resistance of all components in contact with the dust

- Transparent doors for easy charging of surrounding special gaskets
- Dust collecting section below the test chamber
- Wiper
- Vertical airflow

Technical specifications include

Test space volume: 1 m^3
Test room lighting
Mobile design
Ports: 50 and 100 mm Ø

5.5.6.2 Dust Test Chamber Type ST 2000. According to DIN VDE 0470 T1 or EN 60529 for testing for the protection classes IP 5 X and IP 6 X, the ingression of dust into enclosures is tested. The system can be equipped with an underpressure device so that the dust not only settles on the entire specimen but also can ingress into the specimen.

Main Features

- Waste air exhaust via dust filter
- Abrasion resistance of all components in contact with dust
- Transparent doors for easy charging of surrounding special gaskets
- Dust collecting section below the test chamber
- Wiper
- Vertical airflow

Technical Specifications

Test space volume: 2000 L
Test room lighting
Mobile design
Ports: 50 and 100 Ø

Envirotronics Inc. designs and produces the D Series chambers that are available in models D24 and D60. These chambers test the component's resistance to a dust-filled environment defined in SAE Dust Test Specifications J-575, subparagraph "G." The product is placed on the specimen rack and dust is agitated throughout the chamber by injecting compressed air into each of the dust troughs. The dust then settles, covering the product under test.

The D Series chamber consists of 16-gauge sheet metal with an air-dried lacquer-sprayed finish. The hinged lid gives full access to the workspace and is available with the option of viewing windows. In the larger D60 model, 8" to 10" fan blades are mounted in the chamber to keep the dust in an agitated condition. The pressure relief vent provided prevents pressure buildup within the cabinet as well as the escape of dust material.

A compressed air pressure regulation system with a filter and trap provides for connection to the air supply. Air supply air must be at a minimum pressure of 80 psi and 10 CFM. The D Series' control timers allow the following to be preset: the agitation, the number of agitation intervals, and the number of cycles with automatic shutdown at the completion of the preset number of cycles.

Standard features include the following:

- The total access chamber lid stays open at a 60° angle.
- The chamber lid has gaskets with a closed-cell foam material.
- The lid is equipped with quick-release latch hardware.
- Pressure relief vent with filter
- Removable stainless steel with grate specimen platform
- Removable spray nozzles
- Compressed air pressure regulation system with filter and trap
- Two adjustable interval digital timers and one counter timer
- Compressed air dual tower air drier system that dries air to a –85°C dew point level
- 115–1–60 V power-in with an amp draw of a 5 A
- Interior light with exterior switch

Settling Dust Chambers. Easy Loading Clamshell Design

The ESPEC EDC dust chambers (Figs. 5.69–5.71) provide a ready-made solution to common dust test standards for automotive and electronic cabinet requirements. Often specified for use with "Arizona fine dust," these units can also be used with talcum or concrete powder. The chamber has a "clamshell"-type door/lid for easy front-loading, compared with top-loading chambers. Optional reach-in glove ports allow rotating the test sample without opening the door.

Test Standards Met

- SAE J575 (Rev. NOV2006), Section 4.5
- ISO 20653 (2006-08-15), Section 8.3.1
- DIN 40050, Part 9, IP Codes 5 K and 6 K, vertical dust flow
- IEC EN 60529 (IEC 529), IP-5 and IP-6, vertical dust flow (with optional underpressure mode)
- IEC EN 60068-2-68 (IEC 68-2-68), LA2, vertical dust flow

Features

- Stainless steel interior and exterior
- W-shaped bottom for improved dust collection and dispersion

Figure 5.69. Dust test chamber (ESPEC).

Figure 5.70. Dust test chamber EDC-54-ESPEC.

- Easy-access clamshell door with viewing window and optional glove ports
- Mesh product shelf
- Two-inch cable port
- Timers for agitation and settling periods

Figure 5.71. Dust test chamber in use (ESPEC).

Options

- Two reach-in glove ports
- Dry air purge to maintain dryness
- Evacuation (underpressure) system for DUT, as required by IEC 60529
- ISO 12103-1 Arizona test dust

EDC-54 Cubic Foot Setting Dust Chamber Specifications

Item	Value
Product family	Setting Dust
Interior volume	54 ft^3 1500 L
Interior dimensions ($W \times D \times H$)	1800 × 900 × 750 mm

5.5.7 Ozone Test Chambers

Ozone in the air leads to the formation of cracks in the rubber and other elastomers that reduce the reliability and durability of the product. Antioxidants and ozone waxes protect most types of rubber used for this purpose.

Therefore, multi-environmental testing includes ozone test chambers that include the simulation and the measure of rubber aging. Companies in the United States, Germany, India, China, and others produce test chambers for this purpose.

In USA Inc. [183] designs and manufactures a line of ozone test chambers, including a high-concentration ozone analyzer, an off-gas ozone analyzer, an

Figure 5.72. Model OTC-1: ozone test chambers [183].

ozone safety analyzer, and a dissolved ozone analyzer. Each ozone test chamber is available for different applications and is able to measure ozone in both gas and liquid phases. Single and multichannel ozone models are available, and all models use the proven UV technique to make measurements. One can program ozone test chambers to run automatically and unattended with user-specified parameters for ozone concentration, test start time, and test duration.

USA Inc. produces ozone test chambers for various purposes (e.g., Fig. 5.72):

- Weathering of rubber and cables
- Colorfastness testing of textiles
- Image performance testing of photographic paper
- Other testing of environmental exposure

The ozone test chamber, OTC-1, consists of the following modules:

- Ozone generating module
- Ozone monitoring module
- Test chamber module
- Ozone scrubber
- Safety interlocks module

The ozone monitoring and ozone generating modules work together. Typical features include the following:

- Closed-loop servo control ensures a highly stable ozone concentration
- Simple menu-driven display
- Unprecedented uniform exposure of all specimens using a rotating integrated ozone destructor with built-in safety interlocks
- No separate exhaust hood required
- Ozone-resistant wetted materials

- Analog and digital output capability
- Calibration traceable to the National Institute of Standards (NIST)
- Accurate UV-based ozone analyzer for stable ozone concentration readings

Argentox Ozone Technology GmbH [184] designs, constructs, supplies, and installs complete ozone test cabinets to meet their customer requirements.

In the past, the coulometric constant-flow electrolyte method was frequently used to determine and check ozone concentrations. Presently, the UV absorption method is used.

Specifications of the Argentox ozone test chambers include

- Temperature range: –5 to +90°C
- Humidity range: 15–90% relative
- Ozone concentration
  - Test ranges: 0–100 pphm
  - 100–1000 pphm
- Test chamber volume
  - 140 L 500 × 440 × 600 mm
  - 300 L 900 × 540 × 600 mm
  - 450 L 1100 × 650 × 700 mm

Ozone Technologies produces ozone test chambers for

- Accelerated aging studies
- Product quality control and quality assurance
- Research and development

Ozone test chamber features include

- Ozone ranges: 25, 50, 100, and 200 pphm
- Temperature ranges to +90°C, UV absorption analyzer or semiconductor-based analyzer
- Static and dynamic test setup

Anseros Klaus Nonnenmacher GmbH developed and produced the series ozone test chambers. The chamber SIM 6300 has the following testing facilities (Fig. 5.73):

- Static test with or without rotation
  - Six to eight sample locations
  - Variable, adjustable strain on each location
  - Rotating carrier, one revolution every 2 minutes

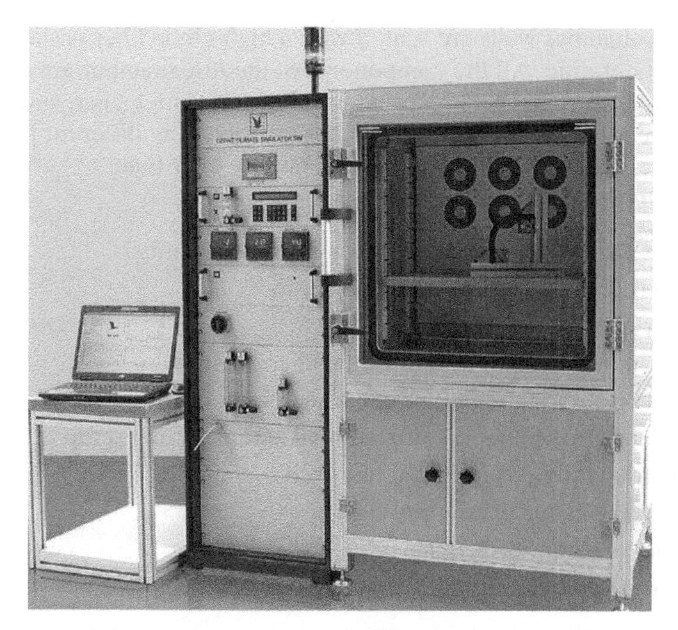

Figure 5.73. Ozone test chamber SIM 6300 [190].

- Dynamic test with or without rotation
    - Eight sample locations
    - Adjustable strain 5, 10, 15, 20, 25, 40, and 60%

Ozone Test Chamber Control Unit

- Complete control and monitoring of ozone test sequence
- Computer interface via Rs-185
- Customized control, reporting, and monitoring
- Ozone measurement by absorption analyzer
- Safety system via interlocking circuitry
- Chemical free operation

Special Features of the Ozone Test Chamber

- PLC and PID operation
- Automatic changeover testing steps like countdown
- Test run and exhaust
- Adjustable operational and working parameters
- Digital ozone output indicator

Stainless chamber walls are seated with a high-grade fiber gasket to assure leak-proof operation. All the components of the test chamber are in one self-contained unit. A strip chart recorder, a solid-state data logger, and/or a computer can be attached to the test chamber for recording different parameters: temperature, ozone concentration, and humidity over time.

The components of an ozone test chamber include

- Ozone generation and distribution into test chamber
- Desired temperature maintaining setup
- Proper air circulation setup
- Ozone concentration monitoring and correction
- Chamber environmental air exhaust system
- Chamber environmental air sampling and analyzing setup
- Sequential operation of total test cycle
- Static or dynamic testing provisions
- Sample rotation mechanism
- Safety interlocks and indications

ANSEROS Klaus Nonnenmacher GmbH produces also the ozone test chamber SIM 7500 (Figs. 5.74–5.77). SIM 7500 is especially designed for cable and seal testing: automatic PLC control with digital and analog data acquisition as well as control of temperature, humidity, and ozone concentration. Different tools help to carry out static and dynamic tests.

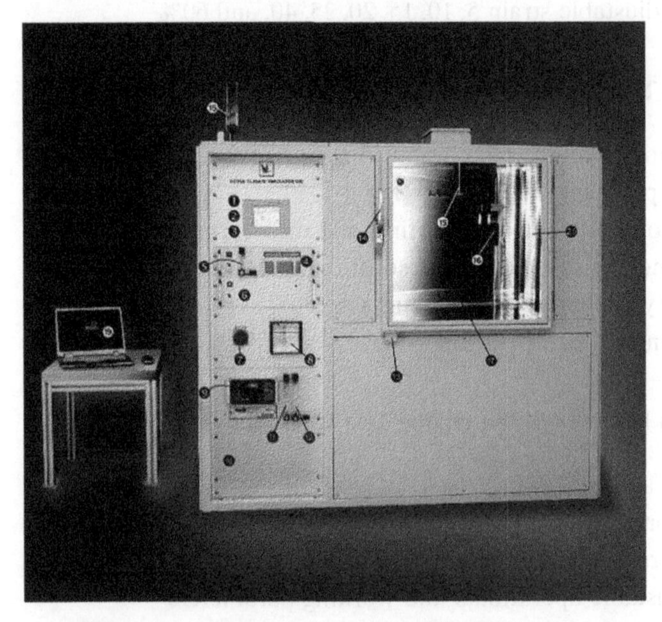

Figure 5.74. Ozone test chamber SIM 7500 [190].

Figure 5.75. Area for application ozone test chamber SIM 750 (aerospace) [190].

Figure 5.76. Area for application ozone test chamber SIM 750 (automotive) [190].

ANSEROS also produces the ozone test chamber SIM 8000 (Fig. 5.78) with the following features:

- Ozone climate control
- Optional test equipment like UV radiation and fogging
- Built-in rotation and dynamic tools
- Inside coating in stainless steel, welded
- Wide tight doors (with safety system) for movement of heavy goods

Figure 5.77. Area for application ozone test chamber SIM 7500 (powertrain) [190].

Figure 5.78. Ozone test chamber SIM 8000 [190].

- Remote control of all test ranges
- Full automatic operation with synthesizing and synchronization
- Software for data storage, internet compatible

Specifications include

- Approvals: DIN 53509, ASTM D1149, and ISO 1431
- Volume of chamber: 8 m^3
- Ozone range: 25 . . . 200 pphm (or on request)
- Temperature range: –35 . . . 130°C (or on request)
- Humidity range: 10 . . . 95% RH, 100% with fogging equipment
- Software: AMACS 2.1
- Recorder: OPTION
- Climate: chiller
- Inside dimensions of the chamber
  - Width: 2000 mm
  - Height: 2000 mm
  - Depth: 2000 mm
- Outside dimensions of the chamber
  - Height: 2380 mm
  - Depth: 2530 mm/3500 mm (with chiller)
  - Width: 1400 mm
- Dimension of door
  - Height: 2000 mm
  - Weight: 900 kg

Mast/Keystone [185], a manufacturer and distributor of ozone test chambers, produces several models of ozone test chambers: Model 700 (700-10LTA, 700-10LTB, and 700-14LTB), as well as others. The 700-10 LTA ozone test chamber provides continuous, uniform temperature and ozone control. The measurement and control system consists of the STAT-1900 Single Set point temperature controller for uniform thermal performance under all loading conditions and the Mast Development 727-#A Ozone Monitor with power flow control assembly to ensure uniform and constant control of ozone levels.

Applications include accelerated and continuous-duty ozone testing of rubber and plastic components.

Specification of the ozone test chamber Model 700-10LTA (Fig. 5.79) includes the following:

- Temperature range: +15°C above ambient to +343°C (+650°F)
- Temperature uniformity: ± 1% of set point

Figure 5.79. 700 ozone chamber [185].

- Ozone control: ± 4% of set point
- Ozone concentration range: 0–10 ppm by volume (conservatively rated using atmosphere level)
- Air replacement rate: 1–8 ft^3/min
- Interior dimensions: 20" $W \times 20$" $D \times 25$" H
- Exterior dimensions: 45" $W \times 36$" $D \times 69$" H
- Voltage: 208/240 V, single phase, 50/60 Hz

Model 700-14LTB has a temperature range from +3 to +93°C (+200°F) and an RH range from 40 to 95%.

5.5.8 Accelerated Weathering Testing

Weathering is a collective term for the processes whereby rock at or near the earth's surface disintegrates and decomposes due to the actions of atmospheric agents, water, and living organisms.

Some of these processes are mechanical, like the expansion and contraction caused by sudden large changes in temperature, the expansive force of water freezing in cracks, the spitting caused by plant roots, and the impact of running water. Other processes are chemical, like the oxidation, hydration, carbonization, and loss of chemical elements by the solution of water. Weathering is important because it aids in the formation of soil and prepares materials for degradation (erosion).

There are two types of weathering testing: outdoor natural weathering testing and accelerated weathering testing in special test chambers. Accelerated weathering testing is considered a goal in this book.

More companies are turning to accelerated weathering due to the increased demand for quickly obtaining quality decision-making data. Nevertheless, no accelerated weathering program can be complete without the confirmation from and correlation with natural weathering. Natural weathering provides the data you need to ensure that your product has protection against costly liability issues.

As an example of accelerated weathering testing, let us show the accelerated aging and weathering testing for plastics and polymerics that Intertek PTLI provides. The weathering laboratory exposes the plastic samples to cycles of intense light, heat, and water. Accelerated testing does not provide the information to predict exactly the plastic aging performance in actual outdoor exposure, but the controlled tests do provide meaningful material comparisons that help engineers make informed choices. Intertek PTLI also provides performance evaluations before and after exposure tests.

Accelerated weathering testing includes	Aging performance evaluations before and after exposure include
• Xenon arc exposure ASTM D2565, ASTM D4459, ASTM DG155 ISO 4892, SAE J1885, SAE J1960 • Highly predictive accelerated weathering Xenon Arc HPAW • Fluorescent light exposure accelerated weathering • UV radiation, moisture, and heat QUV ASTM D4329, D4587, ISO 4892, SAE J2020 • and more	• Optical tests • Gloss ASTM D2457, ASTM D523 • Refractive index ASTM D542, ISO 489 • Haze and luminous transmission ASTM D751, ASTM D1044 • Three-dimensional color measurement to XYZ or LAB scale ASTM D6290, ASTM E1347 • Gray scale per AATCC • Yellowness index, ASTM E313 • Mechanical properties and testing

Figure 5.80. Ci4000 Weather-Ometer with touch screen interface [191].

Atlas MTT LLC offers a complete line of weathering testing instrumentation, as well as laboratory accelerated testing services.

To establish a scientifically designed weathering test program, this company developed the Ci4000 Weather-Ometer (Figs. 5.80 and 5.81). It represents a huge advantage in the application of state-of-the-art digital and optical methodologies to large-capacity laboratory weathering instrumentation. It also establishes new standards of performance for reproducibility, repeatability, and operating efficiency.

The Ci4000 simulates solar radiation using xenon lamps and a filter system. This system simulates UV, IR, and visible components of sunlight. The tailored interchangeable glass filters enable the xenon light spectrum to match light conditions in a product's end-use environment.

The digital control system makes access to its most sophisticated features available to all operators. This system includes new icons that make getting to the necessary information fast and easy; 14 factors are preprogrammed for test methods, multilingual capability, and automated two-paint irradiance calibration. The electric sensor provides measurement of the RH and enables automatic control at the specimen level. The Smart Dampler™ reduces test variability in chamber temperature and humidity and compensates for changes in ambient laboratory conditions.

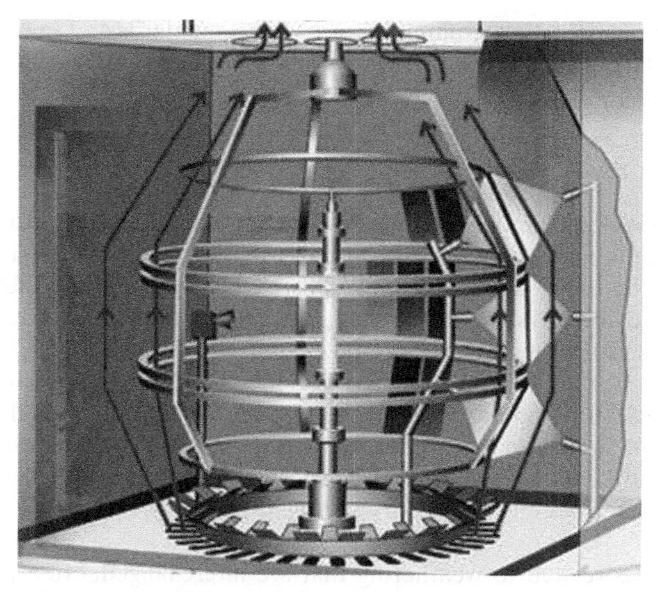

Figure 5.81. The rotating sample rack within the test chamber of the Ci4000 maximizes exposure uniformity over all specimens [191].

This chamber is appropriate for testing products such as auto materials, plastics, inks, paints, coatings, and packaging.

In 2008, Atlas introduced two new Ci5000 Xenon Arc Weather-Ometers with enhanced digital control to give operators more flexibility and control for accelerated weathering testing.

Some of the most notable changes for both units include

- A modern, full-color TFT touch screen display with intuitive icons for easier interpretation of operating parameters and warnings
- An embedded control system that replaces the PLC controller of the previous generation
- A robust digital network (replacing analog control circuits) that is more reliable, provides control that is more accurate, and enhances monitoring of the data
- A significantly increased memory capacity allows additional features and functions previously managed by external controllers, such as water resistivity and lamp water temperature.
- The ability for operators to more efficiently analyze instrument performance, to monitor test parameters, and to perform calibration procedures

The other features and benefits of the new control system are

- Subcycle repeat programming for copying standards and saving them as templates
- Full-color trend plot screen with a large memory
- The streaming data output is formulated for compatibility with modern laboratory information management systems

The low-voltage model will use the same high-quality 12-kW water-cooled lamp, but redesign also enables the unit to use the same incoming line voltage as the discontinued Ci65A series Weather-Ometer. For these older instruments, the control system is becoming obsolete or hard to source. Atlas has made it easier than ever to replace those units.

In 1989, a group of U.S. domestic automotive OEMs and their suppliers developed SAE J1885 and SAE J1960 weathering test methods that qualify interior trim and exterior material for weather durability respectively [186].

J1885 and J1960 are instrument-specific standards because they require the use of specific models of laboratory testing instruments, such as the Atlas Ci35 and Si65.

Two terms related to weathering that are interchanged, often incorrectly, are the parameter set point and the specimen condition. Parameter set points are the levels at which an instrument controls test parameters, such as temperature, RH, or irradiance during operation. They are stipulated in current test methods and are set or programmed by an operator. Specimen conditions are the microconditions of irradiance, temperature, and moisture that exist at, or about a specimen but are not set directly by the operator.

Though the two are typically correlated, they are seldom equal. Perhaps more importantly, the parameter set point and the specimen condition will have a different relationship with one another in different instruments, particularly if the instruments are of different geometric designs.

In performance-based specifications, only the test parameters and their associated tolerances are specified as the nominal spectral power distribution requirement along with liberal tolerances to accommodate various manufacturers' lamps.

Because equal sets of test control parameters in different types of instruments do not necessarily result in identical test conditions on samples, the specification of set points alone is not enough to guarantee similar test results.

The most meaningful parts of the qualification tests are based on the requirements to successfully test standard reference materials (SRMs). By definition, SRMs provide the ability to determine objectively if test conditions in various instruments are identical.

5.6 EQUIPMENT FOR ELECTRICAL TESTING

It is necessary to select from different types of electrical testing equipment depending upon the test subject:

1. Battery testers and fuel cell test equipment that consist of specialized test stations, stands or systems, monitors, and component modules for performance or endurance testing. Test stands or systems may include load banks, test controllers, gas or fuel supply modules, temperature control modules (e.g., ovens and chillers), and data acquisition units. Battery testers, monitors and analyzers, and diagnostic systems monitor overall or cell voltage, charge, amperage, DC resistance, or temperature to indicate battery condition or performance.
2. Burn-in test equipment uses elevated voltages, temperature, and power cycling to evaluate high-power chips, circuit boards, or products. The burn-in process accelerates failures normally known as "infant mortalities" in a device. The burn-in process tests the quality of the semiconductor device before incorporation into a finished device, ensuring that integrated circuit (IC) chips and other microprocessors with latent defects are weeded out. Burn-in test equipment accelerates potential failures in substandard products. Devices that survive a burn-in period are usually free of early failures and other operational problems. Burn-in test equipment is also known as accelerated life testing equipment. This equipment uses specific machines to test memory modules, logic modules, linear components like voltage regulators, mixed signal components such as analog switches and multiplexers, and discrete components like diodes and transistors.
3. Current leakage testers measure the amount of current that leaks to a ground conductor. They include a measuring device or probe that connects to a conductive point, a voltmeter that displays root mean square (RMS) values, and a circuit with specific resistance and frequency characteristics. Current leakage testers operate during both normal and single-fault conditions to determine whether electrical devices pose a shock hazard. For medical devices and other specialized equipment, additional single-fault conditions may be required. Manual, automatic, and semiautomatic current leakage testers are available. Manual devices require the operators to set and change test parameters. Automatic devices are programmable, fully automated, and can often perform an entire series of electrical safety tests in succession. Semiautomatic current leakage testers combine features from both manual and automatic devices.
4. EMC testers are electronic devices used to test or monitor parts and products for EMC. Typically, EMC testers test, monitor, and measure levels of ESD. ESD is the sudden and monetary flow of electric current between two objects at different electrical potentials. Static electricity often causes ESD. In some cases, ESD damage from tribocharging of electrostatic induction may produce visible or audible sparks. On others, ESD subjects semiconductor materials to high voltage. ESD can destroy sensitive ICs in instruments without proper protection. ESD is a common

factor in everyday life since even movement on a carpet can generate significant voltages that can damage equipment.

In addition, short-duration overvoltages on electrical power lines, caused by a variety of conditions, can seriously damage unprotected equipment. Poor insulation within a product can allow damaging short circuits that may appear only after the component experiences other environmental conditions.

One can check the product to determine its resistance to these and other electrical dangers, as well as determine how well the product functions after other tests. This also may affect

- Connectors
- Dielectric materials
- Insulation resistance
- And others

EMC testers can also simulate lighting strikes or short-circuit fault currents that can generate high-level, short-duration magnetic fields.

Ground testers are electrical safety test devices that perform electrical compliance tests. A variety of standards compliance bodies mandate ground testing on all electrical and electronic devices to ensure that people contacting the devices will not be injured by shock or electrical discharge. The two most common ground bond tester types test for earth and ground continuity.

Ground bond testers for earth continuity verify the integrity of the ground connection between the power cord and exposed metals in high current devices and applications. This is achieved by measuring the resistance between the exposed metal parts and the ground blade of the power cord.

Transformer test equipment consists of specialized test modules or systems used to test and/or monitor the electrical and mechanical parameters of transformers and other related devices. Transformer test equipment performs the maintenance of basic loss measurement systems to complete production line testing. Common types of transformers include frequency analysis equipment, loss management equipment, applied potential testing, megohmmeters, turns ratio testing, and other transformer testing applications.

EXERCISES

5.1 Describe the negative aspects of different testing expositions in a situation with equipment for ART/ADT.

5.2 Identify combined equipment for multi-environmental testing, corrosion testing, and vibration testing.

5.3 Can one change a natural medium to an artificial medium? What are the principles for making accurate changes?

5.4 Describe five basic steps of the methodological process to substitute a natural medium for an artificial medium for a waste applicator.

5.5 Why does the simulation of input influences for ART/ADT often need the substitution of artificial media for natural media?

5.6 Where was the first combined test equipment developed for ART/ADT? Show an example of this combined equipment.

5.7 Show a design of combined equipment for an electronic product.

5.8 Describe an example of Weiss Technik's combined test equipment as a combination of different types of equipment for ART/ADT.

5.9 Show the scheme of a climate test chamber with the equipment for dynamometer testing and its specifications.

5.10 Describe the solutions of ESPEC North America Co. as an example of combined test equipment.

5.11 Describe and show the schemes of combined test equipment for mechanical testing.

5.12 Describe the current situation in vibration testing.

5.13 Why does most of the equipment for vibration testing fail to predict how vibration influences the product in real life?

5.14 Compare the trends in the development of vibration testing in the laboratory and proving grounds.

5.15 Describe the basic aspects for the development of electrohydraulic (servohydraulic) vibration testing.

5.16 Show the stages of basic VTE development.

5.17 Describe the specifics of VTE development in the 21st century.

5.18 Show typical specifications for new vibration equipment.

5.19 Describe the basic advantages of new vibration equipment from the standpoint of current vibration equipment.

5.20 The advantages of the new equipment are reflected through specifications. Show the basic specifications of the new vibration technology testing equipment for products (1) with wheels and (2) without wheels.

5.21 Describe the working heads for new vibration equipment.

5.22 Describe and show the structural scheme of a control system for drum vibration equipment.

5.23 Which components comprise the SVC described in this book?

5.24 Describe the principles of operation of dynamometer brake testing.

5.25 Describe how to use the dynamometer for engine testing.

5.26 Describe the types of sweep tests used in dynamometer engine testing.

5.27 What are the specifics of a passenger car's inertia-type brake dynamometer?

5.28 Show the types of chassis dynamometers.

5.29 What is AMET? Describe the current situation in this area.

5.30 Describe the basic reasons slowing the technical progress in AMET equipment.

5.31 Why does the current situation in environmental testing not provide initial information for accurate reliability and durability prediction?

5.32 Show examples of the specifications of current test chambers as an illustration of undeveloped AMET that is used for reliability/durability prediction.

5.33 Show the characteristics of chambers for JIS 0203 Spray Bar Assembly (System #1 and System #2).

5.34 Show the plan of a test chamber for the simultaneous combination of temperature, humidity, pollution, and radiation. Why is this test chamber suitable for ART/ADT?

5.35 Why does not the test chamber for the simultaneous combination of temperature, humidity, pollution, and radiation use a high acceleration coefficient?

5.36 Show the specification of a combined test chamber that Weiss Technik designed and produced.

5.37 List the characteristics of the cyclic corrosion test chamber. What prevents it from being used for the accurate prediction of the product corrosion process in the field?

5.38 Why is the standard salt fog test equipment (Fig. 5.46) unable to provide initial information about the real corrosion in the field?

5.39 What is the purpose of using noxious gas test chambers?

5.40 Show the specifications of the noxious gas test chambers.

5.41 What is the basic goal of wind tunnel testing?

5.42 How can one use wind tunnel testing for reliability/durability prediction?

5.43 Describe the short history of wind tunnels.

5.44 Wind tunnels, particularly large wind tunnels, are used for ART/ADT. How is this done?

5.45 For what purpose does one use solar radiation test chambers?

5.46 Give, as an example, the test chamber SC2000 MHG specifications.

5.47 What is thè goal of dust and sand chambers?

5.48 Give and describe the purposes of the specifications of dust and sand chambers.

5.49 Give examples of the features of sand and dust chambers produced by different companies.

5.50 What can one simulate in the ozone test chambers?

5.51 What is the ozone chamber's role in multi-environmental testing?

5.52 For what purpose, other than multi-environmental testing, can one operate ozone chambers?

5.53 Describe the typical features of ozone test chambers.

5.54 Give an example of the specifications of any ozone test chamber.

5.55 List the components of any one of the available ozone test chambers.

5.56 Why is weathering so important?

5.57 Describe two basic types of weathering testing.

5.58 Describe an example of accelerated weathering testing.

5.59 Show the basic components of the Atlas Ci4000 Weather-Ometer.

5.60 Describe the basic types of electrical testing equipment.

5.61 What is the objective for each type of electrical testing equipment?

Chapter 6

Accelerated Reliability and Durability Testing as a Source of Initial Information for Accurate Quality, Reliability, Maintainability, and Durability Prediction and Accelerated Product Development

6.1 ABOUT ACCURATE PREDICTION OF QUALITY, RELIABILITY, DURABILITY AND MAINTAINABILITY

During the design and manufacturing process, one evaluates the quantitative characteristics of product quality, reliability, durability, and maintainability.

Necessary tests are selected to verify and validate the characteristics of these parameters. The tests are planned to control, review, and document the following requirements:

- Test plans or specifications identify the product being tested and the resources being used to define test objectives and conditions, to identify the parameters to be recorded, and to determine the relevant acceptance criteria.

- Test procedures describe the method of operation, the performance of the test, and recording of the results.
- The correct standard configuration of the product is submitted for the test.
- The requirements of the test plan and the test procedures are observed.
- Acceptance criteria are met.

Purchasing information describes the product including

- Requirements for the approval of the product, procedures, processes, and equipment
- Requirements for the qualification of personnel
- The product name or other positive identification plus issues of all applicable specifications, drawings, process requirements, and inspection instructions
- Requirements for the design, test, examination, inspection, and related instructions for acceptance by the company
- Requirements for test specimens (e.g., production method, number, and storage conditions) for design approval, inspection, investigation, or auditing
- Requirements for the supplier to notify the company of changes in product and/or process definition and to obtain company approval when required
- Right of access by the company, their customer, and regulatory authorities to all facilities involved in the order and to all applicable records

Measurement, analysis, and improvement include

- Design verification (including reliability, durability, maintainability, and safety)
- Process control
- Selection and inspection of key characteristics
- Process capability measurements
- Statistical process control
- Inspection—matching the sampling rate to the criticality of the product and to the process capability
- Failure modes and effects analysis

Industrial companies usually provide verification of reliability, maintainability, and especially quality for a short time during design and manufacturing [218–220, 240, 241]. However, they do not provide a method for accurate prediction of these parameters during longer periods including the warranty period and service life.

The quality control process consists of the measurement of parameters during design and manufacturing (usually for only a short time), comparison with drawing requirements, evaluation of maxima and minima of measured values, calibration, and the determination of criteria and methods needed to ensure that both operation and control of the processes are effective.

However, this process rarely provides the information and methodology necessary for industrial companies and other companies to correctly predict the quality, reliability, durability, and maintainability losses during usage time.

Currently, predictions may be made through computer modeling (simulation).

The accuracy of computer modeling depends on the quality of the algorithm and the input and on how accurately this algorithm is able to simulate the real field situation.

First, usually the simulation does not sufficiently represent the field situation, especially for safety and human factors in combination with the whole complex of input influences. Second, this simulation does not correctly simulate the real test subject. As a result, the predicted maintainability of the product is not accurate.

Analogous to reliability prediction, accelerated reliability testing (ART) can provide the possibility of obtaining initial information to accurately predict the quality, durability, and maintainability losses during usage time.

Therefore, one can provide accurate initial information about the actual cost of maintenance, the necessary measurements, the necessary volume of maintenance operations, tools, and many other particulars during usage time. Similar to reliability prediction, this process can provide practical accurate durability, maintainability, and quality prediction for a given time (warranty period, service life, etc.).

6.2 THE STRATEGY FOR ACCURATE PREDICTION OF RELIABILITY, DURABILITY, MAINTAINABILITY AND QUALITY, AND ACCELERATED PRODUCT DEVELOPMENT

6.2.1 Introduction

The strategy for prediction and development consists of three basic phases (Fig. 6.1). The first phase is to obtain the initial information necessary for accurate prediction and accelerated product development.

This phase includes four basic steps (Fig. 6.2). Initial information can be accurate or inaccurate depending upon the quality of the components in this step. As one can see, each approach depends on the quality of information provided by the corresponding steps. The steps for studying the field situation can be complete (full) or partial (not full) depending upon whether all of the influencing factors in the real-world operating environment (field situation) are addressed and how accurately they model the field conditions.

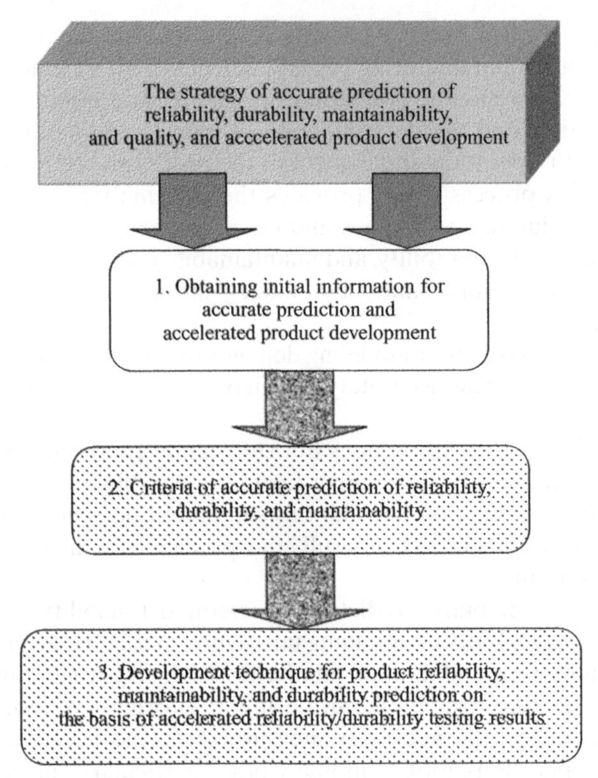

Figure 6.1. The basic scheme of the strategy for accurate prediction of reliability, maintainability, quality and durability.

Full studying is the study of all the parameters of the field situation that have an influence on the product quality, reliability, durability, and maintainability. One can find examples in Chapter 2 of this book. In Section 2.3, "The Collection and Analysis of Failure and Usage Data from the Field," one can read how to conduct Step 1. Currently, industrial companies rarely provide a complete study of the field situation for a product or an accurate simulation of the field situation.

Companies want to save a few pennies. In fact, their loss is much greater than the meager savings achieved by not conducting an accurate simulation of the field situation for the product as required in Step 2. In Section 3.1, this book fully describes the essence of an accurate simulation of the field situation. As for Step 1, most industrial companies provide inaccurate simulation of the field situation (Step 2 of the second column in Fig. 6.2).

Step 3 provides ART and accelerated durability testing (ADT). As was considered in Section 2.1, "Current Practice in Reliability, Maintainability, and Quality," and in Section 1.2, "The Current Situation in AT," industrial companies very infrequently provide these types of testing. They provide many other

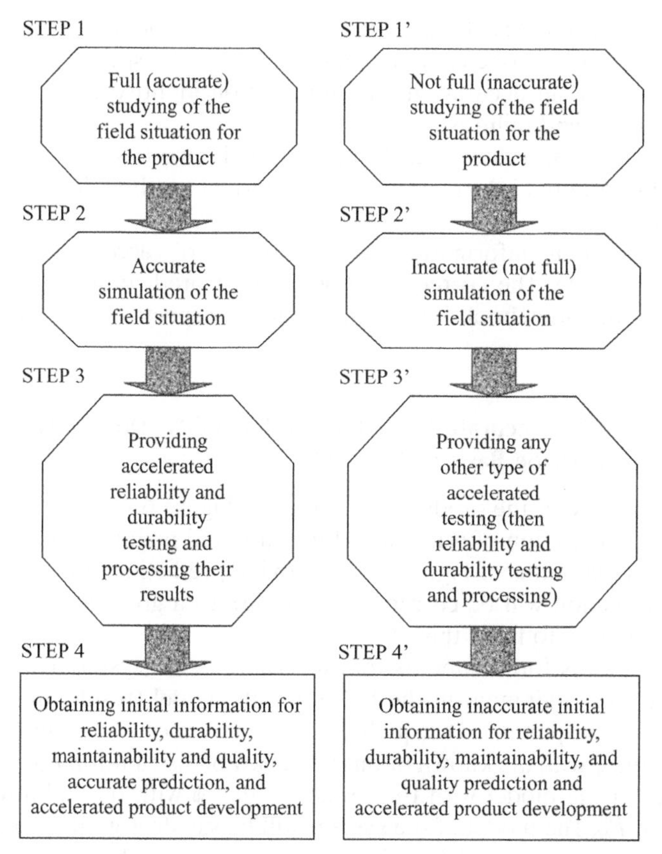
STEP 1
Full (accurate) studying of the field situation for the product
STEP 1'
Not full (inaccurate) studying of the field situation for the product
STEP 2
Accurate simulation of the field situation
STEP 2'
Inaccurate (not full) simulation of the field situation
STEP 3
Providing accelerated reliability and durability testing and processing their results
STEP 3'
Providing any other type of accelerated testing (then reliability and durability testing and processing)
STEP 4
Obtaining initial information for reliability, durability, maintainability and quality, accurate prediction, and accelerated product development
STEP 4'
Obtaining inaccurate initial information for reliability, durability, maintainability, and quality prediction and accelerated product development

Figure 6.2. Scheme of two basic ways of obtaining initial information for reliability, durability, maintainability, and quality accurate prediction, and accelerated product development.

types of testing (Step 3, Fig. 6.2) that are sometimes incorrectly called "ART" or "ADT." We have shown why this is wrong in Chapter 1.

Section 1.4 describes the common principles of ART and ADT. One can read how to correctly do ART/ADT. Chapters 4 and 5 are the largest chapters in this book, and they describe the methodology of ART and ADT performance and equipment for accelerated reliability (durability) testing performance. As a result of using other types of testing, industrial and service companies began to make inaccurate predictions, almost from the start, for product reliability, durability, quality, and maintainability during Step 4: obtaining the initial information for this prediction.

The basic specific direction of Chapters 4 and 5 is the consideration of how one must provide ART and ADT technology to obtain the initial information necessary for accurate reliability, durability, maintainability, and quality prediction.

The above combines directly with the process for accelerated product development because one cannot rapidly decrease the speed of the degradation process without accurate initial information identifying the reasons for the degradation and failures.

To provide an accurate prediction, one needs two basic strategic components: accurate initial information and the correct methodology for accurate prediction.

Accurate initial information is the result of accelerated reliability (durability) testing. The correct methodology of prediction can be found in the author's previous book [18]. Some strategic aspects of this methodology are given next.

6.2.2 Criteria for Accurate Prediction of Reliability, Durability, and Maintainability on the Basis of ART/ADT

The second phase of the prediction strategy in Figure 6.1 uses the criteria for the accurate prediction of reliability, durability, and maintainability.

Using both columns of Figure 6.2 to provide predictions, one must be sure that the prediction will be correct (if possible), to a given accuracy. Use the following solution to meet this objective.

The problem is formulated as follows: There is the system (the results of use the current equipment in the field) and its model (results of ART/ADT for the same equipment). The quality of the system can be estimated by the random value φ with the known or unknown law of distribution $F_S(x)$. Estimate the quality of the model with the random value ϕ with the unknown law of distribution F_M. The model of the system will be satisfactory if the measure of divergence between F_S and F_M is less than a given limit Δ_g.

The model results yield the realization of random variables $\varphi_1^{(1)} \ldots \varphi_1^{(n)}$. If one knows $F_S(x)$, using $\varphi_1^{(1)} \ldots \varphi_1^{(n)}$, then one needs to check the hypothesis.

The null hypothesis H_0, the measure of divergence between $F_S(x)$ and $F_M(x)$, is less than Δ_g. If $F_S(x)$ is unknown, it is necessary also to provide testing of the system. As a result of this testing, one obtains random variables $\varphi^{(1)} \ldots \varphi^{(m)}$. For these two samples, it is necessary to check the hypothesis H_0 that the measure of divergence between $F_S(x)$ and $F_M(x)$ is less than a given Δ_g. If the hypothesis H_0 is rejected, then the model needs updating. One must look for a more accurate way to simulate the basic mechanism of machinery use to perform ART.

Estimate the measure of divergence between $F_S(x)$ and $F_M(x)$ using a multifunctional distribution. The practical use of this criterion depends on the type and forms of this functional distribution. Obtaining an exact distribution of the statistics to test the correctness of hypothesis H_0 is a complicated and unsolvable problem in the theory of probability. Therefore, the author has shown in Reference 43 that if the limits for the studied statistics and their distributions are high and increase the level of the values, then explicit discrepancies can be detected.

6.2.2.1 Solution. Research developed the following solutions for the above-mentioned criteria:

a. The engineering version of the solution obtained follows.

   Develop the upper estimate for the statistical criteria of correspondence, for some measures between the distribution functions of the studied characteristics of reliability, maintainability, and durability in the ART conditions and field conditions. Use this for reliability, durability, and maintainability prediction as well as for solving other engineering problems such as accelerated reliability development and improvement.

b. The mathematical version of the solution obtained follows.

   Approximate criteria, as modifications of the Smirnov and Kolmogorov criteria [194, 195], by divergence ($\Delta_g < 0$) were obtained for the comparison of two empirical functions of the distribution by the measurement of the Smirnov divergence:

$$\Delta[F_S(x), F_M(x)] = \max_{(x)<\infty}[F_M(x) - F_S(x)],$$

and the Kolmogorov divergence,

$$\Delta[F_S(x), F_M(x)] = \max_{(x)<\infty}|F_S(x) - F_M(x)|.$$

In Smirnov's criterion by the null hypothesis,

$$\max_{(x)<\infty}[\tilde{F}_M(x) - F_m(x)] < \Delta_g.$$

By the alternative hypothesis,

$$\max_{(x)<\infty}[F_M(x) - \tilde{F}_m(x)] > \Delta_g.$$

If $\Delta_g = 0$, then we have Smirnov's criterion. An analogous situation exists with Kolmogorov's criterion.

The difference between the two versions is that in the measure using Smirnov's criterion, one takes into account only regions (the oscillogram of loadings) where $F_S(x) > F_M(x)$ and one looks for the maximum of the differences only for those values. In measuring with Kolmogorov's criterion, one takes into account the maximum of differences on all regions by modulus. The consideration of both criteria makes sense because

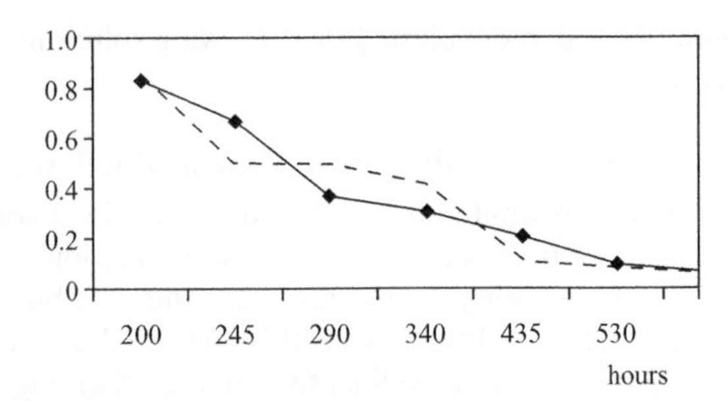

Figure 6.3. The correspondence between functions of distribution of the time to failures of a truck's transmission details in the field (----) and during accelerated reliability/durability testing conditions (____).

Smirnov's criterion is easier to calculate but does not give the full picture of divergences between $F_S(x)$ and $F_M(x)$; Kolmogorov's criterion gives a fuller picture of the above-mentioned divergence but is more complicated to calculate. As a result, one can choose the better criterion for a specific situation if the dependence on specific conditions of the problem is solved.

For example, in the field, 102 failures ($m = 102$) of parts of truck transmissions occurred. As a result of laboratory testing, 95 failures were obtained ($n = 95$); Δ_g is 0.02.

For the field situation, one builds the empirical function of distribution of the time to failures $F_m(x)$ by the intervals between failures. For the ART/ADT conditions, one builds by intervals between failures of the empirical function of distribution time to failures $F_M(x)$.

If we align the graph $F_M(x)$ (Fig. 6.3) and the graph $\tilde{F}_m(x)$, then we will find the maximum difference between $F_M(x)$ and $\tilde{F}(x)$. If we draw the graph $F_m(x)$ on transparent paper and also overlay it, then it is simple visually to find the maximum difference $D^{+}_{m,n} = 0.1$. We obtain $\lambda_0 = 0.99$:

$$\text{The } k = \frac{m}{n} \approx 1.$$

therefore,

$$F_x(x) = 1 - e^{-2x2}[1 + x\sqrt{2\pi \cdot \Phi(x)}].$$

For our situation, we obtain $F_x(0.99) = 0.55$. And $1 - F_x(0.99) \approx 0.4$. Therefore, $1 - F_x(0.99)$ is not small and the hypothesis H_0 can be accepted.

As a result, the divergence between the actual functions of the distribution of time to failures for the failed transmission parts for the truck that was tested in actual field conditions and during ART/ADT conditions by Smirnov's measure is within the given limit $\Delta_g = 0.02$. Meeting this statistical criterion shows that accurate reliability and maintainability prediction of the truck's transmissions can be achieved using the results of this ART/ADT.

6.2.3 Development Techniques for Product Reliability, Durability, and Maintainability Prediction Based on ART/ADT Results

The third and last phase of this strategy (Fig. 6.1) is the development of techniques for the prediction of product reliability, durability, and maintainability based on ART/ADT results. Let us consider seven possible variants (A, B, C, D, E, F, and G).

6.2.3.1 Reliability Prediction. The prediction of product reliability/durability has to be accurate. To achieve this accuracy, one has to follow the procedures shown in Chapters 3–5 to conduct ART/ADT with combined test equipment. The typical situation in engineering practice, especially during the design process, is to provide a small number of specimens (from 3 to 10 specimens of each component) with only two to five possible failures included from testing. It is assumed that the failures of the system (equipment) components (parts and units) are statistically independent in this situation.

The proposed approaches are very flexible and useful for many different types of products such as electronic, hydraulic, pneumatic, electromechanical, mechanical, and others.

The best strategy for reliability prediction cannot be useful if it is not interconnected with a system to obtain accurate initial information for this prediction. As was mentioned earlier, this system is described in this book as the technology of ART/ADT. The strategy for accurate reliability prediction also includes the following:

- Build an accurate model of real-time performance.
- Use this model for the simulation of three basic groups of field situations in the laboratory to test the product and, as a result, to study the physics-of-degradation or chemical degradation (failure) mechanisms during the time and compare them with real field degradation mechanisms of this product. If these degradation mechanisms differ by more than a fixed limit, one must correct the model's real-time performance.
- Make real-time performance forecasts for reliability prediction using these testing results as initial information

Each of the aforementioned items can be performed for different situations, depending on the test subject details, but the reliability can be predicted accurately if the engineers use the proper ART/ADT technology, as well as correct prediction technology.

Execute the first step when it is possible to analyze the real-life situation for the dependence of the product reliability on the combination of three basic groups of the field situation (input influences, human factors, and safety) that are interconnected. As a result, the simulation of the real-life situation will be as complicated as it is in real life. For example, for a mobile product, one needs to use multiaxis and not single-axis vibration.

In order to solve the second step, one has to describe the input influences and output variables, the physics-of-degradation mechanisms for the product, and the parameters of these mechanisms. The results of the product degradation mechanisms include data on the integrated complexes of electrical, mechanical, chemical, thermal, radiation, and other effects of the combined influences. For example, the complex of parameters of each mechanical degradation mechanism includes deformation, corrosion, vibration, crack, wear, creep, and others.

The parameters of the degradation mechanism in the field and during reliability/durability testing must be similar. In addition, one has to remember that in real life, different processes of degradation act simultaneously and in combination. These interacting combinations have to be present for reliability testing conditions.

Therefore, accurate ART includes the simultaneous combination of different types of testing (multi-environmental, electrical, and mechanical) with the assumption that the failures are statistically independent.

In order to solve for the third real-time performance forecasts, the current reliability prediction techniques must have been developed. One can see these developments for different circumstances in Reference 18.

The author briefly shows the basic components of this development in the following discussion. The solution for one specific situation that is published separately in Reference 196 will also be shown in this chapter.

6.2.3.2 *Prediction of the Reliability Function without Finding the Accurate Analytical or Graphical Form of the Failure Distribution Law.*

The author solved this problem for two types of conditions:

(a) The prediction consisting of point expressions for the reliability function of the system's elements

(b) The prediction for the reliability functions of the system with a predetermined accuracy and confidence area

Problem (a) can be solved with graphic-analytical methods based on failure hazard or frequency if we have the graph $f(t)$ of empirical failure frequency. Guided by the failure frequency graph, one can discover the reliability function [18] and [196]

$$p(t) = 1 - \int_0^t f(t)\,dt = 1 - S_f,$$

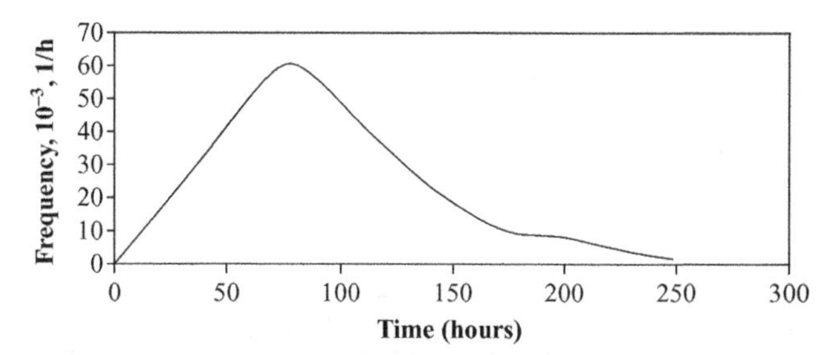

Figure 6.4. Graph of frequency failures of the belts.

where

$$\int_0^t f(t)\,dt = S_f$$

is the area under the curve $f(t)$ that was obtained as a result of ART.

The reliability function of the system that consists of different components (parts) is

$$P(t) = \Pi P_j(t) = \Pi(1 - S_{fj}).$$

For example, as a result of ART on the harvester's belts, $t = 250$ (Fig. 6.4), the area is $S_f = 1.12$, and probability $P(t) = 0.82$.

In the variant (b), one needs to calculate the accumulated frequency function and the values of the confidence coefficient found in the following equations:

$$\underline{Y}(x) = \sum_{m=k}^{n} C_n^m p^m (1 - P)^{n-m}$$

$$\bar{Y}(x) = \sum_{m=0}^{k} C_n^m p^m (1 - P)^{n-m}$$

and evaluate the curves that are limited to the upper and lower confidence areas.

The $C_n^m p^m (1 - P)^{n-m}$ is the probability that the expected event will occur m times during n independent experiments.

Figure 6.5 shows the area of reliability function $\boldsymbol{P(t)}$ of the belts for $\lambda = 0.95$. If one uses this type of graph, it is possible to evaluate the actual reliability of parts, units, and machines (equipment) for whatever time and accuracy is required.

There is also a methodology of prediction in the field based on the mathematical models with an indication of the dependence between product

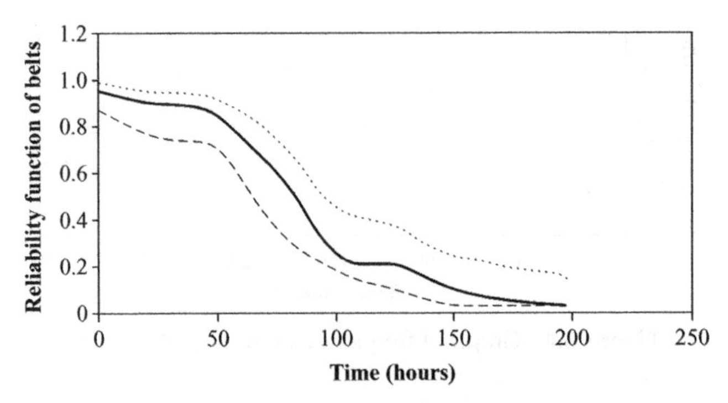

Figure 6.5. Area of possible value $P(t)$ of the belt reliability. ___ is mean of reliability function; ..... is upper confidence level; ---- is the lower confidence level.

TABLE 6.1 The Results of Prototypes of Self-Propelled Spraying Machine Short Field Testing

Index	Prototype of Ro Gator 554		Prototype of John Deere 6500	
	Prestige Farms (Clinton, North Carolina)	Continental Grain Co. (New York)	Prestige Farms (Clinton, North Carolina)	Continental Grain Co. (New York)
Mean time to failure (hour)	104	73.80	104	171.10
Mean time for maintenance (hour)	1.04	1.47	1.82	1.97

reliability and different factors of manufacturing. Those interested can find this in Reference 18.

Here is a practical example of using this methodology.

6.2.3.3 Practical Example. As a result of short field testing for the new self-propelled spraying machines Ro Gator 554 and John Deere 6500, the mean time to failure and the mean time for maintenance were obtained (Table 6.1).

Let us take the prototypes of the Finn T-90 and T-120. The results of field testing of four units of these machines can be seen in Table 6.2. The values of normalized coefficients α_k, β_k, and q_k (Table 6.3) were obtained to correspond to the mean specific parameters of the most important manufacturing and field factors using the author's methodology. The unknown parameters $\alpha_l^{(\tau)}$ and $\alpha_{ij}^{(\tau)}$ (Table 6.4) were obtained using the above-mentioned methodology and Tables 6.2 and 6.3. The coefficients used for new machines were obtained (Table 6.5) by using Table 6.4.

TABLE 6.2 Testing Results of Studied Machine Prototypes

Conditions of Machines Used, No. of Model	Time to Failure (hour)	Mean Time for Maintenance (hour)
1	2	3
Murphy Family Farms		
Finn T-90		
No. 287	21.9	3.02
No. 261	29.04	2.70
No. 290	53.62	2.87
No. 291	47.81	2.24
Mean	37.92	2.71
T-120		
No. 0.59	49.32	2.90
No. 0.30	56.20	1.86
No. 063	67.41	3.81
No. 218	58.27	3.17
Mean	57.80	2.94
Carroll & Foods		
Finn T-90		
No. 316	40.92	2.83
No. 358	1.72	3.20
No. 1001	37.67	1.95
No. 1005	58.21	2.61
Mean	39.63	2.65
T-120		
No. 714	58.72	4.12
No. 1105	80.54	2.80
No.4516	62.98	3.64
Mean	67.41	3.52

TABLE 6.3 Normalized Coefficients with Correspondence of the Most Important Manufacture and Field Factors

Normalized Coefficients	1	2	3	4	5	6
ρ_k	0.2325	0.2225	0.2125	0.175	0.1575	—
α_k	0.1920	0.1940	0.1970	0.206	0.2110	—
q_k	0.2325	0.2250	0.2075	0.130	0.1125	0.0925
β_k	0.1540	0.1550	0.1590	0.174	0.1780	0.1820

TABLE 6.4 Unknown Parameters $\alpha_l^{(\tau)}$ and $\alpha_{ij}^{(\tau)}$

	Finn T-90			T-120		
Unknown Parameters	Marphy Family Farms	Caroll & Foods	Prestige Farms	Marphy Family Farms	Carrol & Foods	Prestige Farms
$\alpha_l^{(\tau)}$	0.42	0.47	0.445	0.19	0.23	0.21
$\alpha_{ij}^{(\tau)}$	0.45	0.75	0.60	0.56	0.80	0.68

TABLE 6.5 Coefficients of Recalculating for Studied Machines

	Ro Gator 554		John Deere 6500	
Coefficients of Recalculating	Prestige Farms	Continental Grain Co.	Prestige Farms	Continental Grain Co.
For the mean time to failure (hour)	0.64	0.66	0.43	0.45
For the mean time for maintenance (hour)	0.41	1.25	1.29	1.18

TABLE 6.6 Predicted Mean Time to Failure and Mean Time for Maintenance of Studied Machines

	Ro Gator 554		John Deere 6500	
Indexes of Reliability	Prestige Farms	Continental Grain Co.	Prestige Farms	Continental Grain Co.
Mean time to failures (hour)	57.12	59.01	59.67	61.61
Mean time for maintenance (hour)	2.17	1.93	33.0	

The mean time to failure and the mean time for maintenance of new machines Ro Gator 554 and John Deere 6500 (Table 6.6) were predicted for the time when they will be manufactured using the author's methodology and Tables 6.1 and 6.5.

6.2.3.4 System Reliability Prediction from Testing Results of the Components. Let us show one possible approach that was published separately in Reference 197.

Introduction. One can experience this problem when testing a complete system that has either a high cost or insufficient time for testing, especially at the beginning of product development.

The author offers the algorithm for calculating the lower confidence bounds (LCBs) at a given confidence level for the system reliability prediction. The accuracy of reliability prediction often depends upon how accurately the ART/ADT was conducted. That depends on how accurately the field conditions were simulated, including simulation of the actual field degradation (failure) process. The results of the product's degradation (failure) process include data on the mechanical, chemical, physical, electrical, and thermal effects.

Accurate ART includes the simultaneous combination of different types of testing (mechanical, multi-environmental, electrical) with the assumption that the failures are statistically independent. Ignoring this fact gives a lower estimate of system reliability. Methodologies that could take into account the failure dependence require additional information.

In this case, the system consists of N components for which the failures are statistically independent. For each of them, the Weibull lifetime distribution is used with scale parameter β_i and shape parameter α_i, $i = 1 \ldots N$. For each component, the test results were obtained using sensors to provide the data.

There are many publications on this subject, including *Statistical Methods for the Estimation of the Reliability of Complex Systems by the Test Results* [198] and Sonkina's article in *IEEE Transactions in Reliability* [199], but none apparently discuss using the multivariate Weibull model.

For these components, well-developed algorithms exist for calculating point and interval reliability estimates and corresponding software such as SuperSMITH/99 that was developed by Wes Fulton.

Only series systems are considered here, but the proposed methodology can be extended to systems with arbitrary series–parallel structures.

The list of assumptions includes the following:

- The system operates until failure.
- Each component used in the system is statistically independent.
- Each component has the Weibull distributed time to failure with different and unknown shape and scale parameters.

There are N independent censored samples for each component as follows:

$$t_{(1)}{}^{(i)} \le t_{(2)}{}^{(i)} \le \ldots \le t_{(ri)}{}^{(i)}, I = 1, \ldots, N, r_i \le n_i.$$

6.2.3.5 *Confidence Bounds for the Simple Weibull Model.* First, consider the simple Weibull model and the algorithm for calculating the LCB of the reliability function $R_i(t)$, $i = 1, \ldots, N$, for each component separately (the index i will be dropped in this section for simplicity).

Use the Weibull model for the lifetime τ of the component to write the reliability index for this component as

$$R(t) = \exp\left\{-\left(\frac{t}{\alpha}\right)^{\beta}\right\}, \quad t > 0, \tag{6.1}$$

where

t is a variable generally given in terms of time or cycles,
α is the scale parameter, and
β is the shape parameter of the Weibull distribution.

The expression (Eq. (6.1)) transforms into the following:

$$R(t)=\exp\left\{-\exp\left(\frac{\ln t-u}{b}\right)\right\},\quad t>0, \tag{6.2}$$

where $u = \ln a$ and $b = 1/\beta$ are the shape and scale parameters of the random variable $x = \ln \tau$ will be held fixed in the future. Assume n units are tested and let $t_1, t_2 \ldots t_\tau$ be the failure data and $t_{\tau+1}, t_{\tau+2} \ldots t_n$ be then censored data (either failure or time censoring).

The methodology for calculating the point and interval estimates of the reliability function $R(t)$ applied to the simple Weibull model has been developed in many publications including [106] and [112] and can be described as follows.

Denote $\bar{a}$ and $\bar{b}$ estimates of the parameters ($u = \ln \alpha$ and $b = 1/\beta$) with the distribution of the ratios

$$B=\bar{b}/b \quad \text{and} \quad U=(\bar{u}-u)/b$$

as independent of the unknown parameters. Here, the distribution of the random variables is determined only by the estimation method and the test plan (i.e., by the numbers n and r). The estimate types include maximum likelihood estimates, the best linear estimates, and the best linear invariant estimates [200], as well as the linear estimates in Reference 201. The tables of the coefficients for calculating the estimates $\bar{u}$ and $\bar{b}$ of the censored samples are included in the indicated papers. For the estimates of the types indicated earlier, one can introduce the function

$$L(\varepsilon,\nu),\ \varepsilon\theta R^1,\text{such that } P\{L(\varepsilon,\nu)<\nu B-U\}=\varepsilon;$$

that is, $L(\varepsilon, \nu)$ is the $(1 - \varepsilon)$th quartile of the distribution of the random variable $\nu B - U$.

The function $L(\varepsilon, \nu)$ decreases with the increasing value of ε (for fixed ν) and increases with the increasing value of ν (for fixed ε). This form is independent of the parameters u and b and is determined only by the form of the estimates $\bar{u}$ and $\bar{b}$ and the test plan (i.e., by the numbers n and r).

Denote

$$L^*(\varepsilon,\nu)=\exp\{-\exp[-L(\varepsilon,\nu)]\}, \tag{6.3}$$

extending its definition by its continuity property to the set

$$\{(\varepsilon, \nu); \varepsilon\theta[0.1], \nu\varepsilon R^1\}, \text{setting } L^*(o, \nu) = 1 \text{ and}$$
$$L^*(1, \nu) = 0 \text{ for all } \nu\varepsilon R^1.$$

The tables of values of $L^*(\varepsilon, \nu)$ for different forms of the estimates $\bar{\mu}$ and b are presented in References 200–203. For example, for the linear estimates by Reference 201 presented in the table of the function

$$L^*(\varepsilon, \nu) \text{ for } \varepsilon = 0.5, 0.75, 0.8, 0.9 \text{ and for censoring } \delta = r/n,$$

which is close to the values of 0.25, 0.5, 0.75, and 1.0. For samples $n > 30$, the normal approximation of the function

$$L(\text{€}, v) \approx v - z_{\text{€}}\sigma/\sqrt{n} \tag{6.4}$$

is presented in the same article. In the last equality, the values of σ_u^2, σ_{ub}, and σ_b^2 can, depending on δ, be determined by Table 6.7.

For maximum likelihood estimates and complete samples of volume $n = 8$, 10, 12, 15, 20, 25, 30, 40, 50, 75, and 100[5] presented in the table of the values of the function L^* as a function of

$$\varepsilon \text{ and } \bar{R} = \exp\{-\exp(-\nu)\}$$

for $\varepsilon = 0.75, 0.9, 0.95, 0.97, 0.98$ and $\bar{R} = 0.5$ (increasing from 0.5 to 0.98 in steps of 0.02).

For the best linear unbiased estimates, the tables of values for the function $L^*(\varepsilon, \nu)$ are presented in Reference 200. A good log χ^2—approximation for the function $L(\varepsilon, \nu)$ is presented in Reference 22 in the form

$$L(\varepsilon, \nu) = -\ln \frac{m\chi_\varepsilon^2(1)}{1}, \tag{6.5}$$

TABLE 6.7 The Component Values of σ for the Given Value of δ

δ	σ_b^2	σ_{ub}	σ_u^2
0.1	9.473	22.183	60.508
0.2	4.738	7.374	16.477
0.25	3.735	4.926	10.497
0.3	3.065	3.438	7.186
0.5	1.716	0.936	2.510
0.6	1.373	0.447	1.612
0.7	1.12	0.145	1.447
0.8	0.928	−0.049	1.253
1.0	0.608	−0.257	1.109

where 1 and m are chosen so that the means and the variations of the quantities $W(\nu) = U - \nu B$ and $\ln[m\chi^2(1)]/1$ are identical.

Having calculated the estimates $\bar{u}$ and $\bar{b}$ by the results of tests of ith-type components ($i = 1 \ldots N$), one can write the expression for the LCB of level q for the component's reliability function $R_I(t)$ in the form

$$R_{iq} = L^*{}_I(q, \bar{\nu}_I), \quad \bar{\nu}_I = \frac{u - \ln t}{B_i}. \tag{6.6}$$

Therefore, to calculate the LCB for the reliability function of any type of component with a given confidence probability q, it is necessary to

- Calculate estimates of the Weibull model's parameters $\bar{u}$ and $\bar{b}$, $I = 1 \ldots N$ (using one of the methods mentioned earlier).
- Calculate the value of $\bar{\nu}_I - (\bar{u}_i - \ln t)/\bar{b}_i$ and the value of the function $L_i(q, \bar{\nu}_I)$, as mentioned earlier.
- Use the expressions in Equations (6.3) and (6.6) for the given values of the confidence probability q.

6.2.3.6 ***Multivariate Weibull Model.*** Let us consider a system with a series connection of N components having independent failures.

If the Weibull model is valid for each N component, then the function of system reliability $R(t)$ for a given value of time t could be represented as

$$R(t) = \exp\left\{-\sum_{i=1\ldots}^{N}\left(\frac{t}{\alpha_i}\right)_{I\beta}\right\} \tag{6.7}$$

or, if we denote $u_i = \ln \alpha_i$ and $b_i = 1/\beta_i$, $i = 1, \ldots N$ as

$$R(t) = \prod_{i=1}^{N} \exp\left\{-\exp\left(\frac{\ln t - u_i}{b_i}\right)\right\}, \quad t > 0. \tag{6.8}$$

Our goal is to give the algorithm for calculating the LCB $R_q(t)$ with confidence level q for system reliability function $R(t)$ if we have N independent censored samples

$$t_{(1)}{}^{(i)} \le t_{(2)}{}^{(i)} \le \ldots \le t_{(ri)}{}^{(i)}, i = 1, \ldots N, r_i \le n_i, \tag{6.9}$$

where n_i is the sample size of the ith-type component and r_i is the number of failures of the ith component.

On the basis of the data (Eq. (6.9)), we can calculate the values $\bar{u}_i$, $\bar{b}_i$ for unknown parameters U_i and b_i, $i = 1, \ldots N$ using one of the methods described in the previous section. It is simple to calculate the point value of the reliability function as follows

$$R(t)=\prod_{i=1}^{N}\exp\left\{-\exp\left(\frac{\ln t-\bar{u}_i}{\bar{b}}\right)\right\}. \tag{6.10}$$

To calculate the LCB $R_q(t)$ with confidence level $q(q = 0.8 - 0.95)$, it is necessary to use the general methodology developed in References 198, 204, and 205. In accordance with Reference 206, the general methodology must be used to solve the next nonlinear external task:

$$R_q(t)=\min_{y\varepsilon S_q}\prod_{i=1}^{N}L^*_I\left[1-e^{-yi},-\ln\ln(I/R')\right],$$

where

$$S_q=\left\{y\varepsilon R^N;\ y\geq 0;\ \sum y_i=\ln\frac{1}{1-q}\right\},$$

and R' is the value of the statistic $R(t)$ determined by the expression (Eq. (6.10)) after calculating the estimated parameters $\bar{u}_i$ and $\bar{b}_i$, $i = 1, \ldots N$.

This is denoted by

$$g_i(y_i)=-\ln L^*_i\left[1-\exp(-y_i),-\ln\ln(1/R(t))\right].$$

As is shown in References 206 and 207, to isolate the two situations most frequently encountered in practice, the expression for $R_q(t)$ can be written in an explicit form.

A. If the functions $g_i(y_i)$ are identical for all $i = 1, \ldots N$ (which is satisfied when the elements are tested according to the same plan (n, U, r) and the estimates of the same type are used for the unknown parameters) and convex upward on $[(0, \ln 1/(1 - q)]$, then

$$\bar{R}_q(t)=\left\{L^*\left[1-(1-q)^{1/N},\ln\ln(1/R'(t))\right]\right\}. \tag{6.11}$$

B. If all the functions $g_i(y_i)$ are convex downward on $[0, \ln 1/(1 - q)]$, then

$$\bar{R}_q(t)=\min_{1\leq I\leq N}L^*_i\left[q,\ln\ln(1/R'(t))\right], \tag{6.12}$$

where $R'(t)$ is the observed value of the statistic $R(t)$ determined by the expression (Eq. (6.10)).

If the convexity conditions are not satisfied for the functions $g_I(y_i)$ on the interval $(0, \ln 1/[1 - q])$, then they can be replaced by convex approximations $g_i(y_i)$ such that the error Δ of Equation (6.12) is no greater than

$$\Delta^*=\exp\left(-\sum_{j=1}^{N}\Delta_i\right),$$

where

$$\Delta_i = \max_y |\bar{g}_i(y) - g_i(y)|, i = 1, \ldots N.$$

Our calculations show that the most practical situations (i.e., for components having a very high level of reliability and relatively small sample sizes) take place in case B.

The calculation of LCB using Equation 6.12 involves calculating the point values of system reliability $R(t)$, then subsequently calculating the LCB of level q for reliability functions of each component of N systems and choosing the minimal value for them. As a corollary of result B, it is not difficult to get the following expression for the LCB $R_q(t)$ if all components of the system have identical (or almost identical) rates of censoring

$$\begin{gathered} \delta_I = r / n_i = \text{constant}, i = 1 \ldots N: \\ R_q(t) = L^*_0 [q, -\ln \ln[1 / R(t)]], \end{gathered} \tag{6.13}$$

where $L^*_0[q, -\ln \ln(1/\subset'(t)]$ is that function from $L^*_l[q, -\ln \ln(1/\subset(t)]$ that corresponds to the minimum sample size (i.e., sample size $n_0 = \min[n_i]$).

6.2.3.7 Examples. **Example 1.** In order to find the value of the LCB at level $q = 0.9$ for the reliability function of an electronic device that consists of $N = 3$ components, and the lifetime of each has a Weibull distribution with unknown parameters, the components were tested according to this plan (n_i, U, r_i), $i = 1, 2, 3$; that is, n_1 specimens of type i are tested until r_i failures appear and the failure specimens are permanent.

For this example, given the following values of the operating times until failure (in hours), the following results were obtained:

$$t_1 = 1820, t_2 = 1960, t_3 = 2172 \text{ (for type 1 specimens);}$$

$$t_1 = 2441, t_2 = 2841, t_3 = 3432, t_4 = 3824 \text{ (for type 2 specimens);}$$

$$\begin{gathered} t_1 = 5329, t_2 = 5682, t_3 = 7016, t_4 = 7919, t_5 = 9566, \\ t_6 = 11,760 \text{ (for type 3 specimens)} \end{gathered}$$

On the basis of these data, the estimates of the unknown parameters found by the method proposed in Reference 203 (i.e., maximum likelihood estimates) are

$$\bar{u}_1 = 7.62, \bar{b}_1 = 0.13, \bar{u}_2 = 8.13, \bar{b}_2 = 0.21, \bar{u}_3 = 8.46, \bar{b}_3 = 0.38.$$

Using Equation (6.10), we can calculate the point estimate for reliability function $R(t)$ as the product of the corresponding point estimates of the component reliability function (for $t = T_0 = 1000$):

$$\bar{R} = \bar{R}_1 \times \bar{R}_2 \times \bar{R}_3 = 0.996 \times 0.997 \times 0.983 = 0.976.$$

To calculate the LCB of level q for device reliability $R(t)$, it is necessary (in the general case) to calculate the LCB at the same level q for each component, assuming that the point estimate of the reliability function of this component is identical to the point estimate reliability function device, that is, to $\bar{R}'(t)$. However, in this particular case ($\delta_1 = \delta_2 = \delta_3 = 0.5$), we can use Equation (6.13), where it is sufficient to calculate the LCB only for the component with the smallest sample size for the first component because $n_i = 6 = \min\{6, 8, 12\}$.

First, determine the value of function $L(q, \nu)$, where $\nu = -\ln \ln \bar{R}' = 3.72$.

Since we used the maximum likelihood estimates for parameters u_i and b_i, now we need to use the tables from Reference 202. For more visualization, use the normal approximation of functions $L(q, \nu)$ by Equation (6.4), although it gives the same uncontrollable error.

From Table 6.7, we can find the values σ_b^2, σ_{ub}, and σ_u^2 corresponding to $\delta = 0.5$ and then calculate $\sigma = (\sigma_u^2 - 2\nu\sigma_{ub} + \nu^2\sigma_b^2)^{1/2} = 4.39$ using Equation (6.4), which gives the following results (since $z_q = 1.28$ when $q = 9$):

$$L(q, \nu) = 3.72 - 1.28 \times 4.39 / \sqrt{6} = 1.424,$$

and Equation (6.2) gives

$$L^*(q, \nu) = \exp\{-\exp(-L(q, \nu))\} = 0.786;$$

that is, the final answer according to Equation (6.13) is $R_q = 0.786$.

For a comparison, we can calculate the LCB of level $q = 0.9$ for each component mentioned earlier.

We obtain the following results:

$$R_{1q} = 0.873, R_{2q} = 0.93, R_{3q} = 0.904.$$

Note that if we multiply these LCBs, we will get only the value 0.73, which is essentially lower.

Remember that the calculation of LCB for the system reliability function requires calculating the conditional LCB R_{iq}* for each components' reliability functions assuming that the point estimate of this component's reliability function is identical to the point estimation of system R'.

Corresponding calculations give the following result:

$$R_{iq} = 0.786, R_{2q}^* = 0.838, R_{3q} = 0.884.$$

The smallest of these values determines the LCB for the device reliability function mentioned earlier:

$$R_q = R_{1q}^* = 0.786.$$

Example 2. Note that min (R_{iq}^*) does not always fall on the component that has the smallest sample size as obtained in Example 1.

Let us slightly modify the initial data in Example 1. For example, we will exchange only the values $n_1 = 12$ and $n_3 = 6$, but the test results of all the components allowed have the same estimate parameters u_i and b_i, $i = 1, 2, 3$.

It is not difficult to ensure that the above-mentioned calculation will give the results below.

The values of conditional LCBs for all components (of the same level $q = 0.9$) are

$$R_{1q}^* = 0.855, R_{2q}^* = 0.838, R_{3q}^* = 0.868.$$

The smallest value of the conditional LCB falls onto the component number 2, which has sample size $n_2 = 8$ (not the smallest!) and the LCB of the device reliability function:

$$R_q = R_{2q}^* = 0.838.$$

The values of LCBs for all components (of level $q = 0.9$) are

$$R_{1q} = 0.914, R_{2q} = 0.93, R_{3q} = 0.89.$$

The product of these values (sometimes used for calculating the LCB for the device reliability function) yields the value 0.75. This is much less than 0.838 obtained by the method suggested earlier.

6.2.3.8 Conclusions from the Above-Mentioned Strategic Methodology. The methodology developed allows the prediction of system reliability for components subjected to different types of the physics-of-degradation processes during the service life. This results from having a separate Weibull model for every type of degradation.

The value of this methodology is that it allows the prediction of the LCB of a system reliability function after testing a few samples of each component of the system during a relatively short time. In addition, it allows prediction of the system reliability at the beginning of development when the testing of the system as a whole either has high cost or may be impossible in a short time.

Moreover, we may use this methodology even when the ART/ADT results of each system's components are known. In this case, we have to transform the testing results found during ART/ADT to the equivalent testing results corresponding to ordinary conditions.

The duration of the testing period is determined by ART/ADT longevity for the system components and can be made sufficiently small in principle. For instance, if we could provide each component in Example 1 with an acceleration coefficient K equal to about 10 (a realistic possibility), then we could obtain the confidence estimate of system reliability function during the required operating time of our system only on the basis of the test results of its components.

There is a possibility of decreasing the new system development time and, consequently, the development cost.

This book considers only series systems, but the proposed methodology can also be extended to systems with arbitrarily series–parallel structures.

The proposed methodology also allows combining service life data with results of ART/ADT.

This methodology is especially important for reducing cost and design time, as well as improving the reliability, durability, safety, and many aspects of quality.

6.2.3.9 The Strategy of Durability Prediction with the Consideration of Expenses and Losses. If one conducts accelerated reliability (durability) testing as described in this book, then it is possible to obtain accurate initial information for durability prediction. During the testing period, the degradation of the test subject as well as the expenses attributable to this subject are approximately identical to the degradation and expenses incurred during the time of use in the field. ART incurs one group of expenses; another is incurred by special field testing as a result of a comparison of test subjects that accomplished different volumes of work during ART.

Test subjects in many areas of industry are connected to the products with which they are working. As a result, the above-mentioned test subjects are related either directly or indirectly to the losses. For example, the degradation and failures of the machinery (including electronic systems of control) for food technology, farm machinery, refrigerators, and many special trucks lead to product losses.

More degradation and failures drive more product losses for the test subject. Therefore, our methodology also considers product losses. If any test subject does not incur product loss, one can more easily measure the results because the components of equations that relate to the product losses will be zero.

Specifics for the methodology of the durability prediction encompass a combination of reliability and usage costs including product losses.

6.2.3.10 Principal Scheme of Accurate Durability Prediction. The total test subject expenses consist of three basic components:

1. Expenses related directly to the work of the test subject
2. Expenses that depend on the loss of the product during the normal work of the test subject
3. Expenses that depend on stopping the test subject work if there are failures or increased degradation of the test subject (increased degradation

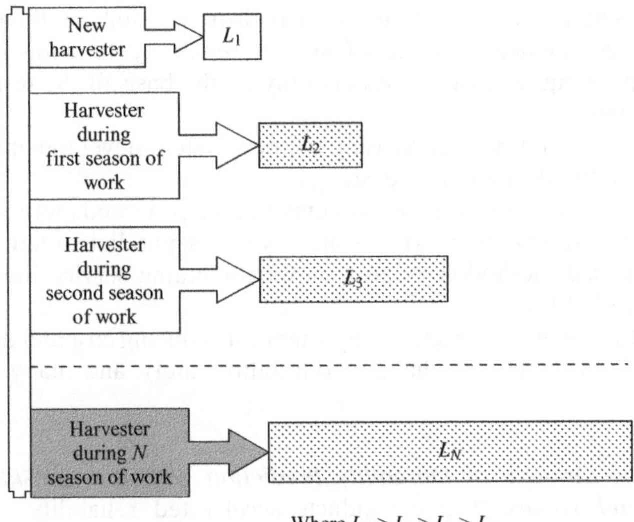

Figure 6.6. Increasing losses during the harvester's work time (seasons).

and stopping work cause increased product loss and incur associated costs)

The total expenses, which influence the total cost of the work, can be expressed as in Chapter 4, notably in the section called "ART/ADT Methodology as a Combination of Different Types of Testing." Figure 4.4 shows that the reliability, quality, and productivity decrease during use (level of wear). The example of a truck is shown. As a result, the losses of the product increase if the test subject interacts with another product that also sustains losses. The losses of both products have a greater influence on the expenses if we consider, for example, farm machinery. During its operation, the wear on the harvester has a direct influence on the loss of grain (Fig. 6.6)

The increase of expenses during normal work can be evaluated as

$$\sum_{i=1}^{n} C_p quant(i) = \left[\sum_{i=1}^{n} Lp(i)\right] \cdot Cu,$$

where

ΣLp is the loss of product during the work of the test subject and
Cu is the cost of the product.

Product quality often influences the degree of deterioration of the test subject resulting in an increase in cost. In this case, calculate the losses due to product cost with the following formula:

$$\sum_{i=1}^{n} C_p quant(i) = \left[\sum_{i=1}^{n} Vt.s.(Cu.n - Cu.d.) \right]$$

where

Cu.n. is the cost of the unit of product that occurs during the work of a new test subject and

Cu.d. is the cost of the unit of product that occurs during the work of a deteriorated test subject.

The examples demonstrated enable one to take into account

- The dynamics of changes in all the basic types of expenses during the lifetime of the product based on initial information obtained by conducting ART
- Losses of quality and quantity of the product with which the test subject is working (depending on whether the test subject is working or not working at all)
- The random character of the losses during the life cycle as a function of test subject reliability

This can be done without taking into account the aging of the test subject. Aging depends upon the length of time on the market before a new more effective test subject appears.

In this case, the optimal durability must be obtained by using current methodologies.

6.2.3.11 *Practical Example.* The optimal durability of a harvester has been predicted by two methods:

- The current method that does not take into account product losses
- The proposed new method that takes into account product losses

The result is shown in Figure 6.7. With the current method, the optimal durability of the harvesters is equal to 7 years, but with the proposed method, it is equal to 14 years. This example illustrates the significance of the difference.

Sometimes, instead of losses, one can take into account a decrease in quality during usage.

6.2.3.12 *About the Basic Strategy of Accurate Maintenance Prediction.* Maintenance is a combination of all technical and administrative actions, including supervision actions, intended to maintain an item in, or restore it to, a state in which it can perform a required function.

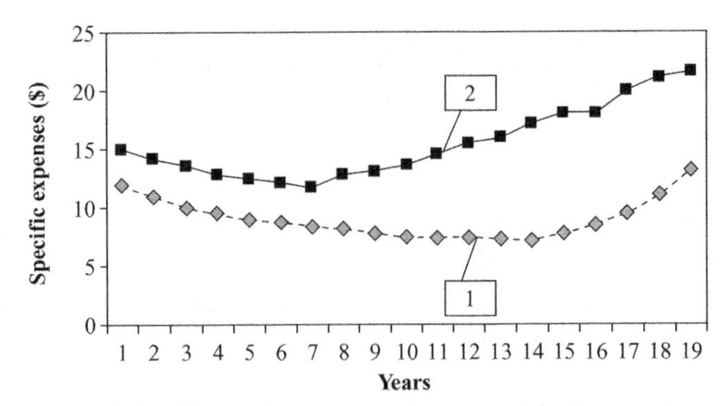

Figure 6.7. Results of counting the accurate durability of the harvester (1—proposed method, 2—current method).

The most important parameters of the maintainability measure are the duration of downtime due to maintenance and the number of repair personnel required. Keep both to a minimum if possible. However, the number of maintenance person-hours required may not be as important as minimizing the time required for repair regardless of the number of men involved or the inefficiency of their utilization. The third important parameter is maintenance cost.

The length of the repair time, for a discrete repair, is the sum of the individual maintenance task times that are required for its completion.

The prediction of the expected number of hours that a device will be in an inoperative state or in a "down state" while it is undergoing maintenance is of vital importance to the user due to the adverse effect that excessive downtime has on mission success. Therefore, once the operational requirements of a device are fixed, it is imperative to predict its maintainability in quantitative terms as early as possible during the design phase. Update this prediction continually as the design progresses to ensure a high probability of compliance with all specified requirements.

The first useful feature of accurate maintainability prediction is that it highlights the areas of poor maintainability justifying product improvement, modification, or a change in the design. Another useful feature of maintainability prediction is that it enables the user to make an early assessment about whether the predicted time, quality of personnel, tools, and inspection equipment are adequate and consistent with the needs of the system's operational requirements.

For some areas of industry, maintenance is the single largest controllable expense for owners/operators. For example, in the aircraft industry [208] (Fig. 6.8),

- Engines account for 50% of maintenance costs
- Operators (specifically airlines) are under pressure to reduce operating costs

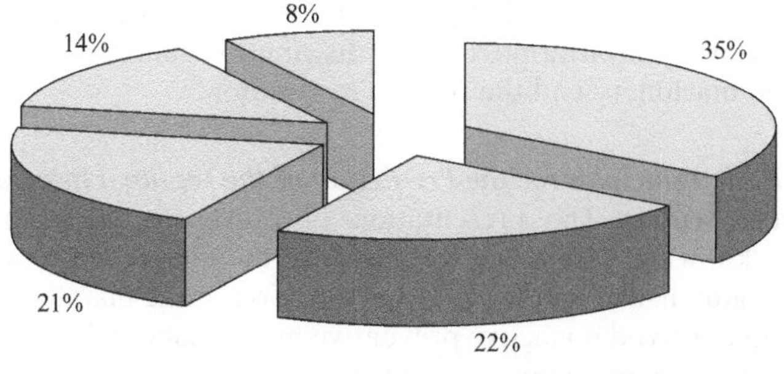

Figure 6.8. Typical aircraft MRO costs [208].

- Maintenance decisions are based only on short-term needs, not long-term impact

The maintainers have had no method to determine what type of maintenance is most cost-effective.

Traditional reliability-centered maintenance (RCM) determines the optimal maintenance for each *component*, not the optimum for the overall system. Optimizing RCM analysis at the system level may significantly reduce long-term operating costs.

One of the basic problems facing RCM is the prediction of the best system strategy, from a reliability perspective, for determining the intervals between maintenance operations, the volume of maintenance work, and the required spare parts. One has to take into account that the maintenance time and the time between maintenance are usually random in character due to the degradation process and the random time to failure.

The practical strategy for developing accurate predictions for maintenance systems could not be solved in the past because it was not possible to obtain sufficient initial information about the frequency of the rise and accumulation of degradation during the service life of the product.

The strategy of ART described earlier provides the methodology to obtain this information. This advancement provides a practical system for optimal maintenance with minimal cost and time, resulting in maximum effectiveness of machinery use.

Theoretical solutions were presented earlier. For some systems, the provision of maintenance should be random. However, for other systems, maintenance must occur at regular intervals, for example, it must not occur only after each failure. This requires a precise maintenance schedule. Thus, the character of the problem has changed. Instead of a typical stationary problem solved in the above-mentioned publications, this is a nonstationary problem that is difficult in view of the arbitrariness of the distribution function for the time to failure of the machinery and the time to restoration.

6.2.3.13 Basic Principles for the Prediction of the Optimal Interval between Maintenance Actions. The presentation of preventive maintenance data requires the known duration and frequency of the maintenance tasks. Active preventive maintenance time is the usual duration of the maintenance action. One may add observed nonactive preventive maintenance when necessary. To aid in maintenance planning, it is desirable to estimate the maintenance person-hours for each task. In addition to the detailed task information, present an overall summary of preventive maintenance times. In the following discussion, we will use the term "maintenance" instead of "preventive maintenance."

We have solved the above-mentioned problem for two situations:

1. As a criterion for the evaluation of the maintenance intervals, one uses the probability of prevention for a given time, T.
2. One uses the maximum availability as an overall criterion.

For the first situation, formulate and solve the problem using the following process.

The distribution of the system time to failures ξ follows the arbitrary law $F(x)$, and the time of restoration ζ follows the arbitrary law $H(x)$. The time between maintenance actions is a constant. Maintenance after each restoration is not essential. Determine the distribution of summary time X_a when the system is in good working order over the interval α under the same condition that began at moment zero.

The interval between maintenance α is given by the equation

$$P\{X_T > T_0\} \geq P_0.$$

Consider the situation with one and with n failures on the interval (o, a) and find the system at the end of the interval in both good and bad conditions. It was ascertained in Reference 18 that the distribution of time X_T, where the system is in good condition over the interval of length $(0, T)$, was shown by the necessary formula.

This function is too cumbersome, so an explicit analytical solution is very difficult. For finding the extremum, use a quantitative method of optimization using a computer; for example, use one of the modifications of the gradient method.

A second solution to the problem uses the method of finding the optimal interval between maintenance actions α by maximizing availability.

One has to take into account the progressive aging process characteristic of most systems. This process can be partially eliminated during the maintenance phase. Our research addressed the aging in one, two, three, and multiple steps. First, let us describe a brief solution to the problem with the simplest variant—one-step aging. In this case, the distribution functions of the times to failure and the times to renewal for the considered period has only one unchanging form.

6.3 THE ROLE OF ART AND ADT IN THE ACCURATE PREDICTION AND ACCELERATED DEVELOPMENT OF QUALITY, RELIABILITY, MAINTAINABILITY, AND DURABILITY

There are three basic elements of ART and ADT [251]:

1. Accurate simulation of the whole complex of the field situation using the given criteria
2. Simulation *simultaneously and in combination* with *each group* of field influences (multi-environmental, mechanical, electrical, and others), as well as safety and human factors
3. Determination of the initial information required for an accurate prediction of reliability, durability, and maintainability

As a result, provide initial information for accelerated quality, reliability, durability, and maintainability development and improvement during the phases of predesign, design, manufacturing, and usage. Moreover, the result is decreasing life cycle costs. Other types of testing cannot provide this information because they do not use an accurate simulation of the field situation and, consequently, the testing results differ from the field results [248–250, 252, 254, 257].

Using accelerated reliability (durability) testing, one can study the dynamics of the product's degradation in the field. As a result, it is easy to identify the reasons for failures, and the losses of quality and maintainability. This results in the rapid elimination of degradation and failures, thereby improving the development of the product.

Figure 6.9 shows the strategic value of ART/ADT as a key factor for the accelerated solution of the interconnected problems of quality, reliability, maintainability, and durability, as well as serviceability, availability, life cycle cost, and warranty length.

These accelerated solutions for development and improvement can be provided through a chain: ART—finding the reasons for the degradation and failures in the field—eliminating these reasons for degradation and failures—checking this elimination—predicting the level of quality, reliability, durability, and maintainability.

The components of this chain can practically achieve an accurate prediction of the quality, reliability, maintainability, and durability values.

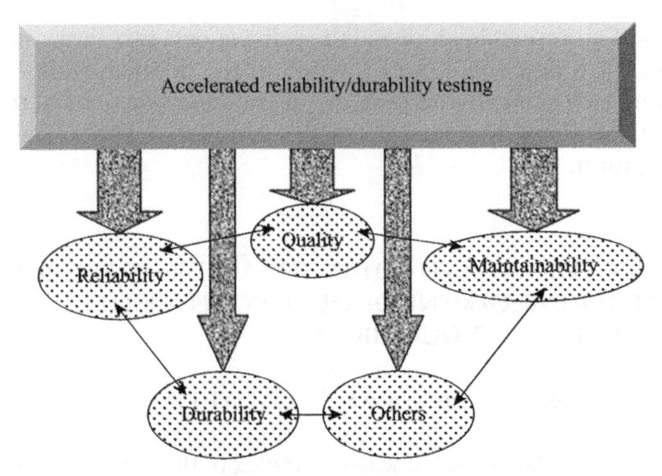

Figure 6.9. Accelerated reliability/durability testing as a key factor for the accelerated solutions of quality, reliability, maintainability, durability, and related problems.

There are many methodologies for this prediction that are suitable. But the second component for an accurate prediction is a problem because the initial information to achieve an accurate prediction is based on currently undeveloped ART/ADT (reliability and/or durability testing). Other types of testing such as highly accelerated life testing (HALT), highly accelerated stress screening (HASS), accelerated aging, thermal shock, mechanical shock, vibration testing, and many other types of laboratory testing as well as field testing and proving ground testing cannot provide accurate initial information for this prediction. They use inaccurate simulations of the field situation during service life. They simulate only a portion of the many field input influences and do not include the human factors and safety problems.

During the current design and manufacturing processes in most industrial companies, ART/ADT is not conducted. In other words, industrial companies do not provide the appropriate information for an accurate prediction.

Therefore, as considered in this book, accelerated reliability and durability testing technology is a key factor for accurate prediction of reliability, durability, maintainability, and availability, and for reducing recalls and life cycle cost of the product. Therefore, it is practically useful.

EXERCISES

6.1 Industrial companies provide verification and validation tests during design and manufacturing. What is the basic negative aspect of these tests for reliability, durability, especially quality, and prediction?

6.2 What is the basic aspect of quality control in the current system for an accurate quality prediction?

6.3 Analyze a computer simulation (modeling) and determine its negative aspects for the accurate prediction of quality, reliability, durability, and maintainability.

6.4 Show the basic scheme of the connection between the ART/ADT technology and accurate prediction of the product's quality, reliability, durability, and maintainability.

6.5 Describe why and how ART/ADT can improve upon the negative aspects for performing accurate predictions.

6.6 Show the scheme of the basic steps of reliability, durability, and maintainability needed for an accurate prediction when ART/ADT is involved.

6.7 Describe what steps are required to obtain accurate initial information for the accurate prediction of quality, reliability, durability, and maintainability.

6.8 Why does the current process to obtain the initial information needed for prediction often make it impossible to accurately predict quality, reliability, durability, and maintainability?

6.9 What criteria are required for an accurate prediction?

6.10 What is the meaning of the divergence between two functions of distribution?

6.11 What is the difference between the author's criteria comparing two functions of distributions from Kolmogorov's and Smirnov's criteria?

6.12 Describe and graph an analytical method of reliability testing. What is the specific component of this method and in what situations should one utilize this method?

6.13 Describe the meaning of the prediction of the reliability function with a predetermined accuracy and confidence level.

6.14 What is the specific method included in this book for system reliability prediction after testing the results of the components?

6.15 Show the basic elements of the above-mentioned system of reliability prediction after testing the results of the components.

6.16 What is necessary to calculate the LCB for the reliability function of any type of component with a given confidence probability?

6.17 Describe the principal scheme of accurate durability prediction.

6.18 What is the specific aspect demonstrated in the book's methodology of accurate durability prediction?

6.19 Describe the basic strategy for accurate maintenance prediction.

6.20 What components are included in the cost of typical aircraft maintenance? What is the typical percentage of each component?

6.21 Describe briefly the basic principles to predict the optimal intervals required between maintenance actions.

6.22 For what two situations did this book solve the maintenance prediction problem?

6.23 For accelerated reliability/durability testing, show the scheme used as a key factor for the accelerated solutions of quality, reliability, durability, and maintainability.

Chapter 7

The Financial and Design Advantages of Using Accelerated Reliability/ Durability Testing

The financial impact of poor reliability and quality on company profits is described in the examples included in Reference 32. Figure 1.10 in Chapter 1 demonstrates this impact.

This book's approach to accelerated reliability testing (ART)/accelerated durability testing (ADT) as a component in the system of systems approach and as the key factor in the "quality–reliability–maintainability–durability–supportability" complex helps to improve this situation. Several industrial companies have partially implemented this approach. An example of the results of this approach, used by companies manufacturing engines and tools, is included in Tables 7.1–7.3.

The following practical difficulties (problems) were evident during the implementation of this technology:

- The methodology's implementation requires an interdisciplinary team for a particular group of products. This team must be composed of highly qualified professionals in the fields of quality, reliability, simulation, testing, safety, and human factors. Yet, the professionals working in these areas and in marketing and sales, plus most suppliers, usually work independently. They have separate requirements (standards) and responsibilities. This is the first problem that all companies need to overcome for the successful implementation of the methodology and technology, including equipment, described in the book.

TABLE 7.1 Dynamics of Engine Complaints for 3 Years

Year	Complaints (%)
First	0.17
Second	0.13
Third	0.03

TABLE 7.2 Comparison of the Complaints for a Reason over 3 Years

	% of Complaints		
	Years		
Reasons	First	Second	Third
1. Design problems	58	43	36
2. Deviation from the instruction procedure	23	31	34
3. Not performed according to drawing	10	10	8
4. Not used according to specification	4	10	15
5. Others	5	6	7
Total	100	100	100

TABLE 7.3 Example of Practical Economic Results of the Proposed Approach for Tools

Year	Sales (Million $)	Complaints (Number)	Rejects (%)
2003	134.3	151	3.2
2004	156.2	144	3.0
2005	196.4	127	2.4
2006	214	114	2.0

- The interdisciplinary team needs a chair who is an outstanding manager who understands the specifics of the different specialties (reliability, durability, maintainability, quality, simulation, testing, safety, and human factors) and their interconnections.
- In the company that first used this proposed technology, a talented quality manager who possessed initiative, understanding, and ability was in charge. He was also experienced in the simulation, testing, reliability, safety, and human factors areas. He proposed to organize an interdisciplinary engineering emergency team composed of the types of professionals listed earlier. With the support of the vice president in development, this team was organized.
- Many difficulties were overcome during the work of this team. Most of the team members realized the necessity of excellent communication skills, good understanding of each other, and the need for coordination across professional specialties by team members.

- This experience reveals that one of the major difficulties in implementing new technology can be resolved by the enthusiastic cooperation of each team member and a willingness to work together. This is required for the implementation of a new technology for ART/ADT.
- The implementation of this new technology revealed that the team members need adequate knowledge in metallic parts, new composite materials and units, and high-level standardization, especially in quality and reliability.

Later, industrial companies producing engines, tools, trucks, and other products implemented this methodology and technology.

Some of the economically beneficial results of implementing this technology are shown in Tables 7.1–7.4 and in Figure 7.1.

The financial results of applying this methodology are impressive. Especially impressive is the increase in the level of profit in these companies over a 3-year

TABLE 7.4 Increased Volume of Sales of Two Products as a Result of the Implementation of the Above-Described Quality Approach

	Product(s), Second Quarter	
Year	First Type of Product	Second Type of Product
2002	23.030	77.201
2004	55.680	108.829

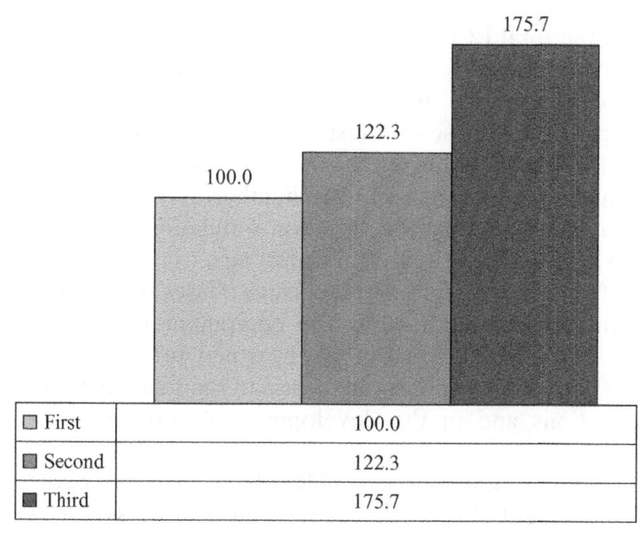

Figure 7.1. Increased volume of sales of instruments after the implementation of the new approach.

period and the substantial increase in the quality and reliability of their products. The market outlook improved so much that Warren Buffet bought one of the companies (in Israel) in 2006. As a result, the product sales increased and the reputation of that company improved in the market (Fig. 7.1 and Table 7.4).

These figures show the effectiveness of this complex analysis and the vital importance of establishing the reasons for potential complaints prior to use and eliminating these faults prior to the completion of the design. Implementing accelerated improvement of the product quality during design and manufacturing yields this desired result.

The process for final quality improvement (Fig. 7.1) coupled with the quality and characteristics of the raw materials used, improvement in the components' quality, improvement of products used by the customers, and the improvements that are specific to each manufacturer yields more marketable products at significant cost savings.

Implementing the above-mentioned approach to ART and ADT and technology in different companies resulted in the following total life cycle cost (LCC) changes.

LCC percentage changes

- Increase the cost of design phase from 4 to 8% (not including the benefits from recalls and maintenance reducing)
- Increase the cost of manufacture phase (not including recalls) from 0 to 1% (not including the benefits from recalls and maintenance reducing)
- Decrease the cost of usage phase (includes recalls and complaints) from 52 to 83%

As a result, the total LCC *decreased* by at least 33–47%.

If the company does not have enough money for full implementation of ART/ADT technology in a single step, then the company can do it step by step: First, implement the use of a test chamber with one type of equipment, for example, for vibration testing. Next, add equipment for the simulation temperature/humidity to the same test chamber, for example. The following step will be to add equipment for pollution simulation, then for other equipment. KAMAZ Inc. (Russia) used this approach.

The cost of the design and manufacturing phases will decrease when the industrial company repeatedly uses the equipment for ART/ADT on the future models of the product and on the next new product designs. The basic parts of equipment will be applicable for use in the following product updates and modernizations and in the development of products with additional savings.

Beginning with the design process, ART/ADT technology could be continued through implementation for the manufacturing process (for quality control and other needs) using the same equipment design. This will dramatically improve the financial results.

EXERCISES

7.1 What kind of practical difficulties occurred during the implementation of ART/ADT technology included in this book?

7.2 Show the dynamics of the complaints made during the implementation of ART/ADT technology in the years after the publication of this book.

7.3 Show an example of increasing a product's sales by the implementation of ART/ADT technology.

Chapter **8**

Accelerated Reliability Testing Standardization

8.1 OVERVIEW AND ANALYSIS

There is a great need for standardization for accelerated reliability testing (ART)/accelerated durability testing (ADT). Standardization will have an important influence on ART and ADT development and implementation. Standardization is needed for the following reasons:

- No standards exist for providing ART/ADT.
- No standards exist for ART/ADT technology.
- Customers do not have nor require ART/ADT standards development and application.
- Lacking standards and requirements, the companies that design and produce testing equipment have less incentive to provide the wide range of equipment necessary for ART/ADT and implementation.

The need for standardization is dominant in this complex of problems.

A standards organization can be classified by its role, position, and the extent of its influence in the company, in the local, national, regional, and global standardization arena.

By geographic designation, there are international, regional, and national standards. By technology or industry designation, there are standards from standards developing organizations (SDOs) and also from standards setting

Accelerated Reliability and Durability Testing Technology, First Edition. Lev M. Klyatis.

organizations (SSOs), which are known as consortia. Standards organizations may be governmental, quasi-governmental or nongovernmental entities.

There are many international standards organizations. Three examples are the International Organization for Standardization (ISO), the International Electrotechnical Commission (IEC), and the International Telecommunication Union (ITU). ISO is composed of the national standards bodies (NSBs), with one member per nation. The IEC is composed of "national committees," with one member per nation. In some cases, the National Committee of the IEC of a country is also the ISO member from that country. Both the ISO and IEC consist of approximately a hundred specific committees. For example, IEC TC56 is the technical committee that updates and develops international standards in dependability (reliability and maintainability). The World Standards Cooperation (WSC) is a cooperative effort between the ISO, the IEC, and the ITU.

Additional independent standards organizations, such as American Society for Testing and Materials (ASTM) International, develop and publish technical standards for international use. Others that set standards within some more specialized context include SAE, IFTF, TAPPI, W3C, Institute of Electrical and Electronic Engineers (IEEE), UPU, and APT. Often these international standards organizations are not based on the principle of one member per country.

Regional standards bodies include the European Committee for Standardization (CEN), the European Committee for Electrotechnical Standardization (CENELEC), the European Telecommunication Standards Institute (ETCI), the European Cooperation for Space Standardization (ECSS), the African Organization for Standardization (ARSO), the Arab Industrial Development and Mining Organization (AIDMO), and others.

Each country or economy has a single recognized standards body. For example, in the United States, this body is the American National Standards Institute (ANSI).

Finally, all middle and large industrial companies, as well as many small companies, have their standards body that develops and updates their company's standards.

The standardization of ART/ADT technology for products remains an unsolved problem. There are only standards for the planning, theoretical (probabilistic and statistical), and reporting aspects of ART (reliability testing). There are no standards for reliability and durability testing technology, that is, how to conduct ART/ADT.

The standardization for conducting ART and ADT is very important. It provides solutions for reliability/durability problems and accurate predictions facilitating solutions for the following issues:

- Maintainability problems
- Minimization of life cycle costs
- Safety aspects of risk assessment

- Need to reduce the time to market
- Accelerated development and improvement of the product
- Minimization of warranty and return costs

One can see the importance of reliability, including reliability testing in the following citation from the U.S. National Research Council to the Department of Defense, Guide 3235.1H [2]:

> The Department of Defense and the military services should give increased attention to their reliability, availability, and maintainability data collection and analysis procedures, because deficiencies continue to be responsible for many of the current field problems and concerns about military readiness.

The current system of standardization, especially in engineering, has positive and negative aspects.

Some basic positive aspects of standardization are the following:

- Standards provide a known level of quality.
- Standards provide a common language for the communication of requirements between suppliers and customers.
- Laws are needed for all the areas involving the activity of people that identify their rights and responsibilities. In engineering, standardization embodies the technical laws that the industry needs.
- Technical laws, in other words, standardization, are necessary to provide the unification of the requirements for the product's technology and quality, and the implementation of advanced solutions of technology and quality (reliability, durability, maintainability, and others) in the practice of design, manufacturing, and usage.

The negative aspects of the current system of standardization, especially of international and national standardizations, are the following:

- Standardization is based on a volunteer base of experts, which may result in the lack of responsibility for the quality and effectiveness of national and international standards.
- Most qualified professionals are not involved in standardization work because they are busy and do not have enough time to volunteer.
- Most qualified professionals who are retired and have enough time to volunteer are financially unable to afford to attend the standards committee meetings.
- Advanced technology and quality solutions of most industrial companies are proprietary; therefore, they are not available to other companies or to any other entity including national and international standardization organizations.

- In national and international standardization, the most interested parties are company producers of a particular product; therefore, their representatives are more involved in the work of standardization committees for their particular products. As a result, these standards are biased to represent the interest and achievements of the producers rather than the customer's requirements. This is especially true for the reliability, maintainability, and durability standards.
- One of the basic requirements of standards, especially national and international standards, is that they must be implemented and approved in practice before they are sanctioned as standards. This resulting delay in implementation is too lengthy and unacceptable. Therefore, these standards represent a past state of affairs, not new solutions. The resulting quality level of these standards is lower than current knowledge.

Let us examine the standards for reliability testing from the world's most powerful standardization organizations.

8.2 IEC STANDARDS

The IEC is the world's leading organization that prepares and publishes international standards for all electrical, electronic, and related technologies, collectively known as "electrotechnology."

On September 15, 1904, the delegates to the International Electrical Congress adopted a report about the "...organization of a representative Commission to consider the question of the standardization of the nomenclature and ratings of electrical apparatus and machinery." As a result, the IEC was officially founded in June 1906, in London, England, where its Central Office was set up. In 1948, the IEC Central Office moved from London to Geneva, Switzerland.

IEC has more than 100 diverse technical committees, including those for capacitors and resistors, semiconductor devices, electrical equipment, medical equipment, maritime navigation, radio communications systems and equipment, lasers, lighting, fiber optics, ultrasonic wind turbine systems, fuel cell technologies, atomic energy, and many others.

IEC Technical Committee TC56 develops and updates the standards in dependability (including reliability and maintainability) including standards for reliability testing. One can see examples of IEC standards for reliability testing in References 209–211.

Standards for Reliability Testing and Statistical Principles

Tools: These standards share the common title "Equipment Reliability Testing":

IEC 60605-1 (1978) Equipment Reliability Testing—Part 1: General Requirements

IEC 60605-2 (1994–2010) Equipment Reliability Testing—Part 2: Design for Test Cycles

IEC 60605–3-1 (1986) Equipment Reliability Testing—Part 3: Preferred Test Conditions. Section 1: Indoor Portable Equipment—Low Degree of Simulation

IEC 60605-3-2 (1986) Equipment Reliability Testing—Part 3: Preferred Test Conditions. Section 2: Equipment for Stationary Use in Weather Protected Locations—High Degree of Simulation

IEC 60605-3-3 (1992) Equipment Reliability Testing—Part 3: Preferred Test Conditions. Section 3: Test Cycle 3: Equipment for Stationary Use in 12 Partially Weather Protected Locations—Low Degree of Simulation

IEC 60605-3-4 (1992) Equipment Reliability Testing—Part 3: Preferred Test Conditions. Section 4: Test Cycle 4: Equipment for Portable and Nonstationary Use—Low Degree of Simulation

IEC 60605-3-5 (1996) Equipment Reliability Testing—Part 3: Preferred Test Conditions. Section 5: Test Cycle 5: Ground Mobile Equipment—Low Degree of Simulation

IEC 60605-3-6 (1996) Equipment Reliability Testing—Part 3: Preferred Test Conditions. Section 6: Test Cycle 6: Outdoor Transportable Equipment—Low Degree of Simulation

IEC 60605-4 (2001–08) Equipment Reliability Testing—Part 4: Statistical Procedures for the Exponential Distribution—Point Estimates, Confidence Intervals, Prediction Intervals, and Tolerance Intervals

IEC 60605-6 (2007–05) Equipment Reliability Testing—Part 6: Tests for the Validity and Estimation of the Constant Failure Rate and Constant Failure Intensity

The following standards share the words "Reliability Testing" in their title:

IEC 61123 (1991) Reliability Testing, Compliance Test Plans for Success Ratio

IEC 61124 (2006–08) Reliability Testing, Compliance Tests for Constant Failure Rate and Constant Failure Intensity

IEC 61650 (1997) Reliability Data Analysis Techniques—Procedures for Comparison of Two Constant Failure Rates and Two Constant Failure (Event) Intensities

For better understanding, note that the above-mentioned standards only mentioned reliability testing, but they are not standards to conduct reliability testing. Examine the contents of these examples.

First Example: International Standard IEC 60605-2
First edition (1994–2009)
Equipment Reliability Testing—Part 2: Design of Test Cycles
Contents

Foreword

The IEC is a worldwide organization for standardization comprising all national electrotechnical committees (IEC National Committees). The object of the IEC is to promote international cooperation on all questions concerning standardization in the electrical and electronic fields. To this end and in addition to other activities, the IEC publishes international standards.

Their preparation is entrusted to technical committees; any IEC National Committee interested in the subject dealt with may participate in this preparatory work. International, governmental, and nongovernmental organizations liaising with the IEC also participate in this preparation. The IEC collaborates closely with the ISO in accordance with conditions determined by the agreement between the two organizations.

The formal decisions or agreements of the IEC on technical matters, prepared by technical committees on which all the National Committees having a special interest therein are represented, express, as nearly as possible, an international consensus of opinion on the subjects.

They have the form of recommendations for international use published in the form of standards, technical reports, or guides and they are accepted by the National Committees in that sense.

In order to promote international unification, IEC National Committees undertake to apply IEC International Standards transparently to the maximum extent possible in their national and regional standards. Any divergence between the IEC standard and the corresponding national or regional standard shall be clearly indicated in the latter.

Introduction

A test cycle is a sequence of different operating and environmental test conditions that are based upon actual conditions of use, as defined, for example, by the relevant product specification. The equipment undergoing reliability testing is normally subjected to repeated test cycles. The number of cycles will depend on the accumulated relevant test time, as required by the selected compliance test plan of IEC 60605-7, or as suitable for determination testing according to IEC 60605-4.

The step-by-step procedure described here is intended for any specific equipment to be tested, when it is considered necessary to simulate closely the real conditions of use of the equipment. It applies fully to laboratory testing but may be applied to field testing, in so far as conditions can be controlled, with respect to operating conditions only (including load and supply).

1. Scope

This part of IEC 60605 provides a general procedure for the design of test cycles, where no applicable preferred test cycles can be found in IEC 60605-3. It applies to the design of operating and environmental test cycles referred to in clauses 8.1 and 8.2 of IEC 60605-1. The resulting test cycle should be included in the detailed reliability test specification.

Tests, which include cycles designed according to this standard, are not intended to replace ordinary tests such as qualification tests, functional performance tests, and environmental tests.

Pre-exposure tests may in some cases be necessary before commencing the test cycles designed by the methods of this standard. The basis of the decision as to whether to include pre-exposure tests is outside the scope of this standard.

Normative References

The following normative documents contain provisions that, through reference in this text, constitute provisions of this part of IEC 60605. At the time of publication, the editions indicated were valid. All normative documents are subject to revision, and parties to agreements based on this part of IEC 60605 are encouraged to investigate the possibility of applying the most recent editions of the normative documents indicated next. Members of IEC and ISO maintain registers of currently valid international standards.

IEC 50(191): 1990, *International Electrotechnical Vocabulary (IEV)—Chapter 191: Dependability and Quality of Service*

IEC 68: *Environmental Testing*

IEC 60605-1: 1978, *Equipment Reliability Testing—Part 1: General Requirements Amendment No. 1 (1982)*

IEC 60721-1: 1990, *Classification of Environmental Conditions—Part 1: Environmental Parameters and Their Severities*

IEC 60721-2, *Classification of Environmental Conditions—Part 2: Environmental Conditions Appearing in Nature*

IEC 60721-3, *Classification of Environmental Conditions—Part 3: Classification of Groups of Environmental Parameters and Their Severities*

Second Example: International Standard IEC 60605-3-1 Equipment Reliability Testing Part 3: Preferred Test Conditions, Indoor Portable Equipment—Low Degree of Simulation

Contents

Preface

Maintainability. The text of this standard is based on the following documents.

The following IEC publications are quoted in this standard.

Safety Requirements for Mains-Operated Electronic and Related Apparatus for Household and Similar General Use

Basic Environmental Testing Procedures—Part 2: Tests, Test B: Dry Heat

60605-1: Equipment Reliability Testing—Part 1: General Requirements

60605-1: Equipment Reliability Testing—Part 2: Guidance for the Design of Test Cycles (in preparation)

8.3 ISO STANDARDS

Founded in February 1947, the ISO promulgates worldwide proprietary industrial and commercial standards. It is headquartered in Geneva.

While ISO defines itself as a nongovernmental organization, its ability to set standards that often become law through treaties or national standards makes it more powerful than most nongovernmental organizations.

ISO is a network of the national standards institutes from 157 countries, one member per country. ISO's main products are the International Standards. ISO also publishes technical reports, technical specifications, publicly available specifications, technical corrigenda, and guides.

There are ISO/IEC Joint Technical Committees, ISO/IEC Joint Study Groups, and jointly issued ISO/IEC standards. There is no special ISO technical committee (unlike IEC) for preparation standards in reliability.

8.3.1 ISO Standards in Reliability Testing

There are three ISO standards in reliability testing [211] called "reliability by testing" for pneumatic fluid power:

ISO/FDIS 19973-1 Pneumatic Fluid Power—Assessment of Component Reliability by Testing—Part 1: General Procedures

ISO/FDIS 19973-1 Pneumatic Fluid Power—Assessment of Component Reliability by Testing—Part 2: Directional Control Values

ISO/FDIS 19973-1 Pneumatic Fluid Power—Assessment of Component Reliability by Testing—Part 3: Cylinders with Piston Rod

8.4 MILITARY RELIABILITY TESTING STANDARDS AND APPROPRIATE DOCUMENTS

8.4.1 Military Standards

MIL-STD-781D Reliability Design Qualification and Production Acceptance Tests [212]

Exponential Distribution

This standard covers the requirements and provides details for reliability testing during the development, qualification, and production of systems and equipment with an exponential time-to-failure distribution. It establishes the requirements for tailoring reliability testing to be performed during integrated test programs specified in MIL-STD-785. Task distributions are provided for reliability development/growth testing (RD/GT) and reliability qualification testing (RQT).

Production reliability acceptance tests (PRATs) and environmental stress screening (ESS) are defined. Specifying any two or three parameters, that is, lower test MTBF, upper test MTBF, or their ratio, given the desired decision risks, will determine the test plan to be utilized. This standard is applicable to six broad categories of equipment, distinguished according to their field service applications.

MIL-STD-721C Definition of Terms for Reliability and Maintainability [213]

This standard defines terms and definitions used most frequently in specifying reliability and maintainability. It provides common definitions for the Department of Defense and for defense contractors.

MIL-STD-2074 Failure Classification for Reliability Testing [214]

This standard establishes criteria for the classification of failures occurring during reliability testing.

This classification into relevant or nonrelevant categories allows the proper generation of MTBF reports. This document applies to any reliability test, including, but not limited to, tests performed in accordance with MIL-STD-781.

MIL-STD-1635 Reliability Growth Testing
DI-RELI-80250 Reliability Test Plan
DI-RELI-80251 Reliability Test Procedures
DI-RELI-80252 Reliability Test Reports
DI-NDTI-81585 Reliability Test Plan
DI-TMSS-81586 Reliability Test Reports

8.4.2 Military Handbooks

MIL-HDBK-781A Reliability Test Methods, Plans, and Environments for Engineering, Development, Qualification, and Production [215]

Scope

This handbook provides test methods, test plans, and test environmental profiles that can be used in reliability testing during the development, qualification, and prediction of systems and equipment.

This handbook explains techniques for use in reliability tests performed during integrated Test programs. Procedures, plans, and environments, which can be used in reliability development/growth tests (RD/GT), reliability qualification tests (RQT), PRATs, ESS methods, and durability/economic life tests are provided.

The data provided in this handbook is typical of reliability test programs and may be specified in Department of Defense contract procurements, requests for proposal, statements of work, and government in-house development that require reliability testing. This handbook is for guidance only. This handbook cannot be cited as a requirement. If it is, the contractor does not have to comply.

When referencing the test methods, test plans, and environmental test conditions, the source is to be cited. The methods in this handbook are applicable to six broad categories of equipment, distinguished according to the field service application of the equipment:

Category 1. Fixed-ground equipment
Category 2. Mobile ground equipment
- A. Wheeled vehicle
- B. Tracked vehicle
- C. Shelter configuration
- D. Manpack

Category 3. Shipboard equipment
- A. Naval surface craft
- B. Naval submarine
- C. Marine craft
- D. Underwater vehicle

Category 4. Equipment for jet aircraft
- A. Fixed-wing
- B. Vertical and short takeoff and landing (V/STOL)

Category 5. Turboprop aircraft and helicopter equipment
- A. Turboprop
- B. Helicopter

Category 6. Missiles and assembled external stores
- A. Air-launched missiles
- B. Assembled external stores
- C. Ground-launched missiles

This handbook is very popular. It includes the technology for reliability testing that

- Explains techniques for use in reliability tests performed during integrated test programs
- Includes the general *procedures, plans, and environments* that can be used in reliability development/growth tests, reliability tests, qualification tests, and PRATs
- Data provided in this handbook are typical of *reliability test programs* and may be specified in the request for proposals, statements of work, and government in-house developments that require reliability testing.

Handbook MIL-HDBK-H 108 Sampling Procedures and Tables for Life and Reliability Testing (Based on Exponential Distribution) [216]

This handbook provides procedures and tables based on the exponential distribution for life and reliability testing. It includes definitions required for the use of the life test sampling plans and procedures, a general description of life test sampling plans, life tests terminated upon the occurrence of a preassigned number of failures, life tests terminated at a preassigned time, and sequential life test sampling plans.

8.5 STANDARDIZATION IN RELIABILITY (DURABILITY) TESTING BY SOCIETIES

Associations

IEEE

American Society for Quality (ASQ)

SAE International—The Engineering Society for Advanced Mobility Land, Sea, Air, and Space

ASTM International (formerly known as the American Society for Testing and Materials)

American Society of Mechanical Engineers (ASME)

And others

Large engineering societies and associations have technical committees and councils in standardization, including those in the reliability and durability areas. For example, SAE International has the G-11 Division—Reliability, Maintainability, and Probabilistic Methods Division, which updates and develops SAE International standards in aerospace. This division has a program of standards development/revision activities. One standard from this program is JA 1009: Reliability Testing Standard, which includes six standards now in development to cover the subject matter:

1. Reliability Testing Standard JA 1009: Glossary of terms and definitions
2. Reliability Testing Standard JA 1009/1: Procedure for the design of reliability testing
3. Reliability Testing Standard JA 1009/2: The strategy of reliability testing
4. Reliability Testing Standard JA 1009/3: Equipment for reliability testing
5. Reliability Testing Standard JA 1009/4: statistical Criteria for a comparison of reliability testing results and field results
6. Reliability Testing Standard JA 1009/5: Collection, calculation, statistical analysis of reliability testing data, development recommendations for improvement of test subject reliability

In another example, ASTM International published the standard ASTM F2477-07 Standard Test Methods for In Vitro Pulsatile Durability Testing of Vascular Stents [217]. As described in the standard, this test method covers the procedure for determining the durability of balloon-expandable and self-expanding metal or alloy vascular stents. The tests are performed by exposing the specimens to physiologically relevant diametric distension levels using hydrodynamic pulsatile loading. Specimens could be deployed into a mock or elastically simulated vessel prior to testing. The test methods are valid for determining stent failure due to typical cyclic blood vessel diametric distension and include physiological pressure tests and diameter control tests. The test apparatus includes a pressure measurement system, dimensional measurement devices, a cycle counting system, and a temperature control system.

The typical duration of this test is 10 years of equivalent use (at 72 beats/min) or at least 380 million cycles. These test methods do not address other models of failure such as dynamic bending, torsion, extension, crushing, or abrasion, as well as test conditions for curved mock vessels or overlapping stents.

General Caveat—This document contains guidance for testing currently carried out in most laboratories. Other testing techniques may prove to be more effective and are encouraged.

Whichever technique is used, it is incumbent upon the tester to justify the use of the particular technique, instrument, and protocol.

As we can see, the above-mentioned techniques for simulation do not correspond to accurate simulation of real-life situations. Therefore, if one uses the aforementioned types of testing, then the durability prediction for the test subject is not accurate and needs improvement. This is described in more detail in Section 1.2.5 of this book.

Conclusion

The above-mentioned analysis and examples demonstrate that no acceptable standards exist for reliability testing technology and durability testing today.

Conclusions

COMMON CONCLUSIONS

As one can see from this book, the technology of accelerated reliability testing (ART) and accelerated durability testing (ADT) consists of the following basic components:

- Accurate simulation of the field conditions
- Complex work during ART and ADT performance
- Accurate prediction of product quality, reliability, durability, maintainability, and life cost cycle during service life (warranty period and any other time) and product improvement

SPECIFIC CONCLUSIONS

1. The current types of stress testing cannot offer useful information for the accurate prediction of product quality, reliability, durability, maintainability, life cycle cost, and supportability because they are based on an inaccurate simulation of real-world conditions.

   ART/ADT can, and does, solve this problem because it is based upon an accurate simulation of real-world conditions.
2. Accurate simulation of real-world conditions requires full simulation of the field input influences integrated with safety and human factors. But

Accelerated Reliability and Durability Testing Technology, First Edition. Lev M. Klyatis.

this proven approach to testing (ART/ADT) is developing too slowly here. This book analyzes the two basic groups of causes for this: cultural and technical.

The cultural group depends upon people, especially top management, which, in turn, depends upon investment in development.

The technical group of causes are

- Insufficient development of accurate simulation
- The lack of ART/ADT theory, strategy, methodology, and equipment, including the study of the degradation (failure) process.

3. The development of ART/ADT requires the development of the following steps:
   - The study of field conditions to determine the important parameters that are to be simulated in the laboratory
   - Develop an accurate (by quality and quantity) simulation of the field conditions
   - Introduce ART/ADT, including the analysis and management of the causes of degradation (failures)
   - Development recommendations for eliminating the causes of failures and degradation
   - Make an accurate prediction of reliability, durability, maintainability, and life cycle costs.
   - Accelerate the development of the product.
4. Each of these steps is complex and must be based upon the analysis of real-world conditions, whereupon recommendations for improvement can be made.
5. This book shows a step-by-step technology of ART/ADT, as well as a way from the simulation of separate input influences to the development of ART/ADT through improvements in both the accuracy of each field component simulation and of combined testing.
6. The accurate prediction of reliability, durability, maintainability, and life cycle cost is the basis for accelerated product improvement.

Glossary of Terms and Definitions

In the European Cooperation for Space Standardization standard "Glossary of Terms," Department of Defense standard "Definitions of Terms for Reliability and Maintainability", the *International Electrotechnical Dictionary*, and the *Oxford English Dictionary*, as well as ISO/IEC Standard for Software Life Cycle Processes-Risk Management, ISO 9000:2000 "Quality Management Systems—Fundamentals and Vocabulary," and other these terms are defined as follows:

Accelerated reliability testing or accelerated durability testing (or durability testing) Is testing in which:

1. The physics (or chemistry) of the degradation mechanism (or failure mechanism) is similar to this mechanism in the real world using given criteria; and
2. The measurement of reliability and durability indicators (time to failures, its degradation, service life, etc.) have a high correlation with these indicators measurement in the real world using given criteria.

Accelerated Reliability and Durability Testing Technology, First Edition. Lev M. Klyatis.

Note 1 Accelerated reliability and durability testing, such as accelerated testing, is connected with the stress process. Higher stress means a higher acceleration coefficient [ratio of time to failures in the field to time to failures during accelerated reliability/durability testing (ART/ADT)], and a lower correlation between field results and ART/ADT results.

Note 2 The basic principles of accelerated reliability and accelerated durability testing (durability testing) are:

1. A complex of laboratory testing and special field testing.
2. The laboratory testing provides a simultaneous combination of a whole complex of multi-environmental testing, mechanical testing, electrical testing, etc.
3. The special field testing takes into account the factors which cannot be accurately simulated in the laboratory, such as the stability of the product's technological process, how the operator's reliability influence on test subject's reliability and durability, etc.
4. Requires accurate simulation of the three integrated complexes (components) of the field situation: whole complex of field input influences, human factors, and safety aspects.

Note 3 ART and accelerated durability testing (ADT) (or durability testing) have the same basis—the accurate simulation of the field situation.

Note 4 Accelerated reliability testing can be for different lengths of time, that is, warranty period, 1 year, 2 years, service life, and others. Accelerated durability testing is provided until the test subject is out of service.

Accelerated testing Is testing in which the deterioration of the test subject is accelerated.

Acceptance The act of an authorized representative of the customer by which the customer for itself, or as an agent of another party, accepts ownership of existing and specified products tendered, or confirms satisfactory performance of specific services, as partial or complete performance of the contract on the part of the supplier.

Accident
An undesired event arising from operation of any project-specific items which result in:

1. human death or injury;
2. loss of, or damage to, project hardware, software or facilities which could then affect the accomplishment of the mission; and
3. loss of, or damage to, public or private property; or detrimental effects on the environment.

Acceptance of risk
Decision to cope with consequences, should a risk scenario materialize.

Accurate prediction is possible
If one has:

1. methodology to incorporate all active field influences and interactions integrated with safety and human factors; and
2. accurate initial information from accelerated reliability/durability testing for accurate prediction calculation.

Accurate simulation of the field input influences
If quality and quantity of the full influences act simultaneously and in mutual combination.

Accurate system of prediction
The system of prediction is accurate if, and only if, the simulation is accurate and ART/ADT is possible.

Accurate physical simulation
Occurs when the physical state of output variables and the physics-of-degradation in the laboratory differs from those in the field by no more than the allowable limit of divergence.

Availability
The ability of an item to be in a state to perform a required function under given conditions at a given instant of time or over a given time interval, assuming that the required external resources are provided.

Note
This ability depends on the combined aspects of the reliability performance, the maintainability performance, and the maintenance support performance.

Assessment
Any systematic method of obtaining evidence from tests, examinations, questionnaires, surveys, and collateral sources used to draw inferences about characteristics of people, objects, or programs for a specific purpose.

Certification Procedure by which a third party gives written assurance that a product, process, or service conforms to specified requirements.

Classification accuracy The degree to which neither false positive nor false negative categorizations and diagnoses occur when a test is used to classify an individual or event.

Common cause failure Failures of multiple items occurring from a single cause that is common to all of them.

Common mode failure Failures of multiple similar items that fail in the same mode.

Common mode fault Faults of multiple items which exhibit the same fault mode.

Confidence interval An interval between two values on a score scale within which, with specified probability, a score or parameter of interest lies.

Configuration control Activities comprising the control of changes to a configuration item after formal establishment of its configuration documents.

Configuration management Technical and organizational activities comprising:

1. configuration identification;
2. configuration control; and
3. configuration status accounting.

Configuration verification Examination to determine whether a configuration item conforms to its configuration documents.

Consequence An outcome of an event.

Note 1 There can be more than one consequence from one event.

Note 2 Consequences can range from positive to negative. However, consequences are always negative for safety aspects.

Note 3 Consequences can be expressed qualitatively or quantitatively.

Corrective action Action taken to eliminate the causes of an existing nonconformity, defect, or other undesirable situation in order to prevent recurrence.

Note 1 The corrective actions may involve changes such as in procedures and systems, to achieve quality improvement at any stage of the quality loop.

Note 2	There is a distinction between "correction" and "corrective action." "Correction" refers to repair, rework or adjustment and relates to the disposition of an existing nonconformity. "Corrective action" relates to the elimination of the causes of the nonconformity.
Note 3	The above term "nonconformance" is equivalent to the term "nonconformity" as used in ISO 8402.
Correlation	The tendency for two measures or variables, such as height, or weight, or other, to vary together or be related for individuals in a group.
Cost (price)	That which must be given or surrendered to acquire, produce, accomplish, or maintain something.
Cost breakdown structure	A systematic decomposition and presentation of the total system cost according to work packages, the nature of the cost element and the organizational elements responsible for the work packages.
Critical item	Any item that introduces risk, which could be unacceptable to the project and requires specific attention or control in addition to that given to items not so categorized.
Customer	Recipient of a product provided by the supplier.
Note 1	In a contractual situation, the customer is called the "purchaser."
Note 2	The customer may be, for example, the ultimate consumer, user, beneficiary, or purchaser.
Data	Information represented in a manner suitable for automatic processing.
Demonstration	A process whereby evidence is produced to provide confidence that the specified requirements are fulfilled.
Dependability	The collective term used to describe the availability performance and its influencing factors: reliability
Note	Dependability is used only for general descriptions in nonquantitative terms.
Deration	Process of designing a product such that its components operate at a significantly reduced level of stress to increase reliability.

Design to minimum risk Design of a product to an acceptable residual risk solely by compliance to specific safety requirements, other than failure tolerance.

Development The process by which the capability to adequately implement a technology or design is established before manufacture.

Note The process may include the building of various partial or complete models of the products and assessment of their performance.

Durability The ability of an item (material, part, unit, or whole machine) to perform a required function under given conditions of use and maintenance, until a limiting state is reached. The measurement of durability is its length of time (hours, months, or years) or its volume of work.

Environment Conditions in which an item exists or is operated.

Estimate at competition The sum of the cumulative costs incurred up to the cut-off date and the estimate to competition from the cut-off date.

Estimate to competition Based on the work completed, approved contract changes and the incurred commitments, the estimate of all costs from the cut-off date required to deliver the product as specified.

Event The occurrence of a particular set of circumstances.

Note 1 The event can be certain or uncertain.

Note 2 The event can be a single occurrence or a series of occurrences.

Note 3 The probability associated with the event can be estimated for a given period of time.

Factor In measurement theory, a statistically derived hypothetical dimension that accounts for part of the intercorrelations among tests. Strictly, the term refers to a statistical dimension defined by a factor analysis, but it is also commonly used to denote the psychological construct associated with the dimension. Single-factor tests presumably assess only one construct; multi-factor tests measure two or more constructs.

Failure The termination of the ability of an item to perform a required function.

Note 1 After failure, the item has a fault.

Note 2	"Failure" is an event, as distinguished from "fault," which is a state.
Note 3	The concept as defined does not apply to items consisting of software only.
Failure mechanism	The physical, chemical, or other processes thathave led to a failure.
Failure mode	The observable effect of the mechanism through which the failure occurs, for example, short-circuit, open-circuit, fracture, and excessive wear.
Note	This term is equivalent to the term "Fault mode."
Fault	1. The state of an item characterized by inability to perform as required, excluding the inability during preventative maintenance or other planed actions, or due to lack of external resources. 2. An unplanned occurrence or defect in an item that may result in one or more failures of the item itself or of other associated equipment.
Note 1	A fault is often the result of a failure of the item itself, but may exist without prior failure.
Note 2	An item may contain a subelement fault, which is a defect that can manifest itself only under certain circumstances (definition 2 above). When those circumstances occur, the defect in the subelement will cause the item to fail, resulting in an error. This error can propagate to other items causing them, in turn, to fail. After the failure occurs, the item as a whole is said to have a fault or to be in a faulty state (definition 1 above).
Fault tolerance	The attribute of an item that makes it able to perform a required function in the presence of certain given subitem faults.
Field test	A test administration used to check the adequacy of testing procedures, generally including test administration, test responding, test scoring, and test reporting.
Harm	Physical injury or damage to health of people, or damage to property or the environment.
Harmful event	Occurrence in which a hazardous situation results in harm.

Hazard	Potential source of harm. In other words, a condition, associated with the design, operation, or environment of a system, that has the potential for harmful consequences.
Hazard acceptance	Decision to tolerate the consequences of the hazard scenarios when they occur.
Hazard analysis	Systematic and iterative process of identification, classification, and reduction of hazards.
Hazard control	Preventive or mitigation measure, associated to a hazard scenario, which is introduced into the system design and operation to avoid the events.
Hazard elimination	Removal of a hazard from a particular hazard manifestation.
Hazard manifestation	Presence of specific hazards in the technical design, operation, and environment of a system.
Hazard reduction	Process of elimination or minimization and control of hazards.
Hazard scenario	Sequence of events leading from the initial cause to the unwanted safety consequence.
Hazard free	Set of hazard scenarios originating from the same set of hazard manifestations.
Hazardous	Property of an item and its environment that provides the potential for mishaps.
Hazardous events	An occurrence arising from the triggering of one (or more) initiator events in the presence of one (or more) hazards, which may lead to undesired consequences.
Hazardous situation	Circumstance in which people, property, or the environment are exposed to one or more hazards.
Human error	The failure of a person to perform an action as required.
Human factors	Is umbrella term for the following areas of (in Europe and other countries—"ergonomics") research: human performance, technology, and human–computer interaction.

Human factors (in general) Is the scientific discipline concerned with the understanding of the interactions between humans and other elements of a system.

Human factors engineering Is the scientific discipline dedicated to improving the human–machine interface and human performance through the application of knowledge of human capabilities, strengths, weaknesses, and characteristics.

Incident An unplanned event that could have been an accident, but was not.

Information Intelligence or knowledge capable of being represented in forms suitable for communication, storage, or processing.

Note Information may be represented, for example, by signs, symbols, pictures, or sounds.

Inspection An activity such as measuring, examining, testing to gauging one or more char-activities of an entity and comparing the results with specified requirements in order to establish whether conformity is achieved for each characteristic.

Item Anything that can be individually described and considered.

Note An item may be, for example:

1. an activity or process a product; an organization, system or person; or
2. any combination thereof.

Life cycle Consists of three basic phases: research and development, production or construction, operation and maintenance.

Life cycle cost The total cost of a system, from "need identification" until disposal. This consists of acquisition cost, ownership cost, and disposal costs.

Maintainability The ability of an item under given conditions of use to be retained in or restored to a state in which it can perform a required function, when maintenance is performed under given conditions and using stated procedures and resources.

Note — The term "maintainability" is also used as a measure of maintainability performance. In this sense, maintainability is "the probability that a given active maintenance action, for an item under given conditions of use can be carried out within a stated time interval, when maintenance is performed under stated conditions and using stated procedures and resources."

Maintainability prediction — An activity performed with the intention of forecasting the numerical values of a maintainability performance measure of an item, taking into account the maintainability performance and reliability performance measures of its subsystems under given operational and maintenance conditions.

Maintenance — The combination of all technical and administrative actions, including servicing actions, intended to retain an item in or restore it to a state in which it can perform a required function.

Mean time between failures — The expectation time between failures.

Model — A physical or abstract representation of relevant aspects of an item or process that is put forward as a basis for calculations, prediction, or further assessment; to create or use such a model.

Multi-environmental complex of field — Consists of temperature, humidity, pollution, *input influences* radiation, wind, snow, fluctuation, and rain. Some basic input influences combine to form a multifaceted complex. For example, chemical pollution and mechanical pollution combine in the pollution complex. Most of these interdependent factors are interconnected and interact simultaneously in combination with each other.

Nonconformance — Nonfulfillment of a specified requirement.

Note — The definition covers the departure or absence of one or more quality characteristics (including dependability characteristics), or quality system elements from specified requirements.

Normative reference — A reference which incorporates requirements from a cited publication into a normative document.

Output variables Are the results of input influences interaction. Output variables can be loading, tension, output voltage, and other. The output variables lead to degradation (deformation, crack, corrosion, vibration, overheating) and failures of the product.

Probability The extent to which an event is likely to occur.

Note 1 ISO 3534-1:1993: the mathematical definition of probability is "a number in the scale 0 to 1 attached to a random event. It can be related to a long-run relative frequency of occurrence or to a degree of belief that the event will occur. For a high degree of belief, the probability is near 1."

Note 2 Frequency rather than probability may be used in describing risk.

Note 3 Degrees of belief about probability can be chosen as classes or ranks, such as rare/unlikely/moderate/likely/almost certain or incredible/improbable/remote/occasional/probable/frequent

Procedure Specified way to perform an activity.

Note 1 In many cases, procedures are documented (e.g., quality system procedures).

Note 2 When a procedure is documented, the term "written procedure" or "documented procedure" is frequently used.

Note 3 A written or documented procedure usually contains the purposes and scope of an activity; what shall be done; what materials, equipment, and documents shall be used; and how it shall be controlled and recorded.

Process Set of interrelated resources and activities which transform inputs into outputs.

Product The result of activities of processes.

Note 1 A product may include service, hardware, processed materials, software, or a combination thereof.

Note 2 A product can be tangible (e.g., assemblies or processed materials) or intangible (e.g., knowledge or concepts), or a combination thereof.

Note 3 A product can be either intended (e.g., an offering to customers) or unintended (e.g., pollutant or unwanted effects).

Product assurance A discipline devoted to the study, planning, and implementation of activities intended to assure that the design, controls, methods, and techniques in a project result in a satisfactory level of quality in a product.

Product state A particular configuration of the product related to the current configuration baseline.

Product tree Hierarchical representation of the system resulting from an orderly or exhaustive identification of its successive levels of decomposition.

Quality Is the ability of the product or service to satisfy the user's needs.

Note 1 In many instances, needs can change with time; this implies a periodic review of requirements for quality.

Note 2 Needs are usually translated into characteristics with specified criteria. Needs may include, for example, aspects of performance, usability, dependability (reliability, availability, maintainability), safety, environment, economics, and aesthetics.

Note 3 The term "quality" should not be used as a single term to express a degree of excellence in a comparative sense, nor should it be used in a quantitative sense for technical evaluations. To express these meanings, a qualifying adjective should be used. For example, use can be made of the following terms:

1. "Relative quality" where entities are ranked on a relative basis in the degree of excellence or comparative sense (not was confused with "grade").
2. "Quality level" in a quantitative sense (as used in acceptance sampling), and "quality measure" where precise technical evaluations are carried out.

Note 4 In some references, quality is referred to as "fitness for use" or "fitness for purpose," or "customer satisfaction" or "conformance to the requirements." These represent only certain facets of quality, as defined above.

Quality assurance All the planned and systematic activities implemented within the quality system and demonstrated as needed, to provide adequate confidence that an entity will fulfill requirements for quality.

Quality control	Operational techniques and activities that are used to fulfill requirements for quality.
Note 1	Quality control involves operational techniques and activities aimed both at monitoring a process and at eliminating causes of unsatisfactory performance at all stages of the quality loop in order to achieve economic effectiveness.
Note 2	Some of the purposes of quality records are demonstration, traceability, and preventive and corrective actions.
Qualitative data collection	Is the collection of softer information, for example, reasons for an event occurring.
Quantitative data collection	Is the collection of data that can be stated as a numerical value
Reliability	Is the ability of an item to perform a required function under given conditions for a given time interval.
Note 1	It is generally assumed that the item is in a state to perform this required function at the beginning of the time interval.
Note 2	The term "reliability" is also used to denote the nonqualified ability of an item to perform a required function under conditions for a specified period of time.
Reliability critical item	An item that contains a single-point failure, with a failure consequence severity classified as catastrophic, critical, or major.
Reliability growth	A condition characterized by a progressive improvement of a reliability performance measure of an item with time.
Reliability testing	Is testing during actual normal service use that offers initial information for the evaluation of the measurement of reliability indicators during the time of provided testing.
Requirement	That which is called for or is demanded: a condition which must be complied with.
Residual risk	The risk remaining in a system after completion of the hazard reduction and control process.
Residual hazard	Hazard remaining after implementation of hazard reduction.

Review Systematic examination of an item for the purpose of assessing the results obtained at a given time in the project, by persons not themselves responsible for the project.

Risk The combination of the probability of an event and its consequence, or a quantitative measure of the magnitude of a potential loss and the probability of incurring that loss.

Note 1 The term "risk" is generally used only when there is a least the possibility of negative consequences.

Note 2 In some situations, risk arises from the possibility of deviation from the expected outcome or event.

Risk acceptance The decision to accept a risk.

Note 1 The verb "to accept" is chosen to convey the idea that acceptance has its basic dictionary meaning.

Note 2 Risk acceptance depends on risk criteria.

Risk assessment Overall process comprising a risk analysis and a risk evaluation.

Risk category A class or type of risk (e.g., technical, legal, organizational, safety, economic, engineering, cost, schedule).

Note A risk category is a characterization of a source of risk.

Risk criteria The terms of reference by which the significance of risk is assessed.

Note Risk criteria can include associated cost and benefits, legal and statutory requirements, socioeconomic and environmental aspects, the concerns of stakeholders, priorities, and other inputs to the assessment.

Risk communication All information and data necessary for risk management addressed to a decision maker.

Risk evaluation Procedure based on the risk analysis to determine whether the tolerable risk has been achieved.

Risk index A score used to measure the magnitude of the risk; it is the product of the likelihood of occurrence and the severity of consequence, where scores are used to measure likelihood and severity.

Risk management The systematic and iterative optimization of the project, resources, performed according to the established project risk management policy.

Risk reduction Implementation of measures that leads to reduction of the likehood or severity of risk.

Risk scenario The sequence or combination of events leading from the initial cause to the unwanted consequence.

Risk trend The evolution of risks throughout the life cycle of a project.

Safety The freedom from unacceptable risk. Safety is one of the aspects of quality.

Safety critical function A function which, if lost or degraded, or which through incorrect or inadvertent operation, could result in a catastrophic or critical hazardous event.

Safety measure Means that eliminates a hazard or reduces a risk.

Set Group of physically or functionally related items that are considered together for technical or administrative reason, but whose association does not increase functionally over that of the individual items.

Severity of safety A classification of a failure or undesired event according to the magnitude of its possible consequences.

Software Programs, procedures, rules, and any associated documentation pertaining to the operation of a computer system.

Stress testing Classified as based on constant stress, step stress, cycling stress, and random stress.

System Set of interdependent elements constituted to achieve a given objective by performing a specified function.

Systems engineering Is a discipline concerned with the architecture, design, and integration of elements that when taken together comprise a system. Systems engineering is based on an integrated and interdisciplinary approach, where components interact with and influence each other. In addition to the technological systems, systems considered include human and organizational systems, where the incorporation of critical human factors with other interacting factors directly affects achieving the enterprise objectives.

Systems of systems Are composed of components that are systems in their own right (designed separately and capable of independent action) that work together to achieve shared goals

Task A specific piece of work to be done.

Test A formal process of exercising or putting to trial a system or item by manual or automatic means to identify differences between specified, expected, and actual results.

Compliance test A test used to show whether or not a characteristic or a property of an item complies with the stated requirements.

Endurance test A test carried out over a time interval to investigate how the properties of an item are affected by the application of stated stresses and by their time duration or repeated application.

Laboratory test A compliance test or a determination test made under prescribed and controlled conditions which may or may not simulate field conditions.

Test development The process through which a test is improved, planned, constructed, evaluated, and modified, including consideration of content, format, administration, scoring, item properties, scaling, and technical quality for its intended purpose.

Test development system A generic name for one or more programs that allow a user to author and edit items (i.e. questions, choices, correct answer, scoring scenarios, and outcomes), and maintain test definitions (i.e. how items are delivered with a test).

Testing techniques Can be used in order to obtain a structured and efficient testing, which covers the testing objectives during the different phases in the life cycle.

Validation Confirmation by examination and provision of objective evidence that the particular requirements for a specific intended use are fulfilled.

Note 1 In design and development, validation concerns the process of examining a product to determine conformity with user needs.

Note 2	Validation is normally performed on the final product under normal operating conditions. It may be necessary at earlier stages.
Verification	Confirmation by examination and provision of objective evidence that specified requirements have been fulfilled.
Note	In design and development, verification concerns the process of examining the result of a given activity to determine conformity with the stated requirements for that activity.
Warning conditions	Conditions in which potentially catastrophic or critical hazardous events have been detected as being imminent and preplanned saving action is required within a limited time.
Whole-life costs (WLC)	Total recourse required to assemble, equipment, sustain, operate, and dispose of a specified asset as detailed in the plan at defined levels of readiness, reliability, performance, and safety.
Note	WLC also includes the costs to recruit, train, and retain personnel, as well as the costs of higher organizations.
Work package	A group of related tasks that are defined at the lowest level within a work breakdown structure.
Workmanship	The physical characteristics relating to the level of quality introduced by the manufacturing and assembly activities."

References

(1) Haq H. Toyota Recall Update: Dealers Face Full Lots, Anxious Customers. *The Christian Science Monitor*. January, 29, 2010.

(2) *What is Technology Education*. UNESCO, 1985.

(3) Guide 3235.1H. Test and Evaluation of Systems RAM. DoD.

(4) Proceedings of the DoD Combined Environment Reliability Test (CERT) Workshop, 2–4 June 1981, Atlanta, Georgia.

(5) Honda Civic Fleet and Accelerated Reliability Testing—July 2005. INL/EXT 06-01262. Energy Efficiency and Renewable Energy. U.S. Department of Energy.

(6) Francfort J, Arguets J, Smith J, Wehrey M. Field Operations Program Toyota RAV4 (NiMH) Accelerated Reliability Testing—Final Report. INEEL/EXT-2000- 00100. Idaho National Engineering and Environmental Laboratory Automotive Systems and Technology Department. March 2000.

(7) Hybrid Electric Vehicle End-of-Life Testing on Honda Insight, Honda Gen 1 Civics and Toyota Gen 1 Priuses. INL/EXT-06-1262.

(8) Birch S. 24 Million Km of Testing for Mercedes C-Class. *Automotive Engineering*. April 2007.

(9) LMS Supports Ford Otosan in Developing Accelerated Durability Testing Cycles. http://www.lmsintl.com/LMS-Ford-Otosan-developing-accelerated-durability-testing-cycles (website URL)

(10) Testing: Durability, Accelerated Vehicle, Component & Systems Durability. http://www.mira.co.uk/Services/VehicleDurability.htm (website URL)

(11) Nevada Automotive Test Center (NATS).

(12) Kyle JT, Harrison HP. The Use of Accelerometer Is Simulating Field Conditions for Accelerated Testing of Farm Machinery. ASAE Paper No. 60–631, Memphis. 1960.

Accelerated Reliability and Durability Testing Technology, First Edition. Lev M. Klyatis.

(13) Most Software Projects Fail to Meet Their Goals. Can This Be Fixed by Giving Developers Better Tools? *The Economist*. November, 27, 2004.

(14) Sensor System for Satellites. *USA News Today*. May, 2003.

(15) Hobbs GK. *Accelerated Reliability Engineering: HALT and HASS*. John Wiley & Sons, 2000.

(16) Capitano JL. Explaining Accelerated Aging. *EE—Evaluation Engineering*. May, 1998.

(17) Yamasaki R, McEvoy JT. *Regional Accelerated Aging Test. Journal of Cellular Plastics*, 39, 4, 2003.

(18) Klyatis LM, Klyatis EL. *Accelerated Quality and Reliability Solutions*. Oxford, UK, Elsevier, 2006.

(19) Stork Materials Technology. http://www.storksmt.com (website URL)

(20) Dmitrichenko CC. Comparative Research of Fatigue of the Caterpillar Elements from Different Steels. NATI Research Institute Proceedings. Vol. 146, Moscow, 1962.

(21) Briskham P, Smith G. Cyclic Stress Durability Testing of Lap Shear Joints Exposed to Hot-Wet Conditions. *International Journal of Adhesion and Adhesives*, 20, 1, 2000.

(22) Bodycote Testing: Environmental Durability Testing. Bodycote Testing Group. 2006.

(23) Tracy CE, Zhang JG, Benson DK, Czanderna AW, Deb SK. *Accelerated Durability Testing of Electrochronic Windows. Electrochemica Acta*, 4, 18, 1999.

(24) Endurates Systems Group. Medical Device. Bose Corporation. 2005.

(25) Medical Devices Testing Services Stent/Graft & Infravascular Testing. Bose Corporation.

(26) ElectroForce® Stent/Graft Test Instruments. ElectroForce Test Instruments. Bose Corporation.

(27) Chu WL. Dynatek Demonstrates First Commercial Stent Tester. In Pharma Technologist.com. Breaking News on Pharmaceutical Technology. 12 October 2005.

(28) Orthopedic and Rehabilitation Devices Panel of the Medical Devices Advisory Committee. November 21, 2002. Draft Questions. Revised 11/12/02.

(29) Non-Clinical Tests and Recommendations for International Stents and Associated Delivery Devices, Issued January 13, 2005. CDRN Guideline.

(30) Anderson O. In Vibro Durability Testing of Coated Stents. Proceedings from the ASM International Materials & Processes for Medical Devices Conference. 2003.

(31) ASTM F2477-07 Standard Test Methods for In Vitro Pulsatile Durability Testing of Vascular Stents.

(32) Dodson B, Johnson E. Hidden Costs of In-Reliability. ASQ's 56th Annual Quality Congress Proceedings. Denver. 2002.

(33) Day JC. Address to the Economic Club in Detroit. 2000.

(34) Testing for Reliability. 1. What is Reliability Testing? TOSHIBA.

(35) Burns MA. Reliability Testing and Product Qualification. IBM Technology Group: Microelectronics Division. IBM Corporation. 2002.

(36) Surridge B, Law J, Oliver B, Pakulski W, Strackholder H, Abou-Khalil M, Bonneville G. Accelerated Reliability Testing of InGaP/GaAs HBTs. GaAs. 2002. Mantech Conference. San Diego, CA.

(37) Khan Z. What You Should Know About Reliability Testing. *Medical Design Magazine*, San Jose, CA. 2008. http://www.medicaldesign.com/articles/ID/12416 (website URL)

(38) Brand O, Feddex GK, Hierold C, Jan Korvink J, Tabata O, Tsuchiya T. *Reliability of MEMS: Testing of Materials and Devices*. John Wiley & Sons, 2008.

(39) Dallas Semiconductors. Product Reliability Testing and Data. MAXIM. 2006.

(40) Solar Simulation Systems. KHS. K.H. Stevernagel Lichttechnik Gmbh.

(41) Shigehary Y. *Concept and Practice of Combined Environmental Test (CERT). Journal of Reliability Engineering Association of Japan*, 24, 5, 2005.

(42) Klyatis LM. *Accelerated Evaluation of Farm Machinery*. Moscow, AGROPROMISDAT, 1985.

(43) Klyatis L, Klyatis E. *Successful Accelerated Testing*. New York, Mir Collection, 2002.

(44) Dzekevich J. Reliability Testing. Raytheon. Reliability Analysis Laboratory. https://www.reliabilityanalysislab.com/tl_jd_0211_ReliabilityTesting.asp (website URL)

(45) Brouwer S, Coddington R. Toyota Recalls Nearly 1 Million Vehicles Worldwide. *USA TODAY*, 5/3/2006.

(46) Nath NS. Mercedes to Recall 50,000 Cars. *All Headline News*, September, 30, 2005.

(47) Cohen P. Apple Recalls Sony-Made PowerBook, iBook Batteries. *Macworld News*. August, 24, 2006.

(48) MS NBC. Associated Press. Apple to Recall 1.8 Million Notebook Batteries. Recall is second-biggest in U.S. history involving electronics or computer. August 25, 2006. http://www.msnbc.msn.com/id/14500443 (website URL)

(49) CNN.com.AUTOS. Ford Recalling 3.8 Million Vehicles. September 9, 2005.

(50) Ackerman D. Another Day, Another Laptop Batteries Recalled. Alfa Blog. September 19, 2006. http://reviews.cnet.com/4531-10921_7-6639289.html (website URL)

(51) Robins B, Ryle G. Beating about the Bushmaster. *The Sydney Morning Herald*, May, 1, 2004.

(52) Smith RJ. Study Faults Army Vehicle. Use a Transport in Iraq Troops at Risk. Internal Report Says. *The Washington Post*. March, 31, 2005.

(53) NASA TestsLounch Abort Parachute System—Releases Crash Photos. Status Report. Data Realised: Tuesday, August 19, 2008. Source: NASA Exploration Systems Mission Directorate—Comments. http://www.spaceref.com/news/viewsr.html (website URL)

(54) Parachute Test Falls for NASA's New Spaceship. By SPACE.com Staff. Posted 21 August 2008. http://www.space.com/news/080821-orion-parachute-test.html (website URL)

(55) Defense Acquisition University (DAU). *Systems Engineering Fundamentals, Supplementary Text*. Virginia, DAU Press, 2001.

(56) Alexander R, Hal-May M, Kelly T. *Computer Safety, Reliability, and Security*. Berlin /Heidelberg, Springer, 2006.

(57) Systems Engineering Fundamentals. DAU. 2001.

(58) Standard IEC 60300-3-2, Ed.2, Dependability management—Part 3–2: Application guide—Collection of dependability data from the field.

(59) IEC 60300-3-7: Dependability Management—Part 3–7: Application guide—Reliability stress screening of electronic hardware.

(60) Chernov YD, Klyatis LM. Reliability of Soil-Cultivating Machinery. Paper AETC-95114. Agricultural Equipment Technology Conference. November 1–4, Chicago. 1995.

(61) Salvendy G (Ed.). *Handbook of Human Factors and Ergonomics*, 3rd edition. John Wiley & Sons, 2006.

(62) Monaghan M. Toyota Turns to Mechanical Simulation, MTS for Driving Simulator Components. AEI Testing and Simulation Technology eNewsletter. SAE International. March 2008.

(63) Alfaro W. Quality, Reliability, and Durability Strategies. General Motors North America Engineering. Presentation at the SAE G-11 RMSL Division Spring 2005 Meeting. Detroit, MI. 2005.

(64) Juran JM, Blanton AG. *Juran's Quality Handbook*, 5th edition. Quality Press, 1999.

(65) Gruna F. *Quality Planning and Analysis: From Product Development Through Use*, 4th edition. Quality Press, 2001.

(66) Ishikawa K. *Guide to Quality Control*. Quality Press, 1986.

(67) Klyatis LM. Analysis of Different Approaches to Reliability Testing. In book: Reliability and Robust Design in Automotive Engineering (SP-1956). SAE International. 2005 SAE World Congress, Paper 2005-01-0820. Detroit, MI.

(68) Klyatis LM. Physical Simulation of Life Processes in the Laboratory. Paper No. 95-4469. Presentation at the 1995 ASAE Annual International meeting. Chicago. 2005.

(69) Klyatis L. Introduction to Integrated Quality and Reliability Solutions for Industrial Companies. ASQ World Conference on Quality and Improvement Proceedings. May 1–3, 2006, Milwaukee, WI.

(70) Klyatis L, Klyatis E. The Strategy to Improve Engineering Concepts of Automotive Reliability, Durability, and Maintainability Integrated with Quality. In book: Reliability and Robust Design in Automotive Engineering 2006 (SP-2032). SAE International. 2006 SAE World Congress. Detroit, MI, April 3–6, 2006.

(71) Chan AH (Ed.). *Accelerated Stress Testing Handbook: Guide for Achieving Quality Products*. Wiley-IEEE Press, 2001.

(72) Boothe RF. Fast Accurate Simulation of Large Shared Memory Multiprocessors. EECS Department. University of California, Berkeley. Technical Report No. UCB/CSD-92-682. 1992.

(73) Akamatsu S, Charles B, Dunleavy L. Accurate Simulation Models Yield High Efficiency Power Amplifier Design. 2006.

(74) Levis P, Lee N, Welsh M, Culler D. TOSSIM: Accurate and Scalable Simulation of Entire TinyOS Applications. University of California, Harvard University. Los Angeles. 2003.

(75) Davis JD, Fu C, Laudon J. The RASE (Rapid, Accurate Simulation Environment) for Chip Multiprocessors. Sun Microsystems.

(76) Corn JA. *Random Processes Simulation and Measurement*. New York, McGraw-Hill, 1996.

(77) Venikov VA. *Theory of Similarity and Modeling*. Moscow, High School, 1996.

(78) Korean Institute of Materials. Introduction to Reliability Assessment Center.

(79) CNN.com—Honda Recalling 699,000 Cars. November 4, 2003. http://www.cnn.com/2003/BUSINESS/11/04/japan.honda.reut/ (website URL)

(80) Lavrinc D. Honda Recalls 270,000 Vehicles in Japan and Elsewhere. Autoblod. Posted October 6, 2006.

(81) Benton J. Toyota Recalls Near 800,000 for July. July 19, 2006. http://Consumeraffairs.com (website URL)

(82) Recall: Suzuki Calls Back 75,000 Renos and Forenzas. Monday, August 6, 2007.

(83) Greenberg K. *Toyota Recall Threatens Quality Reputation*. MediaPost Publications, 2007.

(84) Toyota Recalls 215,000 Cars Worldwide. Recalls/TSBs, Toyota. November 28th, 2007. http://www.autoblog.com/2007/11/28/toyota-recalls-215-000-cars-worldwide/ (website URL)

(85) Ellis L. Honda Recalls 300,000 cars in Japan. *Automotive Business Review*. March, 23, 2007.

(86) Thomas D. Kicking Tires: Honda Recalls 182,756 Civic Sedans. September 14, 2007. http://blogs.cars.com/kickingtires/2007/09/honds-recalls-1.html (website URL)

(87) Engine Problems Force Honda Recall. Monday, March 19, 2007. http://www.sfgate.com/cgi-bin/article.cgi?f=/n/a/2007/03/19/financial/f092251D81.DTL (website URL)

(88) Miersma S. Nissan Recall: Company Must Fix Sensor for 650,000. Centras and Altimas. http://news.windingroad.com/maintenance/nissan-recall-company-must-fix (website URL)

(89) Suzuki Recalls 381,551 Minivehicles. *Japan Today-News*. April, 19, 2007. http://archive.japantoday.com/jp/news/404329 (website URL)

(90) BrightCar Blog: Suzuki Recalls 75,000 Vehicles. Posted by Todd, Tuesday August 7, 2007. http://blog.brightcar.com/2007/08/sizuki-recalls-75000-vehicles.html (website URL)

(91) Lombardi C. Toyota Recalls 1.4 Million Cars. Planerary Gear, January 28, 2009.

(92) Nissan Recalls 2.5 Million Cars (October 30, 2003, June 11 2007). *Consumer News*.

(93) Suzuki Motor Recalls 120,000 Cars. *The Economic Times*. December, 5, 2006.

(94) Odyssey Minivan on Honda Recall List (June 14,2006). http://Consumeraffairs.com (website URL)

(95) Nissan Recalls 92,000 For Faulty Clutch. *The Huffington Post*. June, 7, 2007.

(96) Toyota Recalls Nearly 1 Million Vehicles Worldwide. *USA TODAY*. May, 31, 2006.

(97) Wella M. Toyota Recall Not So Total. *Business Week*. June, 1, 2006.

(98) National Highway Traffic Safety Administration, NHTSA, USA. SunLube Recall Notice. http://www.synlube.com/serv05.htm (website URL)

(99) Klyatis L, Walls L. A Methodology for Selecting Representative Input Regions for Accelerated Testing. *Quality Engineering*, 16, 3, 2004. ASQ & Marcel Dekker.

(100) Lourie AB. *Statistical Dynamics of Farm Machinery*. Leningrad, Kolos, 1970.

(101) Klyatis LM. Step-by-Step Accelerated Testing. The International Symposium of Product Quality and Integrity (RAMS) Proceedings. Washington, DC, January 18–21, 1999.

(102) Chan HA, Parker PT. Product Reliability Through Stress Testing. 1999 Annual Reliability and Maintainability Symposium (RAMS) Tutorial Notes. Washington, DC.

(103) Klyatis LM. Climate and Reliability. ASQ 56th Annual Quality Congress Proceedings. Denver, CO, May 20–22, 2002.

(104) Klyatis LM, Schehtman EM. Physical Modeling of Climatic Conditions Machinery Using. *Journal Mechanization and Electrification of Agriculture Moscow (Russia)*, No. 1, 1970.

(105) Klyatis LM. Effective Accelerated Testing. Reliability Review. *The R & M Engineering Journal*, ASQ. 16, 4, 1996.

(106) Klyatis LM, Teskin OI, Fulton W. Multi-Variate weibull Model for Predicting System Reliability from Testing Results of the Components. The International Symposium of Product Quality and Integrity (RAMS) Proceedings. Los Angeles, January 24–27, 2000.

(107) Fratianne SA, Burrows DC. Integrated Reliability & Maintainability Program-Planning. RAMS Tutorial Notes. Seattle, WA, January 28–31. 2002.

(108) International Electrotechnical Commission (IEC). Standard IEC 60300-3-3, Ed.2: Dependability Management—Part 3-3: Application Guide—Life Cycle Costing.

(109) Baskoro G, Riyvroye JL, Brombacher AC. Developing MESA: An Accelerated Reliability Test. Annual Reliability and Maintainability Symposium Proceedings. January 12–15. Washington D.C. 2003.

(110) Yamamoto T. Guidance for Accelerated Testing and Reliability. ESPEC.

(111) Marcos J, Manuel JL, Pallas L, Fernandez-Gomez S. Reliability Tests on Power Devices. Annual Reliability and Maintainability Symposium Proceedings (RAMS). Seattle, WA, January 28–31, 2002.

(112) Jawaid S, Rogers P. Accelerated Reliability Test Results: Importance of Input Vibration Spectrum and Mechanical Response of Test article. Annual Reliability and Maintainability Symposium (RAMS) Proceedings. Los Angeles. 2000.

(113) Jawaid S, Nesbitt T. Accelerated Reliability Tests: Solder Defects Exposed. Annual Reliability and Maintainability Symposium (RAMS) Proceedings. Washington, D.C. 1999.

(114) Karbassian A, Daren B, Tetsufumi Katakami. Accelerated Vibration Durability Testing of a Pickup Truck Rear Bed. SAE Paper 2009-01-1406. Detroit. 2009.

(115) Ledesma R, Jenaway L, Yenkai W, Shan S. Development of Accalerated Durability Tests for Commercial Vehicle Suspension Components. SAE Commercial Vehicle Engineering Conference. November 2005, Rosemont, IL, USA. Paper 2005-01-3565. SAE International. 2005.

(116) O'Connor P. *Test Engineering*. John Wiley & Sons Ltd, 2001.

(117) McKinney BT. Accelerated Reliability Testing Utilizing Design of Experiments. Report, May- Sepember 1993. Rome Lab., Griffiss AFB, NY. 1993.

(118) Alion Science and Technology. Accelerated Reliability Test plans and Procedures. System Reliability Center, http://src.alionscience.com/consulting (website URL)

(119) Urbancik J. The Evaluation of Accelerated Reliability Tests in Microelectronics. IEEE 26th International Spring Seminar on Electronics Technology. March 2003.

(120) Jepsen R. Sandia National Labs. Superfuge: Re-Entry and Launch Simulation in A Combined Acceleration Axial Spin and Vibration Environment. Development forum of Aerospace Testing Expo 2006. Experts technical presentations.

(121) Keller T. Improved Vibration Simulation Using Multi-Shaker, MIMO, Testing. Development forum of Aerospace Testing Expo 2006. Experts technical presentations.

(122) Klyatis L. A new approach to physical simulation and accelerated reliability testing in avionics. Development forum of Aerospace Testing Expo 2006. Experts technical presentations.

(123) Klyatis LM, Klyatis E. Successful Correlation between Accelerated Testing Results and Field Results. ASQ 55th Annual Quality Congress Proceedings. *Charlot*te, NC, May 7–9, 2001.

(124) Paul PC, Staff CR, Sanger KB, Mahler MA, Kan HP, Kautz EF. Analysiand Test Techniques for Composite Structures Subjected to Out-of-Plane Loads. ASTM International, January 1992.

(125) Vodicka R. Accelerated Environmental Testing of Composite Materials. Technical Report. Accession Number: ADA352718. Defense Science and Technology Organization, Australia. 1997.

(126) Higgins J, Wegner P, Adrian VA, Sanford G. Design and Testing of the Minotaur Advanced Grid-stiffened Fairing. Twelfth International Conference on Composite Structures Proceedings. Vol. 66, 1–4, 2004.

(127) Ohmura A, Wild M. Climate Change: Is the Hydrological Cycle Accelerating? *Science*, 298, 5597, 2002.

(128) Keen K. Global Dimming? Global Warming? What's with the Globe, Anyway? *Grist Magazine*. September, 22, 2004.

(129) Adam D. Goodbye Sunshine or A Very Wildcard. *The Guardian*. December, 18, 2003.

(130) Cost of Corrosion. Report by CC Technologies Laboratories, Inc. to Federal Highway Administration (FHWA), Office of Infrastructure Research and Development. Report FHWA RD-01-156. January 2010. http://www.corrosion cost.com/home.html (website URL)

(131) Report. FHWA-RD-01-156. Cost of corrosion in the USA $276 Billion/year. September 2010.

(132) McDaniel SJ. Failure Analysis of Launch Pad Tubing From The Kennedy Space Center. Microstructural Science, Vol. 25, 1998. ASM International, Materials Park, OH.

(133) Guthrie J, Battat B, Grethlein C. Accelerated Corrosion Testing. Advanced Materials and Processes Technology. *The AMPTIAC Quarterly*, Vol. 6, 3, AMITIAC 19. Material Ease. Rome, NY. DoD USA. 2001.

(134) *Bhopal: Absolutely Liable. Economist*, 308, 560, 1988.

(135) Cassels J. *The Uncertain Promise of Law: Lessons from Bhopal*. University Of Toronto Press Incorporated, 1993.

(136) ASTM G31: Standard Practice for Laboratory Immersion Corrosion Testing of Metals.

(137) Guthrie J, Battat B, Grethlein C. *Accelerated Corrosion Testing. The AMPTIAC Quarterly*, 6, 3, 2002.

(138) Davidson D, Thomson L, Lutze F, Tiburcio B, Smith K, Meade C, Mackie T, McCune D, Townsend H, Tuszynski R. Perforation Corrosion Performance of Autobody Steel Sheet in On-Vehicle and Accelerated Tests. 2003 SAE World Congress. Paper 2003-01-1238. Detroit.

(139) Klyatis LM. Establishment of Accelerated Corrosion Testing Conditions. Reliability and Maintainability Symposium (RAMS) Proceedings. Seattle, WA, January 28–31, 2002.

(140) Golubev AI. *Prediction of Metals Corrosion in Atmospheric Conditions*. Moscow, Russia, GOSNITI, 1961.

(141) Tomashov NF. *Atmospheric Corrosion of Metals*. Moscow, Russia, The Institute of Physical Chemistry, 1968.

(142) Besverhy SF, Jacenko NN. *The Basis of Technology of Special Road Testing and Certification of Automobiles*. Moscow, The Standard Publishing House, 1996.

(143) Baziari A. Accelerated Corrosion Test for Complete Vehicles. U.S. Army RDECOM-TARDEC. AMSRD-TAR-E/ME. Warren, MI 48397. 7/12/2007.

(144) Oh SJ, Hyundai Motor. Co. Development Accelerated Corrosion Test Mode Considering Environmental Conditions. Paper Number 2002-01-1231. 2002 SAE International World Congress and Exhibition. Detroit. 2002.

(145) Munster EG, Rademacher T. Testing of Agricultural Machinery. Landtechnik. #1, 2001.

(146) Klyatis L, Vaysman A. Accurate Simulation of Human Factors and Reliability, Maintainability, and Supportability Solutions. *The Journal of Reliability, Maintainability, Supportability in Systems Engineering*, RMS Partnership. Winter, 2007/2008.

(147) Abdulgalimov YM, Klyatis LM, Tenenbaum MM. Simulated Material for Accelerated Reliability Testing. *Journal Tractors and Farm Machinery*, No. 7, 1989. Moscow.

(148) Formula 1 for Quality direct from the Factory. Weiss Umwelttechnik GmbH. WEISS TECHNIK.

(149) Bus Climatic Wind Tunnel. Seoul Industry Engineering Co., Ltd. R&D Center.

(150) Unholtz-Dickie Corporation. Accelerated Reliability Test Results: Importance of Input Vibration Spectrum and Mechanical Response of Test Article. http://www.udco.com/reliab.htm (website URL)

(151) THERMOTRON. Environmental Stress. Test Simulation and Screening Solutions.

(152) Soft Support System for Ground Vibration Testing at Embraer—Sao Paulo, Brazil. FABREEKA.

(153) Vibration Test Systems. LDS.

(154) T2000 Series Vibration Test System. Umholyz-Dickie Corporation.

(155) Integrated Simulation Solutions For Vehicle Development. MTS Systems Corporation.

(156) MTS Model 353.20 Multi-Axial Simulation Table (MAST). Ground Vehicle Testing No. 2/April 15, 2006.

(157) MTS Multiaxial Elastometer. Test System—Model 836.

(158) Dhillon S. (LDS). Shaken, not heard. Aerospace Testing International. November/December 2007.

(159) ECESO. Dynamometer testing. ECESO.

(160) Dinkel J. *Chassis Dynamometer. Road and Track Illustrated Automotive Dictionary*. Bentley Publishers, 2000.

(161) Committed to Excellence. NSF. Envirotronics.

(162) Environmental Simulation. Noxious Gas Test / Corrosion Test. Technolab (Germany).

(163) Uccelli A. I lobri del volo di Leonardo da Vinci. Hoepli In. 1952.

(164) Moolman V. The Road of Kitty Hawk, Time life Books. 1980.

(165) Albertani R, Hubner P, Ifju P, Lind R, Jackowski J. 2004. Wind Tunnel Testing of Micro Air vehicles at Low Reynolds Numbers. SAE 2004 World Aviation Congress, Paper 2004-01-3090. Reno, Nevada.

(166) Carmichael BH. Low Reynolds Number Airfoil Survey: Vol. 1. NASA CR 165803, November 1981.

(167) Selig MS. Summary of Low Speed Airfoil Data. Vol. 1, 2, and 3 SoarTech Publications.

(168) Riegels FW. *Airfoil Sections*. London Butterworths, 1961.

(169) Laitone EV. (Edited by Mueller TJ). Wind Tunnel Tests of Wings and Rings at Low Reynold Numbers. Fixed and Flapping Wing Aerodynamics for Micro Air Vehicle Applications. *Progress in Astronautics and Aeronautics*, 195, 2005.

(170) Lutz T, Wurz W, Wagner S. (Edited by Mueller TJ). Numerical Optimization and Wind Tunnel Testing of Low Reynolds Number Airfoils. Fixed and Flapping Wing Aerodynamics for Micro Air Vehicle Applications. *Progress in Aeronautics and Astronautics*, 195, 2005.

(171) MIT's Wright Brothers Wind Tunnel.

(172) Christopher H. Road and Runway to Reality. *Aerospace Testing International*. March, 2008.

(173) Suzuki S. Wind Tunnel Test. *Aerospace Testing International*. March, 2008.

(174) High Speed Wind Tunnel. Lockheed Martin.

(175) Large Wind Tunnel for the Aerospace Industry. ONERA. France.

(176) World Most Advanced Wind Tunnel. *Quality Manufacturing Today magazine*. January, 2008. www.windshearinc.com (website URL)

(177) Material Testing Solutions. ATLAS.

(178) Temperature and Climatic Test Chambers. ATLAS.

(179) GAGA Instruments Pte Ltd.

(180) Pacific Laboratory Products—Solar Climatic Test Chambers.

(181) Radiation Test Chambers—EC Plaza. NOGIX Co., Ltd.

(182) USD Series Sand & Dust Chambers. CZZ Cintinnati Sub-Zero.

(183) Ozone Test Chamber. Ozone Detector Instruments and Ozone Generators. In USA Corporation.

(184) Argentox Ozone Technology GmbH (Germany).

(185) Mast/Keystone manufacturers and distributes Ozone Test Chambers, ozone Monitoring. http://www.mastdev.com/700_10ltb_chamber.html (website URL)

(186) Meiler RC, Otto NC, Pielemier WJ, Jeybalan V. The Ford Vehicle Vibration Simulator for Subjective Testing. *Sounds & Vibration*. October, 1996.

(187) AC Engine Dynamometers. MUSTANG Advanced Engineering.
(188) Noxious Gas Climate tests. Weiss Technik.
(189) Dust Test Chambers. WEISS TECHNIK.
(190) Anseros K Nonnenmacher GmbH.
(191) Weathering Testing Guidebook. ATLAS.
(192) Instron Structural Testing Systems. MAST. Multi-Axis Shaking Tables.
(193) The German—Dutch Wind Tunnels. DNW.
(194) Kolmogorov AN. Interpolation and Extrapolation of Stationary Random Sequences. Paper of USSR Academy of Science. Moscow. 1941.
(195) Smirnov IV. Approximation of Distribution Laws of Random Variables by Empirical Data. Papers of the USSR Academy of Science. 1944, 10.
(196) Klyatis LM. Prediction of Reliability and Spare Parts of Machinery. ASQC's 51st Annual Quality Congress Proceedings. Orlando, Fl, May 5–7, 1997.
(197) Klyatis LM, Teskin OI, Fulton JW. Multi-Variate Weibull Model for Predicting Systems Reliability from Testing Results of the Components. Annual Reliability and Maintainability Symposium (RAMS). Los Angeles, CA, 2000.
(198) Pavlov IV. *Statistical Methods for Estimation the Reliability of Complex System by the Test Results*. Moscow, Radio and Svyaz, 1982.
(199) Teskin OI, Sonkina TR. Parametric Reliability-Prediction Based on Small Samples. *IEEE Transactions in Reliability*, 46, 3, 1997.
(200) Mann NR, Chafer RE, Singpurvalla ND. *Methods for Statistical Analysis for Reliability and Life Data*. John Wiley & Sons, 1974.
(201) Johns MV, Lieberman GJ. An Exact Asymptotically Efficient Confidence for Reliability in the Case of the Weibull Distribution. *Technometrics*, 8, 1, 1966.
(202) Engelhardt ML, Bain LJ. Simplified Statistical Procedures for the Weibull or Extreme Value Distribution. *Technometrics*, 19, 3, 1977.
(203) Thoman DR, Bain LJ, Antle SE. Maximum Likelihood Estimation, Exact Confidence Intervals and Tolerance Limit in the Weibull Distribution. *Technometrics*, 12, 2, 1970.
(204) Belyaev JK. Simple Confidence Estimates of the Quintiles of Aging Distributions of the Duration of Failure-Free Operation. *Engineering Cybernetics*, 19, 1, 1981, Moscow.
(205) Gnedenko BV, Belyaev JK, Soloviev AD. *Mathematics Methods in Reliability Theory*. Moscow, Nauka, 1965.
(206) Sonkina TR. Confidence Limits for the Series System Reliability in the Case of A Weibull Law of Failures of Its Components. Reliability and Quality Control, Moscow, No.11, 1982.
(207) Sonkina TR, Teskin OI. Interval Estimation of the Gamma Percent Resource of A System by the Results of Tests of Its Components. *Engineering Cybernetics*, 24, 4, 1986.
(208) Lesmerises A. The Need to Expand The Scope of The Reliability Centered Analysis Process. Presentation at the SAE G-11 Fall Meeting. September 18, 2007. Los Angeles.
(209) List of standards prepared by Technical Committee 56 0f the IEC. International Electrotechnical Commission (IEC). Geneva, Switzerland.

(210) IEC Publication issued by TC-56. Publication. http://www.iec.ch/cgi- in/procgi.pl/www/iecwww.p?wwwlang=e&wwwprog=TCpubs.p&progdb=db1&committee=TC&css_color=purple&number=56 (website URL)

(211) Klyatis LM. Reliability Testing Standardization. SAE International. 2007 SAE AeroTech Congress & Exhibition. Reliability, Maintainability, and Probabilistic Technology G-11 Division Fall 2007 Meeting. Los Angeles, CA. September 19, 2007.

(212) MIL-STD-781D Reliability Design Qualification and Production Acceptance Tests. DoD.

(213) MIL-STD-721C Definition of Terms for Reliability and Maintainability. DoD.

(214) MIL-STD-2074 Failure Classification for Reliability Testing. DoD.

(215) *Guide: MIL-HDBK-781A Reliability Test Methods, Plans, and Environments for Engineering, Development Qualification and Production*. DoD.

(216) *Handbook: MIL-HDBK-H 108 Sampling Procedures and Tables for Life and Reliability Testing (Based on Exponential Distribution)*. DoD.

(217) Stent and Stent-Graft Testing. Test Applications. Medical Device Testing.

(218) Parker P (Ed.). *Accelerated Stress Testing CD-Rom (VHS Tape)*. IEEE Press, 2001.

(219) Full Vehicle Durability. MGA Research Corporation.

(220) Vehicle Durability Tests. IDIADA Automotive Technology. Spain.

(221) Anderson SE. Current Trends in Durability Testing of Vascular Stents. Business Briefing. Global Healthcare Medical Device Manufacturing & Technology. Bose Corporation, Endura TEC Systems Group, Minnetonka. 2005.

(222) Enhancing Life through Testing. Accelerated Durability Specialists. Medical Device Testing.

(223) Toyota's Biggest Recall Affects Cars in Japan, Australia, Malaysia, Singapore and Thailand. AP Worldstream. The Wall Street Journal reported. 10-19-2005.

(224) Chenowrth HB. Reliability Testing—Challenges and Trends. *The Journal of Environmental Sciences*, 30, 1, 1987.

(225) Systems Engineering. Wikipedia encyclopedia.

(226) Medford R. Vehicle Safety at NHTSA. Presentation at the SAE G-11 Meeting. 2005. Detroit, MI.

(227) ECSS-M-00-03A. Space Project Management. Risk management. European Cooperation for Space Standardization (ECSS).

(228) ECSS-Q-40B. Space Product Assurance. Safety. European Cooperation for Space Standardization (ECSS).

(229) ECSS-Q-40-02A. Space Product Assurance. Hazard analysis. European Cooperation for Space Standardization (ECSS).

(230) ISO/IEC. Guide 51. Safety Aspects—Guidelines for Their Inclusion in Standards.

(231) Vaysman A. Road Traffic Medicine in Russia. *Journal of Traffic Medicine*, Supplement, 21, 1, 1993.

(232) Tarasenko A. Simulations for 787 Avionics Testing. Development forum of Aerospace Testing Expo 2006. Experts technical presentations.

(233) Transmissions/Transaxle. MUSTANG Advanced Engineering.

(234) Inertia Brake Dynamometer. GreeningTM.

(235) Innovative Product Support. A World of Capabilities. Dayton T. Brown.

(236) Auto Technology Company Cyclic Corrosion Test Chamber.

(237) Climate Test Chambers for Noxious Gas Testing. Pacific Laboratory Products.

(238) Wind tunnels. NASA Ames Research Center. http://www.windtunnels.arc.nasa.gov (website URL)

(239) Electrical Test Systems. Integrated Assembly & Test Systems. CIRRUS.

(240) Electrical Test Machines for Resistance, Voltage, Current, Power and other tests from TQC. http://www.testmashine.co.uk/electest.htm (website URL)

(241) About Current Leakage Testers.

(242) Kapur KC, Lamberson KC. *Reliability in Engineering Design*. New York, Wiley & Sons, 1977.

(243) Aerospace Standard AS9100. Quality Management Systems-Aerospace-Requirements. 2004.

(244) Baboian R. Corrosion Tests and Standards: Application and Interpretation, ASTM International.

(245) Department of Defense of the USA. Test Method Standard for Environmental Engineering Consideration and Laboratory Tests. MIL-STD-810F.

(246) Klyatis L. Accelerated Reliability Testing as a Key Factor for Accelerated Development of Product/Process Reliability. IEEE Workshop Accelerated Stress Testing. Reliability (ASTR 2009). Proceedings on CD. October 7–9, 2009. Jersey City.

(247) Klyatis LM. Specifics of Accelerated Reliability Testing. IEEE Workshop Accelerated Stress Testing. Reliability (ASTR 2009). Proceedings on CD. October 7–9, 2009. Jersey City.

(248) Wild M, Golgen H, Roesch A, Ohmura A, Long C, Dutton E, Forgam B, Kallis A, Kussak V, Tsvetkov A. From Dimming to Brightening: Decadal Changes in Solar Radiation at Earth's Surface. *Science*, 308, 2005.

(249) Calle LM, Kolody MR, Vinje RD. Electrochemical Investigation of Corrosion in the Space Shuttle Launch Environment. National Aeronautics and Space Administration (NASA), NA-C2-T, Kennedy Space Center, Florida USA ASRC Aerospace Corp., ASRC 15, 2005.

(250) Nelson WB. *Accelerated Testing (Statistical Models, Test Plans, and Data Analysis)*. John Wiley & Sons, 2004.

(251) Klyatis LM, Verbitsky D. Accelerated Reliability/Durability Testing as a Key Factor for Accelerated Development and Improvement of Produt/Process Reliability, Durability, Maintainability, and Quality. Paper 2010-01-0203. SAE 2010 World Congress & Exhibition. Detroit. April 13–15, 2010.

(252) Reliability Prediction and Testing. Percept Technology Labs. http://www.percept.com/reliability.php (website URL)

(253) Taksakulvith P, Jones JA, Warrington L. An Analysis of the Drivers in the Philosophy of Reliability Practice over the Last 50 Years. Annual Reliability and Maintainability Symposium (RAMS). Tampa FL, January 28–30, 2004.

(254) Keelan BW, Cookingham RE, Kane P, Topfer K. *Handbook of Image Quality Characterization and Prediction*. New York, Basel, Marcel Dekker, Inc., 2006.

(255) Jacobs S. Simulate All Flight Conditions. *Aerospace TESTING International.* 2011 showcase.

(256) Van Den Dijssel P. Accessing the Aerospace Trends. *Aerospace TESTING International.* 2011 showcase.

(257) Thiel J. Tubeway Army—Measurement Analysis. *Aerospace TESTING International.* 2011 showcase.

Index

WILEY SERIES IN SYSTEMS ENGINEERING AND MANAGEMENT

Andrew P. Sage, Editor

ANDREW P. SAGE and JAMES D. PALMER
Software Systems Engineering

WILLIAM B. ROUSE
Design for Success: A Human-Centered Approach to Designing Successful Products and Systems

LEONARD ADELMAN
Evaluating Decision Support and Expert System Technology

ANDREW P. SAGE
Decision Support Systems Engineering

YEFIM FASSER and DONALD BRETTNER
Process Improvement in the Electronics Industry, Second Edition

WILLIAM B. ROUSE
Strategies for Innovation

ANDREW P. SAGE
Systems Engineering

HORST TEMPELMEIER and HEINRICH KUHN
Flexible Manufacturing Systems: Decision Support for Design and Operation

WILLIAM B. ROUSE
Catalysts for Change: Concepts and Principles for Enabling Innovation

LIPING FANG, KEITH W. HIPEL, and D. MARC KILGOUR
Interactive Decision Making: The Graph Model for Conflict Resolution

DAVID A. SCHUM
Evidential Foundations of Probabilistic Reasoning

JENS RASMUSSEN, ANNELISE MARK PEJTERSEN, and LEONARD P. GOODSTEIN
Cognitive Systems Engineering

ANDREW P. SAGE
Systems Management for Information Technology and Software Engineering

ALPHONSE CHAPANIS
Human Factors in Systems Engineering

YACOV Y. HAIMES
Risk Modeling, Assessment, and Management, Third Edition

DENNIS M. BUEDE
The Engineering Design of Systems: Models and Methods, Second Edition

ANDREW P. SAGE and JAMES E. ARMSTRONG, Jr.
Introduction to Systems Engineering

WILLIAM B. ROUSE
Essential Challenges of Strategic Management

YEFIM FASSER and DONALD BRETTNER
Management for Quality in High-Technology Enterprises

THOMAS B. SHERIDAN
Humans and Automation: System Design and Research Issues

ALEXANDER KOSSIAKOFF and WILLIAM N. SWEET
Systems Engineering Principles and Practice

HAROLD R. BOOHER
Handbook of Human Systems Integration

JEFFREY T. POLLOCK and RALPH HODGSON
Adaptive Information: Improving Business Through Semantic Interoperability, Grid Computing, and Enterprise Integration

ALAN L. PORTER and SCOTT W. CUNNINGHAM
Tech Mining: Exploiting New Technologies for Competitive Advantage

REX BROWN
Rational Choice and Judgment: Decision Analysis for the Decider

WILLIAM B. ROUSE and KENNETH R. BOFF (editors)
Organizational Simulation

HOWARD EISNER
Managing Complex Systems: Thinking Outside the Box

STEVE BELL
Lean Enterprise Systems: Using IT for Continuous Improvement

J. JERRY KAUFMAN and ROY WOODHEAD
Stimulating Innovation in Products and Services: With Function Analysis and Mapping

WILLIAM B. ROUSE
Enterprise Tranformation: Understanding and Enabling Fundamental Change

JOHN E. GIBSON, WILLIAM T. SCHERER, and WILLAM F. GIBSON
How to Do Systems Analysis

WILLIAM F. CHRISTOPHER
Holistic Management: Managing What Matters for Company Success

WILLIAM B. ROUSE
People and Organizations: Explorations of Human-Centered Design

MO JAMSHIDI
System of Systems Engineering: Innovations for the Twenty-First Century

ANDREW P. SAGE and WILLIAM B. ROUSE
Handbook of Systems Engineering and Management, Second Edition

JOHN R. CLYMER
Simulation-Based Engineering of Complex Systems, Second Edition

KRAG BROTBY
Information Security Governance: A Practical Development and Implementation Approach

JULIAN TALBOT and MILES JAKEMAN
Security Risk Management Body of Knowledge

SCOTT JACKSON
Architecting Resilient Systems: Accident Avoidance and Survival and Recovery from Disruptions

JAMES A. GEORGE and JAMES A. RODGER
Smart Data: Enterprise Performance Optimization Strategy

YORAM KOREN
The Global Manufacturing Revolution: Product-Process-Business Integration and Reconfigurable Systems

AVNER ENGEL
Verification, Validation, and Testing of Engineered Systems

WILLIAM B. ROUSE (editor)
The Economics of Human Systems Integration: Valuation of Investments in People's Training and Education, Safety and Health, and Work Productivity

ALEXANDER KOSSIAKOFF, WILLIAM N. SWEET, SAM SEYMOUR, and STEVEN M. BIEMER
Systems Engineering Principles and Practice, Second Edition

GREGORY S. PARNELL, PATRICK J. DRISCOLL, and DALE L. HENDERSON (editors)
Decision Making in Systems Engineering and Management, Second Edition

ANDREW P. SAGE and WILLIAM B. ROUSE
Economic Systems Analysis and Assessment: Intensive Systems, Organizations, and Enterprises

LEV M. KLYATIS
Accelerated Reliability and Durability Testing Technology

Printed and bound by CPI Group (UK) Ltd, Croydon, CR0 4YY
16/04/2025
14658430-0001